THE FUNCTIONAL FOODS

Nutrient and Health Benefits

Innovations in Plant Science for Better Health: From Soil to Fork

THE FUNCTIONAL FOODS

Nutrient and Health Benefits

Edited by

Megh R. Goyal, PhD, P.E.

Junaid Ahmad Malik, PhD

Anu Kumari, MTech

AAP APPLE ACADEMIC PRESS

First edition published 2025

Apple Academic Press Inc.
1265 Goldenrod Circle, NE,
Palm Bay, FL 32905 USA

760 Laurentian Drive, Unit 19,
Burlington, ON L7N 0A4, CANADA

CRC Press
2385 NW Executive Center Drive,
Suite 320, Boca Raton FL 33431

4 Park Square, Milton Park,
Abingdon, Oxon, OX14 4RN UK

Library and Archives Canada Cataloguing in Publication

Title: The functional foods : nutrient and health benefits / edited by Megh R. Goyal, PhD, P.E., Junaid Ahmad Malik, PhD, Anu Kumari, MTech.
Other titles: Functional foods (Palm Bay, Fla.)
Names: Goyal, Megh R., editor. | Malik, Junaid Ahmad, 1987- editor. | Kumari, Anu, editor.
Series: Innovations in plant science for better health.
Description: First edition. | Series statement: Innovations in plant science for better health : from soil to fork | Includes bibliographical references and index.
Identifiers: Canadiana (print) 20240341406 | Canadiana (ebook) 20240341422 | ISBN 9781774915363 (hardcover) | ISBN 9781774915370 (softcover) | ISBN 9781032684055 (ebook)
Subjects: LCSH: Functional foods. | LCSH: Plant antiviral agents. | LCSH: Probiotics. | LCSH: Nutrition.
Classification: LCC QP144.F85 F86 2025 | DDC 613.2—dc23

Library of Congress Cataloging-in-Publication Data

CIP data on file with US Library of Congress

ISBN: 978-1-77491-536-3 (hbk)
ISBN: 978-1-77491-537-0 (pbk)
ISBN: 978-1-03268-405-5 (ebk)

OTHER BOOKS ON PLANT SCIENCE FOR BETTER HEALTH BY APPLE ACADEMIC PRESS, INC.

Book Series: *Innovations in Plant Science for Better Health: From Soil to Fork*
Editor-in-Chief: Hafiz Ansar Rasul Suleria, PhD

- **Assessment of Medicinal Plants for Human Health: Phytochemistry, Disease Management, and Novel Applications**
  Editors: Megh R. Goyal, PhD, and Durgesh Nandini Chauhan, MPharm
- **Bioactive Compounds from Multifarious Natural Foods for Human Health**
  Editors: Hafiz Ansar Rasul Suleria, PhD, Megh R. Goyal, PhD, and Huma Bader Ul Ain
- **Bioactive Compounds from Plant Origin: Extraction, Applications, and Potential Health Claims**
  Editors: Hafiz Ansar Rasul Suleria, PhD, and Colin Barrow, PhD
- **Bioactive Compounds of Medicinal Plants: Properties and Potential for Human Health**
  Editors: Megh R. Goyal, PhD, and Ademola O. Ayeleso
- **Cereals and Cereal-Based Foods**
  Editors: Megh R. Goyal, PhD, Kamaljit Kaur, PhD, Jaspreet Kaur, PhD
- **Curative and Preventive Properties of Medicinal Plants: Research on Disease Management and Animal Model Studies**
  Editors: Megh R. Goyal, PhD, Junaid Ahmad Malik, PhD, and Ademola Olabode Ayeleso, PhD
- **Health Benefits of Secondary Phytocompounds from Plant and Marine Sources**
  Editors: Hafiz Ansar Rasul Suleria, PhD, and Megh Goyal, PhD
- **Herbal Medicines and Nutritional Supplements for Health Benefits**
  Editors: Megh R. Goyal, PhD, and Anamika Chauhan, PhD
- **Herbs, Spices, and Medicinal Plants for Human Gastrointestinal Disorders**
  Editors: Megh R. Goyal, PhD, Preeti Birwal, PhD, and Durgesh Nandini Chauhan, MPharm
- **Himalayan Medicinal Plants for the Treatment of Depression: A Source of Rich Antidepressant Agents**
  Editors: Abdur Rauf, PhD, Prabhakar Semwal, PhD, and Hafiz Ansar Rasul Suleria, PhD

- **Himalayan Medicinal Plants for the Treatment of Depression: A Source of Rich Antidepressant Agents**
  Editors: Abdur Rauf
- **Human Health Benefits of Plant Bioactive Compounds: Potentials and Prospects**
  Editors: Megh R. Goyal, PhD, and Hafiz Ansar Rasul Suleria, PhD
- **Phytochemical and Pharmacological Investigation of the Family Rutaceae**
  Editors: Abdur Rauf, PhD, Hafiz Ansar Rasul Suleria, PhD, and Syed Muhammad Mukarram Shah, PhD
- **Phytochemicals and Medicinal Plants in Food Design: Strategies and Technologies for Improved Healthcare**
  Editors: Megh R. Goyal, PhD, Preeti Birwal, PhD, and Santosh K. Mishra, PhD
- **Phytochemicals from Medicinal Plants: Scope, Applications, and Potential Health Claims**
  Editors: Hafiz Ansar Rasul Suleria, PhD, Megh R. Goyal, PhD, and Masood Sadiq Butt, PhD
- **Plant- and Marine-Based Phytochemicals for Human Health: Attributes, Potential, and Use**
  Editors: Megh R. Goyal, PhD, and Durgesh Nandini Chauhan, MPharm
- **Plant-Based Bioactive Compounds and Food Ingredients: Encapsulation, Functional, and Safety Aspects**
  Editors: Junaid Ahmad Malik, PhD, Megh R. Goyal, PhD, Preeti Birwal, PhD, and Ritesh B. Watharkar, PhD
- **Plant-Based Functional Foods and Phytochemicals: From Traditional Knowledge to Present Innovation**
  Editors: Megh R. Goyal, PhD, Arijit Nath, PhD, and Hafiz Ansar Rasul Suleria, PhD
- **Plant Secondary Metabolites for Human Health: Extraction of Bioactive Compounds**
  Editors: Megh R. Goyal, PhD, P. P. Joy, PhD, and Hafiz Ansar Rasul Suleria, PhD
- **The Functional Foods: Nutrient and Health Benefits**
  Editors: Megh R. Goyal, PhD, Junaid Ahmad Malik, PhD, and Anu Kumari
- **The Role of Phytoconstitutents in Healthcare: Biocompounds in Medicinal Plants**
  Editors: Megh R. Goyal, PhD, Hafiz Ansar Rasul Suleria, PhD, and Ramasamy Harikrishnan, PhD
- **The Therapeutic Properties of Medicinal Plants: Health-Rejuvenating Bioactive Compounds of Native Flora**
  Editors: Megh R. Goyal, PhD, PE, Hafiz Ansar Rasul Suleria, PhD, Ademola Olabode Ayeleso, PhD, T. Jesse Joel, and Sujogya Kumar Panda

About Senior-Technical-Editor-in-Chief

Megh R. Goyal, PhD, PE

Retired Professor in Agricultural and Biomedical Engineering, University of Puerto Rico, Mayaguez Campus; Senior Acquisitions Editor, Biomedical Engineering and Agricultural Science, Apple Academic Press, Inc.

Megh R. Goyal, PhD, PE, is a Retired Professor in Agricultural and Biomedical Engineering from the General Engineering Department in the College of Engineering at the University of Puerto Rico–Mayaguez Campus. During his professional career, he has worked as a Soil Conservation Inspector; Research Assistant at Haryana Agricultural University and Ohio State University; Research Agricultural Engineer/Professor at the Department of Agricultural Engineering of UPRM; and Professor of Agricultural and Biomedical Engineering in the General Engineering Department of UPRM. He spent a one-year sabbatical leave in 2002–2003 at the Biomedical Engineering Department of Florida International University, Miami, USA.

Dr. Goyal was the first agricultural engineer to receive the professional license in agricultural engineering from the College of Engineers and Surveyors of Puerto Rico. In 2005, he was proclaimed the "Father of Irrigation Engineering in Puerto Rico for the Twentieth Century" by the American Society of Agricultural and Biological Engineers, Puerto Rico Section, for his pioneering work on micro irrigation, evapotranspiration, agroclimatology, and soil and water engineering.

During his professional career of 52 years, he has received many awards, including Scientist of the Year, Membership Grand Prize for the American Society of Agricultural Engineers Campaign, Felix Castro Rodriguez Academic Excellence Award, Man of Drip Irrigation by the Mayor of Municipalities of Mayaguez/Caguas/Ponce and Senate/Secretary of Agriculture of ELA, Puerto Rico, and many others. He has been recognized as one of the experts "who rendered meritorious service for the development of [the] irrigation sector in India" by the Water Technology Centre of Tamil Nadu Agricultural University in Coimbatore, India, and ASABE who bestowed on him the 2018 Netafim Microirrigation Award.

Dr. Goyal has authored more than 200 journal articles and edited more than 130 books. AAP has published many of his books, including *Management of Drip/Trickle or Micro Irrigation; Evapotranspiration: Principles and Applications for Water Management;* ten-volume set on *Research Advances in Sustainable Micro Irrigation.* He has also authored the textbooks *Elements of Agroclimatology* (Spanish) by UNISARC, Colombia, and two *Bibliographies on Drip Irrigation.* Dr. Goyal has also developed several book series with AAP, including Innovations in Agricultural & Biological Engineering, Innovations and Challenges in Micro Irrigation; and Innovations in Plant Science for Better Health: From Soil to Fork.

Dr. Goyal received his BSc degree in Engineering from Punjab Agricultural University, Ludhiana, India, and his MSc and PhD degrees from the Ohio State University, Columbus, Ohio, USA. He also earned a Master of Divinity degree from the Puerto Rico Evangelical Seminary, Hato Rey, Puerto Rico, USA.

About the Editors

Junaid Ahmad Malik, PhD
Lecturer, Department of Zoology, Government Degree College, Bijbehara, Kashmir, Jammu and Kashmir, India

Junaid Ahmad Malik, PhD, is a Lecturer with the Department of Zoology at Government Degree College, Bijbehara, Kashmir (J&K), India, and is actively involved with teaching and research activities. He has more than eight years of research experience. His areas of interest are ecology, soil macrofauna, wildlife biology, conservation biology, etc. Dr. Malik has published over 20 research articles and technical papers in international peer-reviewed journals and has authored and edited books, book chapters, and more than 10 popular editorial articles. He also serves as an editor and reviewer of several journals. He has participated in several state, national, and international conferences, seminars, workshops, and symposia and has more than 20 conference papers to his credit. He is a life member of the Society for Bioinformatics and Biological Sciences.

Dr. Malik received a BSc (2008) in Science from the University of Kashmir, Srinagar, Jammu and Kashmir; MSc (2010) in Zoology from Barkatullah University, Bhopal, Madhya Pradesh, India; and PhD (2015) in Zoology from the same university. He completed his BEd program in 2017 at the University of Kashmir, Srinagar, J&K.

Anu Kumari, MTech

Assistant Professor, Warner College of Dairy Technology, Sam Higginbottom University of Agriculture, Technology, and Sciences (SHUATS), Prayagraj, Uttar Pradesh, India

Anu Kumari, MTech, is working as an Assistant Professor (Dairy Engineering) in the Department of Dairy Engineering, Warner College of Dairy Technology, Sam Higginbottom University of Agriculture, Technology, and Sciences, Prayagraj, Uttar Pradesh, India.

She is the recipient of the Gold Medal (2011) awarded by SHUATS for excellent performance in academics during the graduate program, NDRI fellowship (2011–2013). She secured 2nd position in a dairy product judging contest organized by Anand Agriculture University (AAU), Anand.

She is currently working in the area of development of biocomposites for food and dairy products, process mechanization of dairy equipments, mathematical modeling, image analysis technique, rheology of food and dairy products, by-product utilization of agricultural commodity, and baking of food and dairy products. She has served Dairy Science College, KVAFSU, Hebbal, Bengaluru, as an Assistant Professor. She has guided 15 postgraduate students and co-guided several master scholars in dairy and food technology. She has participated in several national and international conferences and seminars. She has served Birsa Agriculture University, Ranchi; Baba Sahab Dr. Bhim Rao Ambedkar College of Agriculture Engineering and Technology, Chandra Shekhar Azad University of Agriculture and Technology, Kanpur, Etawah Campus; UPPSC (Uttar Pradesh Public Service Commission) as an external examiner and paper setter for dairy engineering courses.

She has received the Best Research Paper Award (2016) by SHUATS, Prayagraj; the Best Paper Award (2019) by ICAR-NDRI, Bengaluru; the Young Scientist Award (2019) by Agricultural Technology Development Society (ATDS), Ghaziabad. She has successfully completed training of 15 days at Feeder Balancing Dairy, Patna Dairy Project (Sudha Dairy), Bihar, in all the activities related to microbiological and chemical analysis and production and storage of milk and milk products; one month at Experimental Dairy of National Dairy Research Institute, Karnal, Haryana in all the activities related to Market Milk section, Cheese and Butter section,

Condensing and Drying Section, and Quality Control Laboratory; four months at Mother Dairy Fruits and Vegetable Pvt. Ltd., Patparganj, New Delhi in all the activities related to Q.A. lab, Dock lab, and Process and Utilities and 21 days CAFT Training on Dairy and Food Process Engineering: Equipment, Processing, and Value Addition organized by Dairy Engineering Division at ICAR-NDRI, Karnal.

She has more than 50 publications including, 13 papers, two books, five book chapters, two popular articles, four articles, and 26 abstracts in compendium of various seminars/conferences. She is a life member of IDEA (Indian Dairy Engineering Association); and Agricultural Technology Development Society (ATDS), Ghaziabad.

She did her graduation (2011) in Dairy Technology from Sam Higginbottom University of Agriculture, Technology, and Sciences (SHUATS), Prayagraj, India; and MTech (2013) in Dairy Engineering from ICAR-National Dairy Research Institute, Karnal, India.

Contents

Contributors

Ashok Kumar Agarwal
Chairman, Department of Medical Research and Development, PetriMed CA, River Engineering Pvt. Ltd., Greater Noida, Uttar Pradesh, India

Tarun Kanti Bandyopadhyay
Assistant Professor, Chemical Engineering Department, National Institute of Technology (NIT), Agartala, Tripura, India

Nabanita Baruah
PhD Research Scholar, Department of Bioengineering and Technology, Gauhati University, Guwahati, Assam, India

Beauty Basumatary
PhD Research Scholar, Department of Bioengineering and Technology, Gauhati University, Guwahati, Assam, India

Anindita Nandi Batabyal
Adhoc Assistant Professor, PG Department of Home Science, Sardar Patel University, Anand, Gujarat, India

Biswanath Bhunia
Assistant Professor, Bioengineering Department, National Institute of Technology (NIT), Agartala, Tripura, India

Rajkrishna Biswas
Assistant Professor, Department of Zoology, Dukhulal Nibaran Chandra College, Murshidabad, West Bengal, India

Savita Budhwar
Assistant Professor, Department of Nutrition Biology, Central University of Haryana, Mahendergarh, Jant Pali, Haryana, India

Manali Chakraborty
PhD Research Scholar, Department of Nutrition Biology, Central University of Haryana, Mahendergarh, Jant Pali, Haryana, India

Soumi Chakraborty
Assistant Professor, Amity Institute of Food Technology (AIFT), Amity University, Noida, Uttar Pradesh, India

Hariswami Das
Assistant Teacher, Department of Life Science, Sovanagar High School, Malda, Murshidabad, West Bengal, India

Sumi Das
PhD Research Scholar, Department of Bioengineering and Technology, Gauhati University, Guwahati, Assam, India

Manab Deka
Professor, Department of Bioengineering and Technology, Gauhati University, Guwahati, Assam, India

Edible S. Erten
Lecturer, Department of Food Engineering, Faculty of Engineering, Aydın Adnan Menderes University, Zafer Mah. Aytepe Mevkii. Efeler, Aydın, Turkey

Tayyibe Erten
Assistant Professor, Department of Nutrition and Dietetics, Faculty of Health Science, Bayburt University, Gencosman Mah., Bayburt, Turkey

Moushumi Ghosh
Professor, Department of Biotechnology and Coordinator for Food Security Center of Excellence, Thapar Institute of Engineering and Technology (TIET), Patiala, Punjab, India

Nupur Ghosh
Researcher, Department of Research and Development, Aseptic Laboratories, Jangambari, Varanasi, Uttar Pradesh, India

Pijush Kanti Ghosh
Director, Aseptic Laboratories, Jangambari, Varanasi, Uttar Pradesh, India

Probir Kumar Ghosh
Head, Department of Research and Development, Aseptic Laboratories, Jangambari, Varanasi, Uttar Pradesh, India

Megh R. Goyal
Retired Professor, Agricultural and Biomedical Engineering, General Engineering Department, College of Engineering at the University of Puerto Rico – Mayaguez Campus, USA

Nishant Gupta
R&D Officer, Department of Medical Research and Development, PetriMed CA, River Engineering Pvt. Ltd., Greater Noida, Uttar Pradesh, India

Poonam Gusain
Scientist-B, Uttarakhand State Council for Science and Technology, Dehradun, Uttarakhand, India

Arif Jaman
PhD Research Scholar, Department of Biosciences and Bioengineering, Indian Institute of Technology Bombay, Mumbai, Maharashtra, India

Jikky Jayakumar
PhD Research Scholar, Department of Food Process Engineering, Agricultural Engineering College and Research Institute, Tamil Nadu Agricultural University, Coimbatore, Tamil Nadu, India

Jayasree T. Joshi
PhD Research Scholar, Agricultural and Food Engineering Department, Indian Institute of Technology (IIT), Kharagpur, West Bengal, India

Kirti Joshi
Assistant Professor, Department of Food and Biotechnology, Jayoti Vidyapeeth Women's University, Vedant Gyan Valley, Village Jharna, Jaipur, Rajasthan, India

Harshpinder Kaur
PhD Research Scholar, Department of Biotechnology, Thapar Institute of Engineering and Technology (TIET), Patiala, Punjab, India

Arya S. Krishnan
Assistant Professor, Department of Food Technology, Amal Jyothi College of Engineering, Kanjirappally, Kottayam, Kerala, India

Suneel Kumar
Professor, Department of Physics and Astrophysics, Central University of Haryana, Mahendergarh, Jant Pali, Haryana, India

Pinki Malick
PhD Research Scholar, Department of Bioengineering and Technology, Gauhati University, Guwahati, Assam, India

V. Manoja
Research Scholar, Department of Food Process Engineering, Agricultural Engineering College and Research Institute, Tamil Nadu Agricultural University, Coimbatore, Tamil Nadu, India

Mallika Mazumder
Independent Researcher, Department of Botany, Dukhulal Nibaran Chandra College, Murshidabad, West Bengal, India

Meghali Nag
PhD Research Scholar, Department of Bioengineering and Technology, Gauhati University, Guwahati, Assam, India

Sebnem Ozturkoglu-Budak
Associate Professor, Department of Dairy Technology, Ankara University, Dıskapı, Ankara, Turkey

Kaninika Paul
Assistant Professor, Department of Food Safety, Quality and Nutrition College of Food Technology, MIT-ADT University, Rajbaugh Loni Kalbhor, Pune, Maharashtra, India

Anita Raisagar
PhD Research Scholar, Department of Dairy Microbiology, SHUATS, Prayagraj, Uttar Pradesh, India

Kandarpa Kumar Saikia
Associate Professor, Department of Bioengineering and Technology, Gauhati University, Guwahati, Assam, India

Anup Kumar Sarkar
Assistant Professor, Department of Botany, Dukhulal Nibaran Chandra College, Murshidabad, West Bengal, India

Anjaly M. Shanker
PhD Research Scholar, Department of Agriculture and Environmental Sciences, National Institute of Food Technology Entrepreneurship and Management (NIFTEM), Sonipat, Haryana, India

Ramesh Sharma
PhD Research Scholar, Bioengineering Department, National Institute of Technology (NIT), Agartala, Tripura, India

Sangeeta Shukla
Assistant Professor, Department of Dairy Microbiology, SHUATS, Prayagraj, Uttar Pradesh, India

Fatima Siddiqua
PhD Research Scholar, Department of Bioengineering and Technology, Gauhati University, Guwahati, Assam, India

Manisha Singh
PhD Research Scholar, Department of Food Technology, Shaheed Rajguru College of Applied Science for Women, University of Delhi, Vasundhara Enclave, Delhi, India

Ranjana Singh
Professor, Department of Food Technology, Shaheed Rajguru College of Applied Science for Women, University of Delhi, Vasundhara Enclave, Delhi, India

P. Sudha
Assistant Professor, Department of Food Process Engineering, Agricultural Engineering College and Research Institute, Tamil Nadu Agricultural University (TNAU), Coimbatore, Tamil Nadu, India

Areeba Tariq
Research Scholar, Department of Dairy Technology, Ankara University, Dıskapı, Ankara, Turkey

Abbreviations

°C	Celsius
μM	micrometer
3D	three-dimensional
5,6-DMBI	5,6-dimethylbenzimidazole
AAS	atomic absorption spectroscopy
ABTS	2,2'-azino-bis(3-ethylbenzothiazoline-6-sulfonic acid)
ACE2	angiotensin-converting enzyme 2
ADA	American Diabetes Association
ADI	acceptable daily intake
ADLCbl	adenosyl cobalamin
ADV	adenoviruses
AFS	absorption and fluorescence spectroscopy
AI	adequate intake
AI	artificial intelligence
ALA	alpha-linolenic acid
AM	additive manufacturing
AMPK	AMP-activated protein kinase
ANS	additives and nutrient sources
Anti-DGP	anti-deamidated gliadin peptides
Anti-tTG	anti-tissue transglutaminase
ASE	accelerated solvent extraction
ATP	adenosine triphosphate
AYUSH	Ayurveda, Yoga and Naturopathy, Unani, Siddha, and Homeopathy
BAMC	British Analytical Methods Committee
BHA	butylated hydroxyanisole
BHT	butylated hydroxytoluene
BM	breastmilk
BSCP	crude polysaccharides of bamboo shoot
BSH	bile salt hydrolysis
$C_{24}H_{28}O_4$	cis-norbixin
$C_{25}H_{30}O_4$	methyl hydrogen 9'-cis-6,6'-diapocarotene-6,6'-dioate
Ca	calcium
CAGR	compound annual growth rate

CATA	check-all-that-apply
CD	coeliac disease
CE	capillary electrophoresis
CFE	cell free extract
CFE	critical fluid extraction
CFS	cell free supernatant
CFU	colony forming units
CHD	coronary heart disease
CL	chemiluminescence
CM	cow's milk
CMOS	complementary metal oxide semiconductor
CNCbl	cyanocobalamin
CNS	central nervous system
CO_2	carbon dioxide
COS	cello-oligosaccharides
COVID-19	corona virus disease-19
COX-2	cyclooxygenase-2
CP	cold pressing
CP	creatine phosphate
CSD	conventional steam diffusion
CV	cultivation variety
CVB1	coxsackie B
CVD	cardiovascular disease
DBS-LDHs	dodecylbenzene sulfonate–layered double hydroxides
DCPIPH	2,6-dichlorophenol indophenol
DENV	dengue virus
DFS	double fortified salt
DH	dermatitis herpetiformis
DHA	docosahexaenoic acid
DKA	diabetic ketoacidosis
DNA	deoxyribonucleic acid
DPPH	2,2-diphenyl-1-picrylhydrazyl
dsRNA	double stranded RNA
DTPA	diethylene diamine pentacetate
E102	tartrazine
E106b	bixin
E110	sunset yellow
E129	Allura red
EBM	exclusive breastmilk

EFSA	European Food Safety Authority
ELISA	enzyme-linked immunosorbent assay
EMA	anti-endomysial antibody
EO	essential oil
EPA	eicosapentaenoic acid
EU	European Union
EV71	enterovirus 71
FAO	Food and Agriculture Organization
FDA	Food and Drug Administration
FFC	functional food center
FOSs	fructo-oligosaccharides
FRAP	ferric reducing ability of plasma
FS	fluorescence spectroscopy
FSSAI	Food Safety and Standards Authority of India
FTIR	Fourier transform infrared
FUFOSE	Functional Food Science in Europe
GA	gluten ataxia
GABA	γ-aminobutyric acid
GC	gas chromatography
GC-MS	gas chromatography–mass spectroscopy
Gd	gadolinium
GFD	gluten-free diet
GI Tract/GIT	gastrointestinal tract
GI	gastrointestinal
GI	glycemic index
GoI	Government of India
GOS	galacto-oligosaccharides
GP	glutathione peroxidase
GRAS	generally recognized as safe
gRNA	genomic RNA
H_9N_2	influenza A Virus
HBV	hepatitis B virus
HC	haptocorrin
hCoV-OC_{43}	human coronavirus OC_{43}
HCV	hepatitis C virus
HD	hydro-distillation
HINBD	4-N, N-di (2-hydroxyethyl) imino-7-nitrobenzo-2-oxa-1,3-diazole
HIV	human immunodeficiency virus

HMO	human milk oligosaccharides
HMW	high molecular weight
HPE	high pressure extraction
HPHT	high-pressure–high-temperature extraction
HPLC	high-performance liquid chromatography
HR	*Hippophae rhamnoides*
HSV	herpes simplex virus
IBD	inflammatory bowel disease
IBS	irritable bowel syndrome
ICPMS	inductively coupled plasma mass spectroscopy
ICR	insulin to carbohydrate ratio
IDF	insoluble dietary fraction
IF	intrinsic factor
IFIC	International Food Information Council
IFN	interferon
IFT	Institute of Food Technologists
IgA	immunoglobulin A
IgE	immunoglobulin E
IgG	immunoglobulin G
IL	interleukin
IL-12	interleukin-12
IL-1β	interleukin-1
IL-4	interleukin 4
IPNI	International Plant Names Index
ISAPP	International Scientific Association for Probiotics and Prebiotics
JAR	just about right
JECFA	Joint FAO/WHO Expert Committee on Food Additives
KA	Karmali agar
KCN	potassium cyanide
kHz	kilohertz
LAB	lactic acid bacteria
LAMP	loop-mediated isothermal amplification
LC	liquid chromatography
LDL	low density lipoprotein
LDPE	low-density polyethylene
LED	light emitting diode
LFA	lateral flow immunoassay
LLE	liquid–liquid extraction

LMW	low molecular weight
LOD	limit of detection
MAE	microwave-assisted extraction
MAHD	microwave-assisted hydro-distillation
MALDI-TOF	matrix-associated laser desorption/ionization-time of flight
MAP	modified atmosphere packaging
MASD	microwave-assisted steam distillation
MBA	microbiological assay
MERS	Middle East Respiratory syndrome
MFGM	milk fat globule membranes
mg	milligram
MHD	microwave hydro-diffusion
MHDG	microwave hydro-diffusion and gravity
MN	micronutrient
MNPs	magnetic nanoparticles
MoA	mode of action
MPa	megapascal
Mpro	main protease
mRNA	messenger RNA
MRS	deMann Rogosa Sharpes medium
MS	mass spectrometry
MSD	microwave steam distillation
MTLCbl	methyl cobalamin
NCD	non-communicable disease
NCGS	non-coeliac gluten sensitivity
NF-κB	nuclear factor-κB
NIR	near infrared
nm	nanometer
NMMP	National Mission on Medicinal Plants
NMR	nuclear magnetic resonance
NO	nitrogen oxide
OHCbl	hydro oxycobalamin
OPD	o-phenylenediamine
OPW	orange peel wastes
OSSO	*Ocimum sanctum* seed oil
OTC	over-the-counter
OTR	oxygen transmission rate
P	phosphorus

PAMPs	pathogen-associated molecular patterns
PBMC	peripheral blood mononuclear cells
PBS	phosphate-buffered saline
PCA	principal component analysis
PCr	phosphocreatine
PCR	polymerase chain reaction
PDMS	plasma desorption mass spectrometry
PDX	polydextrose
PEF	pulse electric field
PET	polyethylene terephthalate
PFU	plaque forming unit
pH	potential of hydrogen
PLE	pressurized liquid extraction
PLL	poly-L-lysine
PLpro	papain-like protease
PMMA	polymethyl methacrylate
pNpG	para nitrophenyl alpha d glucopyranoside
POS	pectic oligosaccharides
POWO	plant of the world online
PRRs	pattern recognition receptors
psi	pound per square inch
PSR	polymerase spiral reaction
qPCR	quantitative PCR
QPS	qualified presumption of safety
RASSF	rapid alert system for food and feed
RDA	radioisotope dilution assay
RDA	recommended daily allowance
RDA	recommended dietary allowance
RdRP	RNA-dependent RNA polymerase
RNA	ribonucleic acid
RP	rapid prototyping
RPA	recombinase polymerase amplification
RS	Raman spectroscopy
RSV	respiratory syncytial virus
RTI	respiratory tract infection
RT-PCR	real time polymerase chain reaction
SA	streptavidin
SARS CoV-2	severe acute respiratory syndrome corona virus
SARS	severe acute respiratory syndrome

SC-CO_2	supercritical carbon-dioxide extraction
SCF	super critical fluid
SCFA	short chain fatty acids
SCFCO_2	supercritical fluid carbon dioxide
SCWE	supercritical water extraction
SD	steam distillation
SDF	soluble dietary fraction
SDS-PAGE	sodium dodecyl polyacrylamide gel electrophoresis
SE	solvent extraction
SERS	surface enhanced Raman scattering
SFM	solvent free microwave
SFME	solvent free microwave extraction
SMPB	state medicinal plant board
SNF	solids-not-fat
sp.	single species
SPC	self-priming compartmentalization
SPE	solid phase extraction
spp.	two or more species
SPR	surface plasma resonance
ss-RNA	single-stranded RNA
T1D	type 1 diabetes mellitus
T2D	type 2 diabetes mellitus
TBHQ	tert-butyl hydro quinone
TNF	tumor necrosis factor
TNF-α	tumor necrosis factor-alpha
TPC	total phenolic content
Trolox	6-hydroxy-2,5,7,8-tetramethylchromane-2-carboxylic acid
UACE	ultrasound assisted Clevenger extraction
UAE	ultrasound assisted extraction
UCNPs	up-conversion nanoparticles
UMPDB	Uttarakhand medicinal plants database
UPLC	ultra-performance liquid chromatography
USFDA	United States Food Safety Authority
UTI	urinary tract infection
UV	ultraviolet
VFA	volatile fatty acids
VHH	variable heavy chain only antibody
Vit B12	vitamin B12
W	watt

WA	wheat allergy
WHO	World Health Organization
XOS	xylo-oligosaccharide
α	alpha
β	beta
γ	gamma
δ	delta

Preface

Food processing has remained one of the most vibrant areas for exploring different frontiers, from the very ancient ages. As such, food science is an integral part of agricultural sciences and as the world aims for nutritional security, food technologists/researchers/scientists/industry people continue to explore further areas of food science.

Functional foods have become more a preferable choice these days within the health and wellness communities. Functional foods, also referred to as nutraceuticals, are extremely nutrient-dense foods and have been linked to a range of potent nutrient and health advantages. For instance, they could shield against illnesses, stop vitamin deficits, and encourage healthy growth and development.

This book details the scope, importance, and nutritional and health benefits of functional foods. The book summarizes the functional activity of plants, and food formulation for diabetic people, sports persons, and old age people. It also introduces some medicinal plants having antiviral, and antidiabetic activity (such as Mushrooms, Tulsi, Medicinal plants, Guava, etc.), the role of probiotic and prebiotic foods on human health, extraction methods for phytochemicals from plants and their derivatives.

There are five key sections to this edited book.

In the first part, the functional activities of plants and their derivatives are discussed. This part focuses on microgreens as functional foods, food formulation for sports persons, functional foods for geriatrics, and medicinal plants and their cultivation methods.

Chapter 1 titled, Microgreens as Functional Food: A Crop of Modern Agriculture, emphasizes the recent research findings on compositional variations and health benefits of various microgreens. Chapter 2 titled, Food Formulation for Sports Persons, elucidates on the recent advancement in sports foods and supplements that include all vitamins, minerals, herbs, amino acids, and enzymes and these can supplement whole foods to enhance the sports performance of an athlete. Chapter 3 titled, Functional Foods for Geriatrics, focuses on the relevance and development of appropriate food products for geriatrics to improve their nutrition, health, and well-being. Geriatrics focuses on the unique requirements of older people, emphasizing functional independence even in the presence of chronic illness. Chapter

4 titled, Cultivation of Highly Demanded Medicinal Plants: Case Study in Uttarakhand, India, elaborates on cultivation, awareness, government subsidies, and traditional uses.

In the second part, antiviral activities of functional foods are discussed (such as against COVID-19). This part emphasizes the importance of Tulsi, functional food formulation, and fortification to combat pandemic situations. Chapter 5 titled, Antiviral Activities of *Ocimum* Species (Tulsi) to Combat RNA Virus-Mediated Pandemic Situations: COVID-19, highlights anti-RNA virus efficacy and the practical significance of their bioactive compounds to combat epidemic and pandemic situations (COVID-19; SARS-CoV-2). Chapter 6 titled, Functional Foods: Formulation, Fortification, and Combating COVID-19, deals with the development of fortified foods based on micronutrients (MNs), dietary fiber, probiotics, and others for diabetic, geriatric, and infant groups. It also emphasizes the formulation of various foods related to sports persons and COVID patients.

In the third part, the therapeutic and antidiabetic activities of mushrooms and guava pulp are discussed. Chapter 7 titled, Potential Therapeutics of Cultivated and Wild Mushrooms Against Diabetes, enlightens on the importance of mushrooms as anti-diabetic agents along with future prospects, the obstacles and limits that can impact the development of mushroom-based anti-diabetic drugs. Chapter 8 titled, Antidiabetic and Nutritional Properties of Guava (*Psidium guavaja* L.) Pulp (Fermented with *Lactococcus lactis*)" deals with the functional and nutritional properties of guava pulp with reference to diabetics.

The fourth part of this edited book elaborates on the role of probiotics as a functional food for our health. This part describes the role of probiotics on gut microflora, types, mechanism of action, and application of postbiotics, health effects of probiotic, prebiotic, and synbiotic dietary supplements, extraction methods, and health effects of plant-based phytochemicals. Chapter 9 titled, Role of Probiotics on Gut Microflora, focuses on the history and concept of probiotics and the therapeutic values of probiotics. Chapter 10 titled, Probiotics, Prebiotics, and Synbiotics as Dietary Supplements: A Healthy Approach, discusses the potential role of dietary supplements, probable product formulation, processing, and preservation of products. Chapter 11 titled, Postbiotics: Types, Mechanism of Action, and Applications, provides an in-depth review of postbiotics including their types, methods of preparation, their functionality or utilization, and their role of action at the target site. Chapter 12 titled, Extraction Methods for Essential Oils From Citrus Peels, studies the emerging oil extraction techniques from citrus peels and compares these with conventional techniques with respect to extraction duration, efficiency, and biocomponents of essential oil. Chapter 13 titled, Extraction and Role of Vitamin B12 in Food Products: Health Effects, explores extraction, estimation, and stability analysis of vitamin B12 with the application of food-grade microorganisms for *in-situ* production of vitamin B12. Chapter 14 titled, Extraction Methods for Phytochemicals from Annatto (*Bixa orellana* L.): Natural Food Colorant, focuses on the nature of annatto, its chemical composition, and the enhanced technologies for extraction of bixin and norbixin from annatto.

Quality aspects of functional foods, the fifth part of this book, provides an overview of foodborne pathogens and food-related microorganisms. It also describes the importance of gluten-free foods and the introduction of legume-based gluten-free bakery food products and their quality aspects. Chapter 15 titled, Foodborne Pathogens and Food-Related Microorganisms: An Overview, collates the recent research findings on foodborne pathogens and discusses their impact on our present understanding of microbial utilization and risk assessment. It also explores foodborne diseases and related microorganisms. The replacement of food gluten is essential for the development of gluten-free foods. To achieve this, a variety of flours (corn flour, chestnut flour, and buckwheat flour) have been investigated by food scientists. Chapter 16 titled, Sensory Analysis and Consumer Acceptance of Legume-Based Gluten-Free Bakery Products, deals with the recent developments in legume-based gluten-free bakery products.

—Editors

PART I

FUNCTIONAL ACTIVITIES OF PLANTS AND THEIR DERIVATIVES

CHAPTER 1

Microgreens as Functional Food: A Crop of Modern Agriculture

Arya S. Krishnan, Jayasree T. Joshi, and Anjaly M. Shanker

ABSTRACT

Microgreens are relatively small, tender, immature, and nutrient-dense plants that can be easily cultivated indoors and are easily available throughout the year. They offer higher nutritional value than the sprouts and other mature parts as well as distinct sensory characteristics. They contain greater concentration of nutritive compounds including Vitamin C, tocopherols, phylloquinone, and carotenoids than in their mature plant parts. Shelf-life of fresh-cut microgreens is significantly influenced by both pre-harvest and post-harvest interventions. Additionally, microgreens' packaging has an impact on their post-harvest quality and shelf-life. These have anti-cancer, anti-inflammatory, anti-microbial, anti-hyperglycemia properties, which can be increased by their inclusion in both *in vitro* and *in vivo* studies. Because of their distinctive sensory characteristics (such as flavor, tender texture, and sparkling colors), they have become increasingly popular in the culinary world. Despite the presence of diverse chemical conformation, lack of *in vivo* and clinical trials limits the application of microgreens and requires further research studies to boost their use. A review of recent research findings on the nutritional composition and health benefits of various microgreens is presented in this chapter.

1.1 INTRODUCTION

In response to the rapid urbanization of the world's population, there is an ever-increasing demand for more sustainable, reliable, accessible,

---

The Functional Foods: Nutrient and Health Benefits.
Megh R. Goyal, Junaid Ahmad Malik, and Anu Kumari (Eds.)

convenient, and wholesome foods. The cultivation of plants in a closed environment under controlled lighting and irrigation is one of the most desirable techniques for indoor farming. Microgreens are among the most espoused crops of controlled environment agriculture. Moreover, current experimental evidence supports the health-protective effects of a diet high in plant-derived foods. Vegetable families (such as the Apiaceae, Amaranthaceae, etc.) have exhibited health-promoting properties due to the functional bioactive compounds that are present in edible parts of the plant [22].

This chapter emphasizes the recent research findings on compositional variations and health benefits of various microgreens.

1.2 SPROUTS AND MICROGREENS

When compared to sprouts and other mature parts, microgreens have higher nutritional content and distinct sensory attributes [33]. Numerous crops (such as cereals, legumes, oilseeds, pseudo-cereals, herbs, and vegetables) can be used to produce sprouts and microgreens [66]. Microgreens can be consumed raw, incorporated into a wide range of dishes, or blended into juices and smoothies. They contain greater concentrations of bioactive constituents, such as carotenoids, ascorbic acid, phylloquinone, and tocopherols than the mature plant parts [70]. Microgreens are typically harvested seven to 14 days after germination [62]. Depending on the plant species, they are generally between 3 and 10 cm in height at the time of harvest [48]. Table 1.1 summarizes some of the most common microgreens and their growth conditions.

1.3 PRE- AND POST-HARVEST TREATMENTS

Fresh-cut microgreens are highly respiring products, as they are made up of delicate tissues, and their deterioration is linked to a stress-induced reaction rather than natural senescence [33]. Pre-harvest and post-harvest interventions have huge effects on the shelf-life of microgreens. Physiological and metabolic activities of microgreens are affected by pre-harvest surroundings (such as intensity of light, oil salinity, nutrient fortification), and presence of other environmental substrates [74].

Light treatment has proven to increase seed germination and microgreen biosynthesis [68]. Illumination can increase the bioactive components and antioxidative potential of chia microgreens significantly [42]. Dark storage

TABLE 1.1 Some of the Commonly Used Microgreens and Their Growing Conditions

Microgreen	Growing Conditions	Harvest Period	Key Findings	References
Amaranth	Greenhouse cultivation Average temperature: 26±2°C.	9 days	Higher amounts of provitamins A, α-carotene, and β-carotene observed in microgreens than in sprouts. Compared to sprouts and microgreens, matured amaranth showed increased protein, Iron, carotene, ascorbic acid, and lutein levels.	[16]
Broccoli	Temperature: 20, 25, 30°C Photoperiods: 8 hours dark and 16 hours light	7 days	With sprout development, glucoraphanin content and sulforaphane production decreased.	[19]
Buckwheat	Grown in germination equipment; ambient conditions	10 days	High amounts of carotenoids, and flavonoids, α-tocopherols observed.	[23]
Carrot	Greenhouse cultivation Average temperature: 22°C.	7 days	Lighting increased the amount of vitamin C and chlorophyll in microgreens.	[2]
Fennel	Average temperature: 25 ± 5°C Relative humidity: 65 ± 10% Photoperiod 11.5 hours Light intensity: 2,500–4,400 lux.	12–14	Has high protein content (4.4 g per 100 g), dietary fiber, calcium, potassium, sodium, and selenium.	[18]
Flaxseed	Average temperature: 20°C Photoperiod: 16 hours	14 days	Microgreens exhibited 62.6% higher chlorophyll content, 24.4% higher carotenoid, and 24.4% higher phenol content compared to sprouts.	[50]
Lettuce	Greenhouse cultivation Minimum temperature (Avg.): 7.6°C Maximum temperature (Avg.): 20.5°C Average photoperiod: 11–16 hours	2 weeks	Microgreens have greater levels of calcium, iron, magnesium, manganese, selenium, zinc, and molybdenum than mature lettuces.	[48]

TABLE 1.1 *(Continued)*

Microgreen	Growing Conditions	Harvest Period	Key Findings	References
Radish cress	Temperature: 18/22 ± 2°C Relative humidity: 60–65% Light intensity: 50 μmol m^{-2} s^{-1}	7 days	Elevated amount of flavonoids and anthocyanins observed.	[25]
Spinach	Temperature: 16–19°C Radiation: 125–159, 153 W/m	17 or 18 days	More ascorbic acid than mature spinach (71.2 mg/100 g). Tocopherol and -carotene levels are similarly higher.	[8]
Sunflower	Temperature: 25 ± 5°C	7 days	Fibers, ascorbic acid, protein, α-tocopherol, β-carotene, lutein, and chlorophyll were found in high concentrations.	[13]
Wheat	Average temperature: 19°C Relative humidity: 66%	16 days	Amount of total carotenoids in the microgreens increased significantly over time, peaking on day 16.	[44]

of radish microgreens resulted in greater carotenoid retention and hydroxyl radical scavenging than storage under light [71]. Microgreens treated with calcium before harvest has been proven to improve sensory and microbiological quality [38]. Calcium additionally plays a significant role in plant growth and development. Pre-harvest calcium treatment was able to increase microgreen biomass and reduced senescence in broccoli microgreens [31].

In wheat microgreens, salt stress has shown to enhance the amounts of bioactive compounds and antioxidant activity. Treatment with NaCl (12.5 mM) enhanced the amounts of carotene, flavonoids, phenolic acids, and vitamin C in wheat microgreen extracts. Furthermore, salt-stress generated by 25 mM NaCl treatment resulted in highest levels of anthocyanin and radical-scavenging activity of DPPH (2,2-diphenyl-1-picrylhydrazyl) [21].

Post-harvest treatments for improving the shelf-life includes disinfecting with chlorine, ozone treatment, coating, and modified atmosphere packaging (MAP) [52]. In the fresh-cut vegetable industry, chlorine-based solutions are commonly used as sanitizing and disinfecting agents. The microbiological quality of buckwheat and radish microgreens was improved after disinfecting with chlorine solution [29, 73]. In recent years, there has also been greater concern on the possible risk of undesired disinfection by-products generated, when chlorinated products get contacted with organic matter [3]. As a result, researchers have started working on alternative techniques.

Dalal et al. [12] investigated the effects of organic acids and ethanol treatment on the post-harvest characteristics of sunflower microgreens. At the end of storage, the treatments with citric acid and ascorbic acid alone and in combination retained the highest amounts of total phenols, ascorbic acid, and DPPH scavenging activity with the lowest microbial load. Treatment with citric acid solution (5% w/v) followed by an ethanol spray (50% v/v) had improved the microbiological quality of Chinese cabbage microgreens [11].

1.4 PACKAGING AND SHELF-LIFE OF MICROGREENS

The microgreens continue to respire even after being harvested. If harvesting techniques are not properly followed, then it can result in negative effects due to excessive respiration rates. One of the most important challenges for the microgreen industry is due to their low yield and short shelf-life [29]. This limits the production and marketing of microgreens on a large scale [8]. The imperative factors that affect the shelf-life of microgreen include temperature, relative humidity, packaging material, gas composition, and initial microbial load. Low-temperature storage coupled with proper packaging can prevent quality loss and can enhance the shelf-life of perishables by slowing the rate of respiration, senescence, and development of spoiling microbes [46, 60]. The storage temperature is one of the most crucial components that can impact the quality attributes of the product during storage, and it requires precise control [58].

The influence of temperature on the storage life of microgreens has been the focus of many researchers [6, 31, 46]. The effect of temperature on the shelf-life and quality of buckwheat microgreens was studied by Kou et al. [29], who found that the microgreens need to be stored at 5°C to keep their quality and shelf-life.

Packaging is another important parameter, which influences the postharvest quality and shelf-life of the microgreens. Respiration rate of the product, mechanical and barrier properties of packaging film, storage temperature, and relative humidity are key elements that influence the product package atmosphere, and thus the final quality [55]. The best strategy to protect the product quality and increase shelf-life is to use a packing film with proper oxygen transmission rate (OTR) to match the product respiration rate [73].

Modified atmospheric packaging, in which preservation is done by lowering oxygen and raising carbon dioxide (CO_2) concentrations, is one of the most often used methods for enhancing the shelf-life of microgreens [27]. Passive MAP using films with relatively high OTR has shown improved retention of postharvest quality parameters of microgreens [33]. A similar pattern was seen in radish microgreens [18, 73]. It is also necessary to select proper packaging films with minimal qualitative and quantitative changes.

Dalal and Siddiqui [12] investigated the effectiveness of using polystyrene trays for the packaging of sunflower microgreens; and noticed that the longevity and nutritional quality of the microgreens filled in the polystyrene trays were superior to those packaged in low-density polyethylene (LDPE) bags.

Edible coatings are viable alternatives for postharvest storage of microgreens. In edible coating, a thin layer of film is produced on the food surface thus preventing moisture and gas migration across the film. This slows the decay of food and improves its quality [34]. There are only few studies on the application of the edible coating on microgreens to date. Ghoora et al. [18] studied the effectiveness of spray- and dip-coating techniques for edible coating of radish and roselle microgreens. They used aloe vera gel as the coating material. Following this, the treated samples were stored in macro-perforated packaging, LDPE self-seal bags, and PET (polyethylene terephthalate) clamshell containers. Coated microgreens showed decreased postharvest degradation and greater ascorbic acid content than the untreated microgreens. In comparison to LDPE self-seal bags, PET clam-shell container samples demonstrated decreased physiological weight loss, rate of respiration, microbial counts, electrolyte leakage, and overall acceptability. The lower electrolyte leakage is attributed due to improved cell integrity [1].

1.5 NUTRITIONAL CHARACTERIZATION OF MICROGREENS

Growing number of people are becoming familiar with microgreens due to their easy cultivation methods and high nutritional content. Pesticide applications in the environment, population growth, and the need to ensure food security continuously support indoor farming of microgreens. The high nutritional profile of microgreens and their ability to prevent disease as well as health promoting characteristics make them part of the super food category. In comparison to mature plants, microgreens provide a higher level of vitamins, minerals, carotenoids, and glucosinolates.

The results of previous studies have shown that microgreens contain high amounts of polyphenolic compounds, which exhibit substantial antioxidant properties. Although wide range of microgreens are currently being cultivated, yet scientific data about their direct potential to affect human physiological functions are not widely available. Herbs and legumes can be grown as microgreens based on their nutritional content and health-promoting characteristics. Commonly cultivated taxonomic families include: Brassicaceae (broccoli, mizuna, cabbages, radishes), Fabaceae (fenugreek, sweet pea, alfalfa), Apiaceae (carrot, parsley, celery) Asteraceae, Amaranthaceae, and many others. This section focuses on discussion of the different nutritional components of microgreens, such as Vitamins, Carotenoids, Phenolic Compounds, Mineral Nutrients, Antioxidant Activity, and Glucosinolates.

1.5.1 VITAMINS

A principal class of nutrients adding nutritional value to microgreens is vitamins and their precursors. Research have revealed significant levels of tocopherol (vitamin E), carotene (provitamin A), ascorbic acid (vitamin C) and phylloquinone (vitamin K1) in microgreens when compared to their matured counter parts. Phylloquinone, which is also known as vitamin K, is widely present in microgreens. This vitamin helps make a wide range of proteins essential for blood clotting and bone development. According to the studies, phylloquinone concentrations ranged from 0.66 to 4.1 g per kg of fresh weight. Phylloquinone concentrations in broccoli microgreens were 3.2 times higher than those in mature broccoli. This is in agreement with a study that focused on the phytochemical concentrations and antioxidant capacity of Brassicaceae microgreens [74]. A high percentage of phylloquinone is present in a number of plants belonging to the Amaranthaceae, Brassicaceae, and Lamiaceae families.

Vitamin C (Ascorbic acid) is another prominent vitamin in microgreens, and it plays several biological functions, such as wound healing, collagen synthesis, and immune system regulation. High amounts of vitamin C are found in the Malvaceous, Brassicaceae, or Cucurbitaceae families. Around 10 microgreens were grown from seeds of spinach, carrot, mustard, radish, roselle, onion, fenugreek, fennel and French basil; and the nutritional profile was studied by Ghoora et al. [18]. Vitamin C content was varied from 29.9 to 123.2 mg/100 g; and of all the nutritional components analyzed, vitamin C was the most abundant. Highest concentration of particular vitamin was observed in case of Roselle, trailed by French basil, fenugreek, radish microgreens and with onion microgreens having the lowest concentration. Results revealed that the selected microgreens had 120%, 127% and 119% higher vitamin C content than their mature complements. It has been noticed that microgreens contain corresponding amounts of ascorbic content compared to citrus fruit that are acknowledged to be wealthy assets of these nutritional compounds [26].

Vitamin E (called tocopherol) is another predominant vitamin, which might be distinguished in microgreens; and is vital for the right functioning of many organs in our body. Among β-, γ-, and δ-tocopherols, γ-tocopherol is the most popular in plants and especially, Brassicaceae and Apiaceae families of microgreens are found to be rich source of these. Varietal conditions can affect the phytochemical constituents of microgreens. Xiao et al. [74] conducted a study analyzing the nutritional properties of 30 varieties of Brassicaceae microgreens; and among these microgreen varieties, higher values of alpha-tocopherol content was observed in comparison with portions of γ-tocopherol. Microgreens from ruby radish had the highest concentration of tocopherol (1.55 mg/100 g fresh weight) and those from mizuna or red kale had the lowest (0.1 mg/100 g fresh weight).

Ghoora et al. [18] reported analogous results in a study to determine the nutritional composition of some culinary microgreens. In terms of amount of α-tocopherol in fresh weight, sunflower or radish were the most abundant sources (58.6 and 48.7 mg/100 g), while mustard and spinach had the highest amounts of α-tocopherol (15-fold and 8-fold, respectively), compared to their mature counterparts. The α-Tocopherol makes up 90% of the vitamin E in human tissues and acts as an antioxidant to prevent the chain reaction of free radicals that produce more free radicals. Moreover, they can protect against oxidative damage in many tissues, as well as can influence muscle or immune functions [64].

1.5.2 CAROTENOID CONTENT

Carotenoids are phytonutrients, which are located inside the cells of huge variety of plants, algae, and bacteria. Moreover, carotenoids have an antioxidant ability that can scavenge free radicals – unstable atoms that can impair cells by reacting with other molecules. Fruits and vegetables are greater source of carotenoids, which are fat-soluble pigments. Studies conducted to reveal the nutritional profile of microgreens suggest that microgreens are excellent source of carotenoids. Some carotenoids act as precursors of Vitamin A, e.g., β-carotene.

Microgreens contain three major groups of carotenoids: β-carotene, lutein/zeaxanthin and violaxanthin. A study conducted to analyze the nutritional profile of Brassicaceae microgreens reveals that concentrations of β-carotene, lutein/zeaxanthin and violaxanthin fluctuated by 2.3, 7.9, and 5.5-folds, respectively. It has been noted that most of the Brassicaceae microgreens examined in the previous study are rich sources of carotenoids (10 mg/100 g of edible portions) with overall β-carotene, lutein/zeaxanthin and violaxanthin concentrations of 10.4–18.8 mg/100 g fresh weight.

Similar findings were reported by Podsędek [49]; and Saini et al. [54]. From the previously reported studies [15, 47], it is observed that varietal differences can affect the carotenoid concentrations of plants. Besides varietal differences, carotenoid concentrations are also affected by climatic and geographical factors, pre- and post-harvest interventions, harvesting methods, etc.

1.5.3 PHENOLIC COMPOUNDS

Phenolic compounds are another category of phytonutrients that are present in plant materials exhibiting several health benefits. There are many types of phenolic compounds, such as flavonoids, phenolic acids, tannins, stilbenes, and lignans, all of which exhibit antioxidant properties. Research has limited results on the nutritional profile and antioxidant content of microgreens; but according to available research, Brassicaceae, Apiaceae, Lamiaceae, or Malvaceae families are more popular for their phenolic compounds. Based on the evidence, coriander had the highest total polyphenol content (5,920 mg/g dry weight) followed by green basil and tatsoi (3,506 and 2,645 mg/g dry weight, respectively). The highest antioxidant capacity was confirmed in the Brassicaceae family, followed by those of the Lamiaceae family [32].

According to Paradiso et al., similar findings were observed in a study carried out to analyze the nutritional characteristics of packaged microgreens. The health-promoting phytoconstituents found in mature Brassicaceae include kaempferol, quercetin, caffeic acid, ferulic acid, isorhamnetin, and hydroxycinnamic acid with its derivatives. The recent research on microgreens suggests they have a broader polyphenol profile, strong antioxidant potential, and may be a better choice for human diets because of their bioavailability [32, 37].

1.5.4 MINERAL NUTRITION

Among the essential mineral elements, principal nutrients for human health are those that are acquired primarily through diet. Mineral nutrients are generally classified into: Macronutrients (such as Ca, Mg, P, K, and Na) and MNs (also known as trace elements) Fe, Zn, Cu, and Mn). Both of these classes of nutrients are crucial for several biological processes in plants and animals, and deficiencies can result in metabolic disorders and other acute and chronic diseases.

According to recent studies, microgreens contain excellent mineral supplies, and 90% of the microgreens possess excessive concentrations of macro- and micro-elements compared to their mature plants. A recent study reveals that fennel microgreen is a valuable source with highest concentrations of calcium, potassium, and sodium; and spinach microgreens contain highest concentrations of Mg; and roselle microgreens are richest source of P, Zn, and Se [18].

1.5.5 GLUCOSINOLATES

Glucosinolates are major class of plant secondary metabolites with considerable health benefits and are usually predominantly observed in the family

Brassicaceae. Myrosinase, an enzyme found in unprocessed plant parts, hydrolyzes glucosinolates in the digestive tract to different metabolites, such as isothiocyanates, nitriles, oxazolidine-2-thiones, and indole-3-carbinols. Myrisinase enzymes get inactivated during the processing of cruciferous plants. Numerous factors can affect the bioavailability of glucosinolates, such as storage conditions (Storage time and temperature), packaging conditions and environmental factors. Glucosinolates are known as nitrogen-sulfur derivatives and the prime class comprises of aliphatic, aromatic, and indole derivatives [35]. It has become apparent from recent investigations that microgreens contain high levels of glucosinolates compared to their matured counterparts.

According to Huang et al. [20], red cabbage microgreens contain two times more glucosinolates than mature plants. Recent investigations also exhibit that glucosinolates possess a vital part in treating cardiovascular, lung, and cancer diseases, besides their antioxidant, antidiabetic, and anti-inflammatory properties [40].

1.5.6 ANTIOXIDANT POTENTIAL

Several investigations have documented the presence of bioactive compounds, minerals, vitamins, and phytochemicals; these have affirmed that microgreens are abundant source of bioactive components than their matured counter parts. Research has focused on the existence of antioxidants to combat free radicals and to reduce oxidative stress. Carotenoids, ascorbic acid, phenolic compounds, and some minerals (Cu, Zn, and Se) in microgreens act as antioxidants and protect against diseases caused by high oxidative stress. Microgreens and their mature counterparts have been compared with respect to their antioxidant content and capacity. Several microgreens contained higher antioxidant concentrations, but the results of the study cannot be generalized [74].

Copper, zinc, and selenium act as cofactors for superoxide dismutase (an antioxidant enzyme) and perform a major role in the antioxidant defense structure, so that these minerals are known as antioxidant enzymes. Dietary deficiencies of antioxidant minerals may reduce the activity of antioxidant enzymes. Microgreens contain high level of macro- and microminerals, compared to other vegetables [69]. Phytochemicals and vitamins in microgreens provide them with a strong antioxidant capacity and are effective at regulating plasma lipoprotein and cholesterol metabolism, suggesting a possible role in curing chronic diseases.

1.6 HEALTH BENEFITS

With the litheness of sourcing small scale and indoors for the production of microgreens, they are enticing customers owing to their nutritional and health benefits. Opulence of certain vitamins, phytochemicals including phenolic compounds and carotenoids, and minerals validates the prevailing requirement of microgreens among the populations. Anti-cancer, anti-inflammatory, anti-microbial, anti-hyperglycemia, and other properties enriched by the inclusion of microgreens are widely reported by *in vivo* and *in vitro* studies by certain authors.

1.6.1 ANTI-CANCER POTENTIAL

Jeopardy allied with fatal tumors is accumulative, with cancer being one of the leading bases of death among humans. Upsurge in incidence embarks the importance of prevention of disease and success of studies related to the modification of diets with phytochemical intake [41]. In this framework, the likelihood of microgreens with significant constituents (such as polyphenols, vitamins, carotenoids, etc.) as preventive cure is something that must be studied vastly. Assessment of these potentials has been done by different authors to comprehend the importance of these plant clusters. Such an evaluated potential of *Brassicaceae* family comprising cabbage, cauliflower, etc., has been documented in detail by Singh et al. [59].

The occurrence of glucosinolates metabolites in these plant products contributes progressively by consenting the cells to guard against potential carcinogens and activating enzymes accountable for the metabolism. Higher glucosinolate fractions were reported in broccoli microgreens compared to their florets and mature leaves, which justify their ability [38]. Competence of glucosinolates is related to the hydratable forms of their functional constituents, which has anti-cancer potential and can notably stimulate the molecular and signaling trails [22].

Estimations on cell viability and mechanisms related to proliferative effect explains that the bioaccessible fractions of broccoli, kale, mustard, and radish microgreens were operative in reducing proliferation of tumor cells, with broccoli microgreens exhibiting less effect equated to others. Constituent variations like low ascorbic acid content and antioxidant potential could be the reasons of low activity in the assessment. The cytotoxicity against types of carcinoma cells including that of lungs, colon, prostate, and colorectal was exhibited by broccoli microgreens, which bounce an optimism from the prevention standpoint [35].

Analogous to the bioactive phytochemical profile of broccoli microgreens, the usability of microgreens of radish in controlling and reducing the presence of cancer cells in colon and bladder were reported by Baenas et al. [4].

Apart from the functionality, the other apprehension related to microgreens is the effect of handling, preservation, processing, and storage methods on the potentiality of their phytochemicals. Despite the fact that there are only limited studies in this perspective, different methods like low temperature storage, processing techniques, etc., implied on these products is not found to affect their final outcome. In case of broccoli and radish sprouts, the refrigerated storage at 4–5°C was suggested, which helps in maximum retention of the phytochemicals and other functional constituents [4]. Even introduction of non-thermal plasma or cold plasma treatment was able to enhance the anticancer properties of mustard green microgreens for better health allusions [53]. Intonations of these pathways remain ambiguous and there are only limited studies backing the efficacy of these constituents for prevention. To have a better understanding, it is important to have more studies on the benefits of these products on prevention of proliferative action of carcinoma cells.

1.6.2 ANTI-MICROBIAL ACTIVITY

Reviews on anti-microbial potential of microgreens are still in burgeoning phase confining to the applicability and potential perception. There are only limited studies focusing on the detrimental effects of microgreens on pathogenic bacteria and other microbial populations. Assessment of kale and red spinach microgreens counter to particular pathogenic bacteria has uncovered the likelihood of using them as a reserve of synthetic antimicrobial agents [56].

Kale and red spinach microgreens were more effective in controlling the *B. subtilis* and *E. coli* populations in comparison with the mature greens. Similar results were reported by Le et al. [36] about the potential of broccoli on antimicrobial properties. With minimum inhibition concentration of around 0.39 and 0.78 mg/mL, the extracts were most effective against *Bacillus subtilis* and *Salmonella Typhimurium*, *Staphylococcus aureus* and *E coli*.

Sharma & Gupta [57] analyzed the antimicrobial activities of microgreens garnered from *Pennisetum typhoides* (Pearl millet). Evaluations showed that the capability of pearl millet microgreens contrary to *Staphylococcus aureus* with least inhibition concentration of around 0.125 mg/ml. Generalization of these outcomes is unfeasible as this domain requires more and more scientific proof. Backing of scientific proof can even build up a pathway counting the usage of these microgreens as a substitute for synthetic antimicrobial agents.

1.6.3 ANTI-DIABETIC ACTIVITY AND ANTI-OBESITY ACTIVITY

Changes in dietary patterns and intake have tailored the populations imperiling them with odds of obesity, a world health care challenge piloting to adverse metabolic issues (such as cardiovascular diseases (CVDs), diabetes, cholesterol, etc.). Health benefits allied with microgreens are important considerations for regulating blood glucose levels, control of body mass index, accumulation, and concentration of adipose tissue. The antidiabetic potential of microgreens of red cabbage, radish, and kohlrabi with higher levels of glucosinolates and phenolic compounds has opened up the prospects of incorporating these in the diet pattern of affected populations [65].

Evaluations by several other authors indicate the significance of the compound sulforaphane present in young broccoli sprouts for monitoring the serum insulin concentration and thereby lessening the complications associated with diabetes [5]. Constructive findings on obesity in humans were observed by broccoli microgreens by regulating the lipid metabolism. Thus, the importance and potential of *Brassicaceae* family microgreens must be reviewed on a larger scale to provide benefits to a larger population confronting serious health issues. Apart from the *Brassicaceae* family, fenugreek microgreens exhibit the capability of controlling the blood glucose levels by hindering enzyme regulating the carbohydrate metabolism and upgrading glucose uptake in cells [67]. Elevated levels of total phenolics, flavonoids, and antioxidants have major role in this observed activity.

1.6.4 ANTI-INFLAMMATORY POTENTIAL

Inflammation is considered to be a biological reaction of our body to injury, infection or any other issues [24]. Deterrence of inflammation directly involves the inhibition of inflammatory mediators or proinflammatory cytokines (such as interleukin-1 (IL-1β), interleukin 6 or 8 (IL-6, IL-8) and tumor necrosis factor-alpha (TNF-α)). Microgreens-rich glucosinolates, polyphenols, and other active compounds could impede the action of kinases, thereby influencing and hindering NF-kB (nuclear factor-κB) inflammation signaling pathway [75].

Presence of sulforaphane in broccoli microgreens was effective in regulating neuroinflammatory conditions. Inhibition of proinflammatory cytokines IL-1β, TNF-α, prostaglandin E2 and modification of NF-kB inflammation signaling pathway helped in reducing the inflammatory conditions [61]. Effectiveness of red cabbage microgreens in inhibiting proinflammatory cytokines TNF-α and thereby controlling the inflammation issue was studied by Huang et al. [20] by inducing a high fat diet in a mice-model.

1.6.5 OTHER BENEFICIAL EFFECTS

Conception of eliminating the toxicity and side effects of drugs has augmented the enclosure of bioactive components in our regular diet. Importance of microgreens has amplified with an anticipation that this would decrease the risk associated with chronic diseases. Apart from the debated capacities, their potential towards regulation and control of chronic diseases has been assessed and recorded by several authors. Dietary supplementation with broccoli microgreens was effective in controlling the infirmities linked with placental insufficiency. *In vitro* studies for the inclusion of microgreens during pregnancy and newborn phase has shown positive effects on brain growth thus reducing the chances of cerebral palsy and other disabilities in rats [9].

Lettuce and Chicory microgreens under hydroponic conditions favored low potassium content, which reassures the inclusion of a healthy dietary pattern among patients with damaged kidney function [51]. Possibility of microgreens including broccoli, red mustard, green curly kale and radish under hydroponic conditions included in the dietary recommendations of patients with kidney diseases could be due to their low potassium levels [14].

Elimination of adverse effects of the drug therapy is possible with antinociceptive activity of extracts from broccoli sprouts thus signifying its usefulness for the treatment of visceral and nociceptive pains [4]. Although the presence of diverse chemical conformation provides ground for the wider application of microgreens yet lack of *in vivo* and clinical trials limits the application; and more studies are needed to boost their application.

1.7 SENSORY EVALUATION

The acceptability of microgreens is increasing day-by-day, as they are packed with nutrients including phytochemicals, vitamins, minerals, etc. Culinary professionals have become increasingly attracted to their distinctive sensory characteristics, such as aroma, tender texture, and sparkling colors. Color, taste, and odor of microgreens are significantly affected by bioactive components and nutrients. Each type of microgreen possesses its own set of characteristics.

According to a study conducted on Broccoli microgreens from both local and commercial farms, samples collected from local farms had higher acceptability since they contain more chlorophyll, which increases the sugar production [63]. As phenols tend to taste astringent in vegetables and fruits, microgreens may also have an astringent flavor due to their phenolic content. Furthermore, other than the sucrose content of microgreens, changes in acid and aroma components can also affect the sweetness of microgreens. Compared to microgreens containing higher amounts of glucosinolates (such as China rose radish and Dijon mustard of Brassicaceae family), microgreens rich in phenolic compounds received higher acceptance rates in terms of taste, color, quality, and overall acceptance [72].

In a sensory evaluation test of 12 microgreens species, Caracciolo et al. [10] found that acceptance is highly influenced by the sensory characteristics (such as aroma, astringency, bitterness, heat, sourness, etc. The flavor of microgreens rich in allyl isothiocyanate is pungent and spicy, like mustard and wasabi. Social awareness and acquiescence of microgreens can therefore remain as a significant factor in promoting this new functional food [74].

1.8 SAFETY CONCERNS

After harvest, number of factors may affect the microbial load on microgreens, including contiguity to soil at harvest, moisture content following storage temperature, and methods of washing before packaging. As a result of their gauzy and soft-textured hypocotyls, microgreens are more likely to support microbial growth than mature plants. Washing microgreens with chlorinated water before packaging can effectively reduce the number of aerobic mesophilic bacteria [11]. A study performed to ascertain the impact of processing methods on the overall value of buckwheat indicated that chlorinated-wash treatments could reduce the initial load of aerobic mesophilic bacteria to 100 mg/l. In addition, wash treatments were effective only for the first 7 days after storage, after which bacterial count was increased to 10.3 log CFU/g at 21 days in the buckwheat microgreens washed with water [29].

A similar pattern of microbial escalation has been reported in case of washed broccoli microgreens storage by Kou et al. [30]. Microgreens, whose microbial counts have risen, are also associated with rapid leakage of electrolytic components and waterlogged hypocotyls, resulting in overly wet residuals after the wash treatments [11, 29, 30]. Despite the fact that washing treatments lead to excessive moisture content due to inadequate drying conditions, unwashed microgreens are more microbiologically safe. These results indicate that different washing treatments can only reduce initial microbial loads to a certain extent and despite the initial benefits, waterlogged hypocotyls lead to mechanical damage and shortened shelf-life. Sanitation is the key to the popularization and diversification of ready-to-serve packaged microgreens as well as industrial microgreen production. It is necessary to conduct further research on the efficacy of various sanitation techniques and the effect of drying on the quality parameters and shelf-life of microgreens [33].

1.9 MARKET POTENTIAL AND COMMERCIAL APPLICATIONS

A range of top-notch microgreens is being developed financially and in some cases are developed by people on smaller scale for domestic usage. Variety of microgreens can be grown together for industrial applications and the success of microgreen cultivation lies in having the suitable varieties of microgreens with proper maturity. Standard, loose, and sterile soil can be used for the cultivation of healthy microgreens and the seeding can be done in rows or as broad-cast method. Overcrowding of seeds may lead to elongated stem and increase the chances of various diseases [29].

Although most of the microgreen crops do not require any fertilizer application, yet several kinds of fertilizers especially organic fertilizers can positively modify microgreen yield potential [43, 72]. Microgreens are harvested at different stages of development for different crops, ranging from one to three weeks. Harvested microgreens are prone to spoilage and the microbiological safety can be achieved with suitable washing treatment followed by appropriate drying conditions.

Microgreens are usually packed in polyethylene bags; and after packaging it is recommended to store at suitable temperatures as storage temperature can greatly influence the microbiological safety [31]. Packing, distribution, and storage of microgreens in grocery stores are confined to different materials and storage environments and are usually allied with a light source. Studies have revealed that the loss of vitamin C in Chinese kale during storage can be prevented by low intensity light [45].

Similar results were observed by Xiao et al. [73]; and constant light exposure can enhance the tocopherol and carotenoid contents of baby leaf spinach. It is simple and convenient to grow microgreens because they do not require a lot of equipment or time and can be incorporated into our diet in a variety of ways. The versatile ingredients can be found in sandwiches, wraps, and salads. In addition, they can also be blended into smoothies or juices. Microgreens are nutrients rich compounds and also known as "Vegetable Confetti" because of the high levels of antioxidants, minerals, carotenoids, and phytochemicals. In addition to being beneficial in curing several diseases, they can be grown in a variety of ways to meet consumer preferences.

Renna et al. [51] indicated that cultivated lettuce and celery microgreens with low potassium content are beneficial for patients with impaired kidney function. In their study, patients with kidney diseases, who ate vegetables-based dishes, could reduce their potassium intake by consuming microgreens.

1.10 SUMMARY

Since microgreens are still a fairly new subject, research into their nutritional content and ability to promote health is only at its infancy. In addition to having an enhanced antioxidant profile (carotenoids, glucosinolates, polyphenols, and vitamins) than mature greens, microgreens have greater health benefits, such as preventing chronic diseases. Several treatments can enhance microgreens' nutritional value, including light exposure, salinity stress, nutrient supplementation, and organic fertilizer use. Besides being rich in nutrients, microgreens are also effective in regulating lipoprotein and cholesterol metabolism. However, several pre- and post-harvest factors (such as medium and method of cultivation, harvesting methods, storage atmosphere, and packaging condition) can affect the phytochemical contents in microgreens. Hydroponics and aquaponics are potent cultural techniques for the cultivation of microgreens, particularly under controlled environmental conditions. Moreover, number of limitations are present in the commercial cultivation of microgreens. Therefore, additional analysis is required to enhance the industrial cultivation, yield, and phytochemical composition of microgreens.

KEYWORDS

- **anti-inflammatory potential**
- **anti-obesity activity**
- **bioactive components**
- **glucosinolates**
- **microgreens**
- **post-harvest quality**
- **sprouts**

REFERENCES

1. Abbasi, N. A., Ashraf, S., Ali, I., & Butt, S. J., (2015). Enhancing storage life of bell pepper by UV-C irradiation and edible coatings. *Pakistan Journal of Agricultural Sciences, 52*(2), 403–411.
2. Andrejiová, A., Hegedűsová, A., Mezeyová, I., & Kóňová, E., (2017). Content of selected bioactive substances in dependence on lighting in microgreens. *Acta Horticultura eet Regiotecturae (Journal of Horticulture and Regiotecture), 20*(1), 6–10.
3. Artés, F., Gómez, P., Aguayo, E., & Escalona, V., (2009). Sustainable sanitation techniques for keeping quality and safety of fresh-cut plant commodities. *Postharvest Biology and Technology, 51*(3), 287–296.
4. Baenas, N., Gómez-Jodar, I., Moreno, D. A., García-Viguera, C., & Periago, P. M., (2017). Broccoli and radish sprouts are safe and rich in bioactive phytochemicals. *Postharvest Biology and Technology, 127*, 60–67.
5. Bahadoran, Z., Mirmiran, P., & Azizi, F., (2013). Potential efficacy of broccoli sprouts as a unique supplement for management of type 2 diabetes and its complications. *Journal of Medicinal Food, 16*(5), 375–382.
6. Barba, F. J., Nikmaram, N., Roohinejad, S., Khelfa, A., Zhu, Z., & Koubaa, M., (2016). Bioavailability of glucosinolates and their breakdown products: Impact of processing. *Frontiers in Nutrition, 3*, E-Article ID 24.
7. Berba, K. J., & Uchanski, M. E., (2012). Post-harvest physiology of microgreens. *Journal of Young Investigators, 24*(1), 1–5.
8. Bergquist, S. Å., Gertsson, U. E., & Olsson, M. E., (2006). Influence of growth stage and postharvest storage on ascorbic acid and carotenoid content and visual quality of baby spinach (*Spinacia oleracea* L.). *Journal of the Science of Food and Agriculture, 86*(3), 346–355.

9. Black, A. M., Armstrong, E. A., Scott, O., Jurlink, B. J. H., & Yager, J. Y., (2015). Broccoli sprout supplementation during pregnancy prevents brain injury in the newborn rat following placental insufficiency. *Behavioral Brain Research, 291*, 289–298.
10. Caracciolo, F., El-Nakhel, C., Raimondo, M., Kyriacou, M. C., Cembalo, L., De Pascale, S., & Rouphael, Y., (2020). Sensory attributes and consumer acceptability of 12 microgreens species. *Agronomy, 10*(7), 1043–1053.
11. Chandra, D., Kim, J. G., & Kim, Y. P., (2012). Changes in microbial population and quality of microgreens treated with different sanitizers and packaging films. *Horticulture, Environment, and Biotechnology, 53*(1), 32–40.
12. Dalal, N., Siddiqui, S., & Neeraj, (2019). Effect of chemical treatment, storage, and packaging on physicochemical properties of sunflower microgreens. *International Journal of Chemical Studies, 7*(5), 1046–1050.
13. Dalal, N., Siddiqui, S., & Phogat, N., (2020). Evaluation of effects of chemical treatments on sensory attributes of sunflower microgreens with storage. *Journal of Pharmacognosy and Phytochemistry, 9*(1), 439–443.
14. De La Fuente, B., López-García, G., Máñez, V., Alegría, A., Barberá, R., & Cilla, A., (2020). Antiproliferative effect of bioaccessible fractions of four *Brassicaceae* microgreens on human colon cancer cells linked to their phytochemical composition. *Antioxidants, 9*(5), 368–373.
15. Dhuique-Mayer, C., Fanciullino, A. L., Dubois, C., & Ollitrault, P., (2009). Effect of genotype and environment on citrus juice carotenoid content. *Journal of Agricultural and Food Chemistry, 57*(19), 9160–9168.
16. Ebert, A. W., Wu, T. H., & Yang, R. Y., (2014). Amaranth sprouts and microgreens: A homestead vegetable production option to enhance food and nutrition security in the rural-urban continuum. In: *Proceedings of the Regional Symposium on Sustaining Small-Scale Vegetable Production and Marketing Systems for Food and Nutrition Security (SEAVEG 2014)* (pp. 25–27). Bangkok – Thailand.
17. Ghoora, M. D., & Srividya, N., (2020). Effect of packaging and coating technique on postharvest quality and shelf-life of *Raphanus sativus* L. and *Hibiscus sabdariffa* L. microgreens. *Foods, 9*(5), 653–660.
18. Ghoora, M. D., Babu, D. R., & Srividya, N., (2020). Nutrient composition, oxalate content and nutritional ranking of ten culinary microgreens. *Journal of Food Composition and Analysis, 91*, E-article ID 103495.
19. Guo, L., Yang, R., Wang, Z., Guo, Q., & Gu, Z., (2014). Glucoraphanin, sulforaphane and myrosinase activity in germinating broccoli sprouts as affected by growth temperature and plant organs. *Journal of Functional Foods, 9*, 70–77.
20. Huang, H., Jiang, X., Xiao, Z., Yu, L., Pham, Q., Sun, J., & Wang, T. T., (2016). Red cabbage microgreens lower circulating low-density lipoprotein (LDL), liver cholesterol, and inflammatory cytokines in mice fed a high-fat diet. *Journal of Agricultural and Food Chemistry, 64*(48), 9161–9171.
21. Islam, M. Z., Park, B. J., & Lee, Y. T., (2019). Effect of salinity stress on bioactive compounds and antioxidant activity of wheat microgreen extract under organic cultivation conditions. *International Journal of Biological Macromolecules, 140*, 631–636.
22. Jambor, T., Knizatova, N., Valkova, V., Tirpak, F., Greifova, H., Kovacik, A., & Lukac, N., (2022). Microgreens as a functional component of the human diet: A review. *Journal of Microbiology, Biotechnology and Food Sciences, 2022*, e5870–e5878.

23. Janovská, D., Stocková, L., & Stehno, Z., (2010). Evaluation of buckwheat sprouts as microgreens. *Acta Agriculturae Slovenica, 95*(2), 157–161.
24. Jayarathne, S., Koboziev, I., Park, O. H., Oldewage-Theron, W., Shen, C. L., & Moustaid-Moussa, N., (2017). Anti-inflammatory and anti-obesity properties of food bioactive components: Effects on adipose tissue. *Preventive Nutrition and Food Science, 22*(4), 251–255.
25. Keutgen, N., Hausknecht, M., Tomaszewska-Sowa, M., & Keutgen, A. J., (2021). Nutritional and sensory quality of two types of cress microgreens depending on the mineral nutrition. *Agronomy, 11*(6), 1110–1115.
26. Khosravi, F., & Asadollahzadeh, H., (2014). Determination of ascorbic acid in different citrus fruits under reversed phase conditions with UPLC. *Eur. J. Exp. Biol, 4*, 91–94.
27. Kim, J., Luo, Y., & Gross, K., (2004). Effect of package film on the quality of fresh-cut salad savoy. *Postharvest Bio. Tech., 32*, 99–107.
28. Kou, L., Luo, Y., Park, E., Turner, E. R., Barczak, A., & Jurick, W. M., (2014). Temperature abuse timing affects the rate of quality deterioration of commercially packaged ready-to-eat baby spinach. Part I: Sensory analysis and selected quality attributes. *Postharvest Biology and Technology, 91*, 96–103.
29. Kou, L., Luo, Y., Yang, T., Xiao, Z., Turner, E. R., Lester, G. E., & Camp, M. J. (2013). Postharvest biology, quality, and shelf-life of buckwheat microgreens. *LWT-Food Science and Technology, 51*(1), 73–78.
30. Kou, L., Yang, T., Liu, X., & Luo, Y., (2015). Effects of pre-and postharvest calcium treatments on shelf-life and postharvest quality of broccoli microgreens. *HortScience, 50*(12), 1801–1808.
31. Kou, L., Yang, T., Luo, Y., Liu, X., Huang, L., & Codling, E., (2014). Pre-harvest calcium application increases biomass and delays senescence of broccoli microgreens. *Postharvest Biology and Technology, 87*, 70–78.
32. Kyriacou, M. C., El-Nakhel, C., Graziani, G., Pannico, A., Soteriou, G. A., Giordano, M., & Rouphael, Y., (2019). Functional quality in novel food sources: Genotypic variation in the nutritive and phytochemical composition of thirteen microgreens species. *Food Chemistry, 277*, 107–118.
33. Kyriacou, M. C., Rouphael, Y., Di Gioia, F., Kyratzis, A., Serio, F., Renna, M., & Santamaria, P., (2016). Micro-scale vegetable production and the rise of microgreens. *Trends in Food Science & Technology, 57*, 103–115.
34. Lacroix, M., & Vu, K. D., (2014). Edible coating and film materials: Proteins. In: *Innovations in Food Packaging* (pp. 277–304). London: Academic Press.
35. Le, T. N., Chiu, C. H., & Hsieh, P. C., (2020). Bioactive compounds and bioactivities of *brassica oleracea* sprouts and microgreens: An updated overview from a nutraceutical perspective. *Plants, 9*(8), 946–951.
36. Le, T. N., Luong, H. Q., Li, H. P., Chiu, C. H., & Hsieh, P. C., (2019). Broccoli (*Brassica oleracea* L.) sprouts as the potential food source for bioactive properties: A comprehensive study on *in vitro* disease models. *Foods, 8*(11), 532–537.
37. Li, Z., Lee, H. W., Liang, X., Liang, D., Wang, Q., Huang, D., & Ong, C. N., (2018). Profiling of phenolic compounds and antioxidant activity of 12 cruciferous vegetables. *Molecules, 23*(5), 1139.
38. Lu, Y., Dong, W., Alcazar, J., Yang, T., Luo, Y., Wang, Q., & Chen, P., (2018). Effect of preharvest CaC_{12} spray and postharvest UV-B radiation on storage quality of broccoli

microgreens, a richer source of glucosinolates. *Journal of Food Composition and Analysis, 67*, 55–62.

39. Lu, Y., Dong, W., Yang, T., Luo, Y., & Chen, P., (2021). Preharvest UV-B application increases glucosinolates contents and enhances postharvest quality of broccoli microgreens. *Molecules, 26*(11), E-article ID 3247; online; https://doi.org/10.3390/molecules26113247.
40. Maina, S., Misinzo, G., Bakari, G., & Kim, H. Y., (2020). Human, animal, and plant health benefits of glucosinolates and strategies for enhanced bioactivity: A systematic review. *Molecules, 25*(16), E-article ID 3682.
41. Mir, S. A., Shah, M. A., & Mir, M. M., (2017). Microgreens: Production, shelf-life, and bioactive components. *Critical Reviews in Food Science and Nutrition, 57*(12), 2730–2736.
42. Mlinarić, S., Gvozdić, V., Vuković, A., Varga, M., Vlašiček, I., Cesar, V., & Begović, L., (2020). The effect of light on antioxidant properties and metabolic profile of chia microgreens. *Applied Sciences, 10*(17), E-article ID 5731.
43. Murphy, C. J., Llort, K. F., & Pill, W. G., (2010). Factors affecting the growth of microgreen table beet. *International Journal of Vegetable Science, 16*(3), 253–266.
44. Niroula, A., Khatri, S., Timilsina, R., Khadka, D., Khadka, A., & Ojha, P., (2019). Profile of chlorophylls and carotenoids of wheat (*Triticum aestivum* L.) and barley (*Hordeum vulgare* L.) microgreens. *Journal of Food Science and Technology, 56*(5), 2758–2763.
45. Noichinda, S., Bodhipadma, K., Mahamontri, C., Narongruk, T., & Ketsa, S., (2007). Light during storage prevents loss of ascorbic acid, and increases glucose and fructose levels in Chinese kale (*Brassica oleracea* var. alboglabra). *Postharvest Biology and Technology, 44*(3), 312–315.
46. Oliveira, M., Abadias, M., Usall, J., Torres, R., Teixido, N., & Vinas, I., (2015). Application of modified atmosphere packaging as a safety approach to fresh-cut fruits and vegetables–A review. *Trends Food Sci. Tech., 46*, 13–26.
47. Payyavula, R. S., Navarre, D. A., Kuhl, J. C., Pantoja, A., & Pillai, S. S., (2012). Differential effects of environment on potato phenylpropanoid and carotenoid expression. *BMC Plant Biology, 12*(1), 1–17.
48. Pinto, E., Almeida, A. A., Aguiar, A. A., & Ferreira, I. M., (2015). Comparison between the mineral profile and nitrate content of microgreens and mature lettuces. *Journal of Food Composition and Analysis, 37*, 38–43.
49. Podsędek, A., (2007). Natural antioxidants and antioxidant capacity of *Brassica* vegetables: A review. *LWT-Food Science and Technology, 40*(1), 1–11.
50. Puccinelli, M., Maggini, R., Angelini, L. G., Santin, M., Landi, M., Tavarini, S., & Incrocci, L., (2022). Can light spectrum composition increase growth and nutritional quality of *Linumusita tissimum*: Sprouts and microgreens? *Horticulturae, 8*(2), 98–101.
51. Renna, M., Castellino, M., Leoni, B., Paradiso, V. M., & Santamaria, P., (2018). Microgreens production with low potassium content for patients with impaired kidney function. *Nutrients, 10*(6), 675–679.
52. Riggio, G. M., Wang, Q., Kniel, K. E., & Gibson, K. E., (2019). Microgreens—A review of food safety considerations along the farm to fork continuum. *International Journal of Food Microbiology, 290*, 76–85.
53. Saengha, W., Karirat, T., Buranrat, B., Matra, K., Deeseenthum, S., Katisart, T., & Luang-In, V., (2021). Cold plasma treatment on mustard green seeds and its effect on

growth, isothiocyanates, antioxidant activity and anticancer activity of microgreens. *Int. J. Agric. Biol, 25*, 667–676.
54. Saini, P., Arora, M., & Kumar, M. R., (2016). Poly (lactic acid) blends in biomedical applications. *Advanced Drug Delivery Reviews, 107*, 47–59.
55. Sandhya, (2010). Modified atmosphere packaging of fresh produce: Current status and future needs. *LWT-Food Science and Technology, 43*(3), 381–392.
56. Shamlan, K. A., Yahya, H., Ismail, I. N. A., & Yahya, H. N., (2020). Antibacterial activities of microgreens and mature extract of kale and red spinach against selected pathogenic bacteria. *J. Agri. Life Sci., 3*(10), 337–342.
57. Sharma, R., & Gupta, P., (2021). Nutraceutical potential of *Pennisetum typhoides* microgreens: *In vitro* evaluation of antioxidant and antibacterial activities and in silico *Staphylococcus aureus* FtsZ inhibition. *Food Bioscience, 42*, E-article ID 101151.
58. Sharma, S., Shree, B., Sharma, D., Kumar, S., Kumar, V., Sharma, R., & Saini, R., (2022). Vegetable microgreens: The gleam of next generation super foods, their genetic enhancement, health benefits and processing approaches. *Food Research International*, E-article ID 111038.
59. Singh, N., & Mishra, A., (2019). Cruciferous microgreens: Growing performance and their scope as super foods at high altitude locations. *Progressive Horticulture, 51*(1), 41–44.
60. Spinardi, A., & Ferrante, A., (2012). Effect of storage temperature on quality changes of minimally processed baby lettuce. *Journal of Food, Agriculture and Environment, 10*(1), 38–42.
61. Subedi, L., Cho, K., Park, Y. U., Choi, H. J., & Kim, S. Y., (2019). Sulforaphane-enriched broccoli sprouts pretreated by pulsed electric fields reduces neuroinflammation and ameliorates scopolamine-induced amnesia in mouse brain through its antioxidant ability via Nrf2-HO-1 activation. *Oxidative Medicine and Cellular Longevity, 2019*, E-article ID: 3549274; doi: 10.1155/2019/3549274.
62. Sun, J., Xiao, Z., Lin, L. Z., Lester, G. E., Wang, Q., Harnly, J. M., & Chen, P., (2013). Profiling polyphenols in five *Brassica* species microgreens by UHPLC-PDA-ESI/HRMSn. *Journal of Agricultural and Food Chemistry, 61*(46), 10960–10970.
63. Tan, L., Nuffer, H., Feng, J., Kwan, S. H., Chen, H., Tong, X., & Kong, L., (2020). Antioxidant properties and sensory evaluation of microgreens from commercial and local farms. *Food Science and Human Wellness, 9*(1), 45–51.
64. Thompson, M. D., & Cooney, R. V., (2020). The potential physiological role of γ-tocopherol in human health: A qualitative review. *Nutrition and Cancer, 72*(5), 808–825.
65. Tomas, M., Zhang, L., Zengin, G., Rocchetti, G., Capanoglu, E., & Lucini, L., (2021). Metabolomic insight into the profile, *in vitro* bioaccessibility and bioactive properties of polyphenols and glucosinolates from four *Brassicaceae* microgreens. *Food Research International, 140*, E-article ID 110039.
66. Verlinden, S., (2020). Microgreens: Definitions, product types, and production practices. *Horticultural Reviews, 47*, 85–124.
67. Wadhawan, S., Tripathi, J., & Gautam, S., (2018). *In vitro* regulation of enzymatic release of glucose and its uptake by fenugreek microgreen and mint leaf extract. *International Journal of Food Science & Technology, 53*(2), 320–326.

68. Wang, H., Gui, M., Tian, X., Xin, X., Wang, T., & li, J., (2017). Effects of UV-B on vitamin C, phenolics, flavonoids and their related enzyme activities in mung bean sprouts. *International Journal of Food Science & Technology, 52*(3), 827–833.
69. Wołonciej, M., Milewska, E., & Roszkowska-Jakimiec, W., (2016). Trace elements as an activator of antioxidant enzymes. *Postepy Higieny i Medycyny Doswiadczalnej (Online: Advances in Hygiene and Experimental Medicine), 70*, 1483–1498. doi: 10.5604/17322693.1229074
70. Xiao, Z., Lester, G. E., Luo, Y., & Wang, Q., (2012). Assessment of vitamin and carotenoid concentrations of emerging food products: Edible microgreens. *Journal of Agricultural and Food Chemistry, 60*(31), 7644–7651.
71. Xiao, Z., Lester, G. E., Luo, Y., Xie, Z., Yu, L., & Wang, Q., (2014a). Effect of light exposure on sensorial quality, concentration of bioactive compounds and antioxidant capacity of radish microgreens during low temperature storage. *Food Chem., 151*, 472–479.
72. Xiao, Z., Lester, G. E., Park, E., Saftner, R. A., Luo, Y., & Wang, Q., (2015). Evaluation and correlation of sensory attributes and chemical compositions of emerging fresh produce: Microgreens. *Postharvest Biology and Technology, 110*, 140–148.
73. Xiao, Z., Luo, Y., Lester, G. E., Kou, L., Yang, T., & Wang, Q., (2014). Postharvest quality and shelf-life of radish microgreens as impacted by storage temperature, packaging film, and chlorine wash treatment. *LWT-Food Science and Technology, 55*(2), 551–558.
74. Xiao, Z., Rausch, S. R., Luo, Y., Sun, J., Yu, L., Wang, Q., & Stommel, J. R., (2019). Microgreens of *Brassicaceae*: Genetic diversity of phytochemical concentrations and antioxidant capacity. *LWT, 101*, 731–737.
75. Zhang, Y., Xiao, Z., Ager, E., Kong, L., & Tan, L., (2021). Nutritional quality and health benefits of microgreens: A crop of modern agriculture. *Journal of Future Foods, 1*(1), 58–66.
76. Zhu, F., Du, B., & Xu, B., (2018). Anti-inflammatory effects of phytochemicals from fruits, vegetables, and food legumes: A review. *Critical Reviews in Food Science and Nutrition, 58*(8), 1260–1270.

CHAPTER 2

Food Formulation for Sports Persons

Soumi Chakraborty, Kaninika Paul, and Anindita Nandi Batabyal

ABSTRACT

This chapter elucidates the recent advances in sports foods and supplements. Sports foods have advantages, such as prolonging endurance, enhancing recovery, reducing body fat, increasing muscle mass, minimizing the risk of illness, and enhancing sports performance (known as ergogenic aids). Therefore, food fortification and supplementation with nutrients (or with bioactive compounds) have been practiced for centuries. These modifications of nutrient content in the food are of great importance, especially for formulation of food products for muscle builder foods (creatine and dehydroepiandrosterone), energy booster foods (caffeine, arginine, and ginseng), immunity booster foods (glutamine, vitamin C, and vitamin E), bone and joint protector foods (glucosamine and chondroitin compounds). For approving these dietary supplements to be safe, many trade associations have designed different programs; however, the basic knowledge of choosing the right supplements should be there for an athlete.

2.1 INTRODUCTION

Sports foods or sports supplements include all "vitamins, minerals, herbs, botanicals, amino acids and enzymes;" and these can supplement the whole foods to enhance the sports performance of an athlete. According to the "National Health and Examination Survey (2011–2014)," more than half of the US adults and one-third of American children (younger than 19) regularly consume any one of the sports supplements. It is a general tendency among the athletes that they rely more on supplements or sports foods than

The Functional Foods: Nutrient and Health Benefits.
Megh R. Goyal, Junaid Ahmad Malik, and Anu Kumari (Eds.)

consuming a balanced diet [3, 9, 26, 29, 31]. However, research studies show that female runners particularly are using multivitamins/minerals, calcium, vitamin D, iron, and probiotics as sports supplements, whereas male runners used creatine and sport foods (protein bars, protein gels, etc.) more frequently for the same [42].

Krumbach et al. specifically highlighted that males usually consume less sports supplements because of its high market price; and their consumption is to improve their performance and to build muscles [34].

Now-a-days, several forms of sports foods are available in the market, such as solid sports foods, semi-solid sports foods, sports drinks, and sports nutrition supplements, etc. [15]. Sports foods are categorized based on their function, forms, availability, and nutritional claims. The classification of sports foods is provided in Figure 2.1.

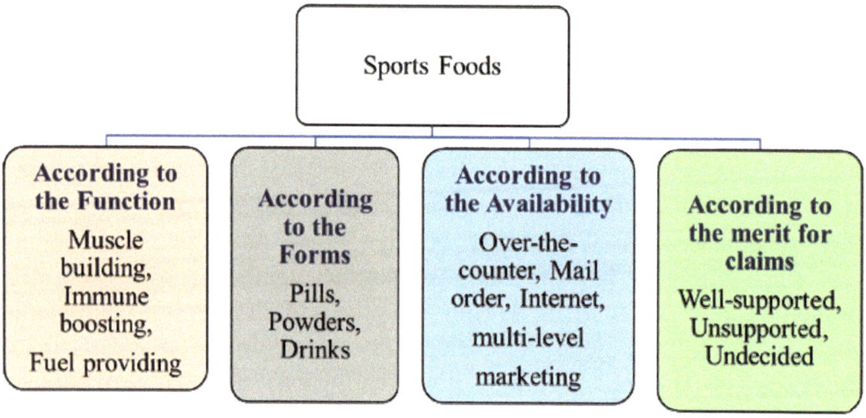

FIGURE 2.1 Classification of sports foods.

It is reported that these sports supplements assist to meet the various physiological and psychological activities in the athlete's body, such as:

1. **Direct Performance Enhancement:** Additionally, few bioactive components from animals and plants are also considered as sports supplements, such as [24]: glucosamine, curcumin, L-carnitine, and lycopene. They directly influence sports performance by improving either of the biochemical pathways (such as galacto-anaerobic, or glycolytic-anaerobic, or aerobic) in our body.
2. **Nutritional Goals:** Sports foods act as an indirect performance enhancer by contributing to the basic nutrients, such as carbohydrate, protein, and fats.

3. **Placebo Effect (Psychological Boost):** Few sports foods do not provide any physiological or ergogenic benefits; however, these foods act only as psychological booster.

It is undoubtedly stated that these foods play an important role in the athlete's diet, but there is no universal system on quality regulation of these sports foods and supplements. In the absence of rigorous government evaluation and quality checking, these food supplements should not be recommended to the athletes. Therefore, only few foods are approved as sports food supplements, as shown below:

- Antioxidants: vitamins C and E;
- Bicarbonate;
- Caffeine;
- Calcium supplement;
- Creatine;
- Electrolyte replacement;
- Glycerol (for hyperhydration);
- Iron supplement;
- Liquid meal supplements;
- Multivitamin/mineral supplements;
- Sick pack (zinc and vitamin C);
- Sports bars;
- Sports drinks;
- Sports gel.

2.2 COMMERCIAL MARKET SCENARIO OF SPORTS SUPPLEMENTS

Recently, the market of supplements/sport foods has been flourishing worldwide. In the worldwide market, North America and Europe regions account for 25% and 30%, respectively. It has been estimated that around 163.1 billion US dollar, with an annual prediction to grow by 6.5% for at least the next five years. Globalizations and the digitalization of our lives are facilitating such a rapid expansion in sport supplementation market [7, 13, 31]. Being one of the world's fastest-growing economies, rising income in India has prompted households to expand budgets and spend more on health and nutrition above essentials such as food [16].

In India, the accelerated growth of the sports nutrition market has been observed in the last few years, due to:

- Huge growth in India's sports industry;
- Increased demand for various health supplements and energy drinks among bodybuilders and athletes;
- Increasing participation of youths in sports and activities that require a high level of fitness and nutrition;
- The busting of myths about supplements; and
- The health-conscious millennial generation.

2.3 RECOMMENDED DIETARY INTAKE AND DAILY VALUE FOR SPORTS PERSON

2.3.1 CARBOHYDRATE

Carbohydrate in our body aids in digestion, assimilation, and elimination. For an average person, minimum 55% of daily energy should be from carbohydrate. According to WHO recommendation, athletes require total carbohydrate to reserve adequate fuel for their events [41].

Usually, the energy provided by fats, carbohydrates, and protein are utilized by the muscle cells. However, protein is not required as a source of energy in sports activities.

In fact, few sports performances require substantial amount of glycogen to be spent during the physical events. If carbohydrate stores are below normal, suitable carbohydrate loading techniques may be employed. However, loading of carbohydrates should be avoided in case the carbohydrate intake is in accordance with the recommended levels, as it might interfere with the intake of other essential nutrients. During physical exercise, muscles utilize its own glucose store. On exhaustion of the glucose in glycogenolysis, gluconeogenesis continues the supply of glucose to the muscles. Food sources, such as, whole grains, vegetables, fruits, milk, and yogurt are good sources of carbohydrates [43].

2.3.2 PROTEIN

The protein requirement for power and endurance athletes is 1.2–1.5 g/kg of body weight since they need extra protein for restoration of the injured muscle fiber. The "International Center for sports Nutrition" approves the protein content in the range of 1.0–1.5 g/kg of body weight/day. In comparison to power athletes, endurance athletes need more protein as they hold some of the protein in the muscles. Few evidence suggests that formation of creatinine from methionine and glycine along with arginine may be advantageous. It is one of the most significant nutrients for maintaining good health and body strength. It plays significant role in the development of body tissues. In general, the protein present in the diet provides about 2–5% of energy required in low or moderate exercise and 10–15% of energy in case of endurance exercise. However, a diet rich in carbohydrate saves protein from being utilized as fuel. Generally, protein consumption is not mandatory for the athletes except in case of gaining muscle mass or endurance sports [41].

2.3.3 FAT

Fat is mainly used as a fuel for an athlete, during low to moderate intensity performance. It is recommended that fat consumption should be moderate with a proper balance between saturated and unsaturated fats [45].

2.3.4 MICRONUTRIENTS (MNS)

Micronutrients (such as vitamins and minerals) are essential for metabolic reactions, immune functions and development of body tissues. Among these, calcium, vitamin D and iron are essential for the athletes [53].

Adequate intake (AI) of calcium is required for proper maintenance of bone health and prevention of osteoporosis. Athletes may lose calcium through sweating and the loss can be replenished through dietary intake of calcium-rich foods and supplements [35].

Low levels of vitamin D may reduce the training and health efficiency of the athletes. Recent studies indicate that athletes might be at risk for suboptimal vitamin D status leading to stress related-fractures, acute illness and low muscle functions. Normal levels of vitamin D vary on the basis of geographical location and race. Generally, athletes from northern latitudes or those who are involved in indoor training (gymnasts, figure skaters) are more prone to vitamin D deficiency [46].

Surveys of several athletic groups have revealed low and mid-range of hemoglobin concentration in both males and females, especially those engaged in strong endurance sports. This is commonly referred to as sports anemia. The low concentration of hemoglobin in most athletes is a result of training, which leads to increased plasma volume that in turn dilutes the red blood cells. The rise in blood plasma is an advantageous adaptation to aerobic exercises and is called dilutional pseudo-anemia. However, athletes may develop true anemia (i.e., deficit in the total amount of hemoglobin or red blood cells). The degree of iron deficiency seems to be higher in athletes than in general population. Owing to this hemodilution, runners generally have lower levels of plasma ferritin than those of general population [41]. Vitamin E or tocopherol is necessary for normal muscle functions. Exercise affects blood flow in skeletal muscle and influences oxidation metabolism in the muscles. Intake of vitamin E will lower the oxidative stress associated with exercise and allow normal functioning of the muscles in athletes [41].

2.3.5 DIETARY FIBER

Dietary fiber is helpful in controlling weight and reducing blood cholesterol and colon disease. It also aids health in regular excretion by reducing glucose absorption along with hemorrhoids and diverticulitis [4, 41].

2.3.6 WATER

Water not only transports nutrients and other desired substances to cells but also eradicates waste materials from cells as well. It reacts with the viscous molecules to produce lubricating fluids for joints and also facilitates easy movement of food in the digestive tract. It helps in maintaining the body temperature. It also maintains the shape of the cells and is one of the basic parts of cell membrane. Water also protects the organs by acting as cushion and helps in preserving body structures [41]. Table 2.1 presents RDA (radio-isotope dilution assay) of food components for a sports person.

TABLE 2.1 Recommended Daily Allowance (RDA) of Food Components for a Sports Person [20,43]

Food Components	Recommended Daily Allowance (RDA)
Carbohydrate	A minimum amount of 300 g/day (when the total calorie intake is 2,000 Kcal).
Dietary fiber	The amount of dietary fiber ranges from 25 to 48 g/day for diets ranging from 3,000–7,000 Kcal/day.
Fat	Total fat should be < 30% of total calories; and saturated fats should be within 10% of total calories.
Micronutrients	Micronutrients including vitamins and minerals are necessary in micro-quantities.
Protein	RDA of protein is 0.8 g/kg of body weight for endurance athletes and body builders.

2.4 SPORTS SUPPLEMENTS, PERFORMANCE ENHANCER (ERGOGENIC AIDS) FOODS, ENGINEERED SPORTS FOODS

2.4.1 *VITAMIN AND MINERAL SUPPLEMENTS*

Vitamin and mineral supplements are highly required for the pre-adolescent endurance runners. Since they are in the critical period of growth (including bone mass accrual), they are at high risk of being malnourished. Due to the increased sweating in sports participation, they can lose significant amount of sodium, potassium, iron, and magnesium. Additionally, it has been reported that vegetarian athletes are at high risk of developing sports anemia owing to the low dietary Fe density and consumption of plant food-based Fe sources. During Aerobic training, a high amount of Fe is lost, which in turn may increase the Fe turnover in the body. Therefore, athletes (specially pre-adolescents) should take the adequate MNs supplements to battle the MN deficiencies [53].

Medically, it is recommended that athletes should consume daily low doses of "multivitamins and/or vitamin-rich carbohydrate/protein supplements," during intensive training. It will help in maintaining an adequate MN level in the body, thereby will help to combat nutrient deficiencies among the athletes. Athletes mostly take multivitamin with minerals, followed by vitamin C, multivitamin (with no minerals), B-complex vitamin and calcium, iron, vitamin E, and zinc [8, 34].

2.4.2 *ANTIOXIDANT SUPPLEMENTATION AND IMMUNITY BOOSTER SPORTS FOODS*

Sports persons frequently use these antioxidant supplementations, such as vitamin C, Vitamin E, coenzyme Q10, beta-carotene. The benefit of using antioxidant is to mitigate the oxidative stress (produced by free radicals), resulted by the increase in oxygen uptake by body tissues. Recent research studies support that antioxidant consumption has no beneficial effects on sports performance. Cui et al. reported that moderate exercise has a positive effect on improving immune function, whereas high and prolonged exercise damage the immune cells [14, 15].

2.4.3 *MUSCLE BUILDER FOODS*

In an athlete's body, skeletal muscle accounts for as much as 45% of the total body mass. Muscles help in movement of the joint during exercise or sports events. The chemical energy required for the muscle movements is provided by the hydrolysis of adenosine triphosphate (ATP). Additionally, creatine phosphate (CP: a high energy compound) is stored in greater amounts (approximately 20 mMol/kg) in the skeletal muscle, which can be broken-down quickly during intense exercise to provide energy for ATP resynthesis. It is evident that athletes have a higher requirement of protein than the normal individual. It is reported that skeletal muscle fiber composition has positive influence on the exercise performance [50].

2.4.4 PERFORMANCE ENHANCER

These foods, besides their general nutritional contribution, also enhance or improve the athletic performance or recovery and are marked as performance enhancers. For example, creatine is used to generate ATP within the muscle during the sports performance, which is highly required for short-term intense activity. It has been observed that sugar cane juice is not considered as performance enhancer, since it has extremely high sugar content that may hamper the performance by slowing gastric emptying and will delay in supply of nutrients and oxygen to the cells. Additionally, sugar-rich foods may trigger the fatigue and hunger pangs in the athletes, by instant secretion of insulin from pancreas (resulted hypoglycemia) [31].

2.4.5 ENERGY BOOSTER FOODS

It has been observed that an athlete can lose 30% of total 24-hour energy within one hour of hard training. Therefore, it is highly recommended to replenish this energy debt by providing a healthy supplement. Anaerobic energy systems provide energy during high intensity exercise of short duration, whereas the aerobic system becomes predominant during prolonged exercise. The major oxidative substrates are carbohydrates and lipids. The body will utilize the muscle glycogen as a source of carbohydrate. Therefore, carbohydrate and fat should be enough in an athlete's diet [10].

For the past few decades, energy drinks are generally marketed to individuals, who are interested in athletics and active lifestyle. The chief ingredient of energy drink is caffeine, which is often used by the sports person as performance enhancer. Along with the caffeine, the drinks also include a combination of other components, such as "sucrose, taurine, guarana, niacin, ginseng, pyridoxine, and cyanocobalamin (CNCbl);" however, the erogenicity of these energy drinks is imparted by the caffeine and glucose therein. It is reported that the use of energy drink with exercise can also reduce body fat. However, it is also stated that addition of pharmacologically active substance with energy drinks are accompanied with undesirable effects. It has been observed among the college-aged male athletes that they often consume this energy drinks in combination with alcohol, which certainly will hamper the erogenicity of the energy drink [37, 52].

Ballard et al. reviewed that the ergogenic effects of caffeine (6 mg/kg body weight) is detected in the aerobic activities (such as running, cycling, and swimming (even for 5 minutes)). At the same dose, caffeine is helpful to speed up the "endurance testing with high-intensity exercise" for 60 to 180s, while studies using the "30-second Wingate test" (cycling test which is used to measure the anaerobic power of the lower body) have shown minimal ergogenic effect with caffeine [5].

Sucrose, vitamin B_3 (niacin), vitamin B_6 (pyridoxine), vitamin B_{12} (cyanocobalamin), and guarana, ginseng is used for making energy drinks. Sugar, in various forms, is used in the energy drinks. Guarana, a usual ingredient of energy drink, contains caffeine and therefore is expected to have the same effects as caffeine. Ginseng is also used by sports individual; however, its health benefits have not been documented till date. Niacin, pyridoxine, and CNCbl are also consumed as ingredients for making energy booster drinks [6].

2.4.6 REHYDRATION FLUID

For an athlete, maintaining proper hydration is extremely challenging for body's metabolic balance. Consumption of sufficient amount of water prior to physical activity is always recommended. During and after exercise to compensate the weight loss, athletes should drink three glasses of water per day. One study found that rehydration in athletes may decrease the risk of developing chronic diseases, such as urolithiasis, impaired renal function and high blood sugar levels, and free radical damage owing to high-intensity exercise. Therefore, it is suggested to keep an "optimal hydration" index (osmotic pressure of 500 milliosmolar concentration/kg) to compensate the daily fluid loss by an athlete [32, 33, 39, 44].

It is important to achieve the proper rehydration after fluid intake. Rehydration drinks should have moderately high level of electrolytes to replace the sweat loss. The volume of fluid should be greater than the volume of sweat loss [39]. For rehydration, a drink with 5–10% carbohydrate and

30–40 mEq of Na has proved to be beneficial. It has been reported that 50 g of carbohydrate per hour should be taken to maximize the glycogen depletion after first two hours of exercise [23].

2.4.7 BONE AND JOINT PROTECTOR FOODS

A diet with proper nutrients helps in the healing process of the injured athletes, such as it improves recovery from muscle or bone injuries or from surgery. The diet of these athletes requires extra energy, nutrients (including protein, vitamins, minerals) and antioxidants [36]. It has been found that pre-adolescent runners consumed dietary supplements on regular basis. Behaviors consistent with dietary control and history of bone stress injury were related to likelihood for supplement use. Table 2.2 indicates health benefits of different sports foods for athletes.

2.5 SAFETY OF FOOD SUPPLEMENTS

There are many challenges in using these sports supplements for their possible side effects. Few studies have reported that dietary behavior is responsible for developing disorders namely, high blood pressure, coronary artery disease and some cancers. Therefore, it is suggested that the performance enhancing, or ergogenic effects of bioactive component has to be confirmed first. It is crucial to confirm whether these effects are "galactic-anaerobic, or glycolyto-anaerobic, or aerobic." Further one has to confirm the manifestation of bioenergetic properties in the athletes, such as in terms of power, capacity or efficiency of anaerobic and aerobic energy conversion, etc.

The consumption of "over-the-counter (OTC)" sports supplements is very common among the athletes. It is noteworthy that the supplements are safe only at their recommended doses. Overconsumption of these foods may lead to sever to fatal health concerns. The following compounds reported to have possible harmful side effects, such as:

- β-Alanine (paresthesia, pruritus, and itching);
- Creatine nitrate (high blood urea nitrogen, impact on lymphocytes and monocytes of those have high resting blood urea nitrogen and weakened immune system);
- Guanidino acetic acid (mild nausea, loss of appetite, abdominal bloating and diarrhea, and higher incidence of elevated serum creatinine);
- Higen amine (moderate dizziness and nausea);

- Nitrate ("endogenous nitrozation and gastric cancer under conditions of low vitamin C and high nitrite levels");
- Omega-3 fatty acids (prolonged bleeding time at high doses).

In addition to these risks, cross-effects could be possible due to unnecessary combining different supplements at the same time [25, 43].

Additionally, it has been found that quality control of sports foods is extremely challenging due to improper handling during manufacturing and storage. The Food and Drug Administration (FDA) found several reports regarding these supplements, such as presence of microbiological contamination undeclared allergens or foreign objects, etc. The contaminants in the products include "stimulants, estrogenic compounds, diuretics, and anabolic agents, including anabolic androgenic steroids, designer steroids and prohormones." Therefore, accurate knowledge and labeling of sports foods with proper information could lead to safe use of sports foods by athletes [25, 43].

2.6 SUMMARY

A range of dietary strategies can be adapted by the athletes, to improve their performance. Nonetheless, recommendations for dietary supplements can vary depending on individual athlete and their sport. This recommendation should be provided by an appropriately qualified professional and therefore optimal performance can be achieved. Athletes should use dietary supplements with caution and as part of an overall nutrition and performance plan. To maximize training outcomes as well as to maintain health, adequate amount of energy should be consumed during the periods of high-intensity and/or long duration training.

Carbohydrate stores in the body are an important source of fuel for both the brain and muscle during exercise which can be manipulated by dietary intake and exercise. Athlete should be provided with first class protein (0.3 g/kg body weight) in post-exercise phase and also continue it regularly [20]. The fat consumption for most athletes depends on their eating habits, however the range is between 20% and 35% of total daily energy intake. The diet of the athletes should meet their "recommended dietary allowance (RDA)/adequate intake (AI)" for all macro- and MNs. In has been suggested prior to an event (before 1–4 hours), fluids and foods should be consumed to help in carbohydrate reserve. It also ensures proper hydration status and maintains gastrointestinal (GI) health throughout the event. Athlete who consumes an appropriate diet providing high energy from a variety of nutrient dense foods, vitamin, and mineral supplements are unnecessary for them. Supplement with "multivitamin/mineral" may be given to the athletes when proper diet is not taken.

Lastly, appropriate use of dietary supplements can support the athletes to achieve proper nutrition and train himself smoothly without having any complication from injury or stress. However, extensive effort and expert knowledge should be there to identify the appropriate products, how to incorporate them into the athlete's sports nutrition plan, and how to ensure optimum benefits. Effectiveness, safety, and risks of the sport nutrition plan should be identified by a well-informed sports nutrition professional with the help of a strict risk-benefit analysis.

KEYWORDS

- **antioxidant supplementation**
- **ergogenic aids**
- **fortification**
- **immunity booster foods**
- **rehydration fluid**
- **sports foods**
- **supplementation**

REFERENCES

1. Alessio, H. M., Hagerman, A. E., Romanello, M., Carando, S., Threlkeld, M. S., Rogers, J., & Wiley, R. L., (2002). Consumption of green tea protects rats from exercise-induced oxidative stress in kidney and liver. *Nutrition Research, 22*(10), 1177–1188.
2. Alford, C., Cox, H., & Wescott, R., (2001). The effects of red bull energy drink on human performance and mood. *Amino Acids, 21*(2), 139–150.
3. American College of Sports Medicine, American Dietetic Association, & Dietitians of Canada, (2000). Joint position statement: Nutrition and athletic performance. *Medicine and Science in Sports and Exercise, 32*(12), 2130–2145.
4. Anderson, J. W., Baird, P., Davis, R. H. Jr., Ferreri, S., Knudtson, M., Koraym, A., Waters, V., & Williams, C. L., (2009). Health benefits of dietary fiber. *Nutrition Reviews, 67*(4), 188–205.
5. Ballard, S. L., Wellborn-Kim, J. J., & Clauson, K. A., (2010). Effects of commercial energy drink consumption on athletic performance and body composition. *The Physician and Sports Medicine, 38*(1), 107–117.

6. Beck, K. L., Thomson, J. S., Swift, R. J., & Von, H. P. R. (2015). Role of nutrition in performance enhancement and postexercise recovery. *Open Access Journal of Sports Medicine, 6*, 259–267.
7. Binns, C. W., Lee, M. K., & Lee, A. H., (2018). Problems and prospects: Public health regulation of dietary supplements. *Annual Review of Public Health, 39*(1), 403–420.
8. Blumberg, J. B., Bailey, R. L., Sesso, H. D., & Ulrich, C. M., (2018). The evolving role of multivitamin/ multimineral supplement use among adults in the age of personalized nutrition. *Nutrients, 10*(2), 248–255.
9. Bragg, M. A., Roberto, C. A., Harris, J. L., Brownell, K. D., & Elbel, B., (2018). Marketing food and beverages to youth through sports. *The Journal of Adolescent Health: Official Publication of The Society for Adolescent Medicine, 62*(1), 5–13.
10. Brotherhood, J. R., (1984). Nutrition and sports performance. *Sports Medicine, 1*(5), 350–389.
11. Burke, L., Cort, M., Cox, G., Crawford, R., Desbrow, B., Farthing, L., Minehan, M., Shaw, N., & Warnes, O., (2006). Supplements and sports foods. In: Burke, L., & Deakin, V., (eds.), *Clinical Sports Nutrition* (Vol 3, pp. 485–579). Sydney – Australia: McGraw-Hill Education.
12. Candow, D. G., Kleisinger, A. K., Grenier, S., & Dorsch, K. D., (2009). Effect of sugar-free red bull energy drink on high-intensity run time-to-exhaustion in young adults. *J. Strength Cond Res, 23*(4), 1271–1275.
13. Catalani, V., Negri, A., Townshend, H., Simonato, P., Prilutskaya, M., Tippett, A., & Corazza, O., (2021). The market of sport supplement in the digital era: Analysis of perceived risks, side-effects and other safety issues. *Emerging Trends in Drugs, Addictions, and Health, 1*, E-article ID 100014.
14. Clarkson, P. M., & Thompson, H. S., (2000). Antioxidants: What role do they play in physical activity and health? *The American Journal of Clinical Nutrition, 72*(2 Suppl.), S637–S646.
15. Cui, P., Li, M., Yu, M., Liu, Y., Ding, Y., Liu, W., & Liu, J., (2022). Advances in sports food: Sports nutrition, food manufacture, opportunities, and challenges. *Food Research International, 157*, E-article ID 111258.
16. Cunningham, S. A., Shaikh, N. I., Datar, A., Chernishkin, A. E., & Patil, S. S., (2021). Food subsidies, nutrition transition, and dietary patterns in a remote Indian district. *Global Food Security, 29*, E-article ID 100506.
17. Deldicque, L., & Francaux, M., (2016). Potential harmful effects of dietary supplements in sports medicine. *Current Opinion in Clinical Nutrition and Metabolic Care, 19*(6), 439–445.
18. Dillard, C. J., Gavino, V. C., & Tappel, A. L., (1983). Relative antioxidant effectiveness of α-tocopherol and γ-tocopherol in iron-loaded rats. *The Journal of Nutrition, 113*(11), 2266–2273.
19. Earnest, C. P., Quindry, J., Panton, L., & Broeder, C., (2015). Effect of deer antler velvet on aerobic, anaerobic and strength performance. *Central European Journal of Sport Sciences and Medicine, 9*(1), 17–26.
20. FAO/ WHO/ UNU, (1985). *Energy and Protein Requirements*. Technical report series no. 724. Geneva – Switzerland: World Health Organization (WHO).
21. Forbes, S. C., Candow, D. G., Little, J. P., Magnus, C., & Chilibeck, P. D., (2007). Effect of red bull energy drink on repeated Wingate cycle performance and bench-press muscle

endurance. *International Journal of Sport Nutrition and Exercise Metabolism, 17*(5), 433–444.

22. Francis, K., & Hoobler, T., (1987). Comparison of peak torque values of the knee flexor and extensor muscle groups using the cybex II® and lido 2.0® isokinetic dynamometers. *Journal of Orthopaedic & Sports Physical Therapy, 8*(10), 480–483.
23. Gisolfi, C. V., & Duchman, S. M., (1992). Guidelines for optimal replacement beverages for different athletic events. *Medicine and Science in Sports and Exercise, 24*(6), 679–687.
24. Grout, A., McClave, S. A., Jampolis, M. B., Krueger, K., Hurt, R. T., Landes, S., & Kiraly, L., (2016). Basic principles of sports nutrition. *Current Nutrition Reports, 5*(3), 213–222.
25. Gunina, L. M., Shustov, Y. B., Belenichev, I. F., Vysochina, N. L., Golovashchenko, R. V., & Morozova, O. V., (2022). Specialized nutrition for athletes: Evaluation of ergogenic action using the principles of evidence-based medicine. *Pharmacia, 69*(1), 37–41.
26. Indoria, A., & Singh, N., (2016). Role of nutrition in sports: A review. *Indian Journal of Nutrition, 3*(2), 147–153.
27. Ivy, J. L., Kammer, L., & Ding, Z., (2009). Improved cycling time-trial performance after ingestion of a caffeine energy drink. *International Journal of Sport Nutrition and Exercise Metabolism, 19*(1), 61–78.
28. Jaughan, R., & Shirreff, S. M., (1997). Recovery from prolonged exercise: Restoration of water and electrolyte balance. *Journal of Sports Sciences, 15*(3), 297–303.
29. Jeukendrup, A. E., (2008). Carbohydrate feeding during exercise. *European Journal of Sport Science, 8*(2), 77–86.
30. Johnson, C. L., Dohrmann, S. M., Burt, V. L., & Mohadjer, L. K., (2014). *National Health and Nutrition Examination Survey: Sample Design, 2011–2014* (p. 45). Atlanta – GA: US Department of Health and Human Services, Center for Disease Control and Prevention, National Center for Health Statistics.
31. Juhn, M. S., (2003). Popular sports supplements and ergogenic aids. *Sports Medicine, 33*(12), 921–939.
32. Kalpana, K., Lal, P. R., Kusuma, D. L., & Khanna, G. L., (2013). The effects of ingestion of sugarcane juice and commercial sports drinks on cycling performance of athletes in comparison to plain water. *Asian Journal of Sports Medicine, 4*(3), 18–25.
33. Kenefick, R. W., & Sawka, M. N., (2007). Hydration at the work site. *Journal of the American College of Nutrition, 26*(suppl. 5), S597–S603.
34. Kenney, E. L., Long, M. W., Cradock, A. L., & Gortmaker, S. L., (2015). Prevalence of inadequate hydration among US children and disparities by gender and race/ethnicity: National health and nutrition examination survey, 2009–2012. *American Journal of Public Health, 105*(8), e113–e118.
35. Krumbach, C. J., Ellis, D. R., & Driskell, J. A., (1999). Report of vitamin and mineral supplement use among university athletes in division I institution. *International Journal of Sport Nutrition and Exercise Metabolism, 9*(4), 416–425.
36. Kunstel, K., (2005). Calcium requirements for the athlete. *Current Sports Medicine Reports, 4*(4), 203–206.
37. Manore, M., & Thompson, J., (2000). Protein and exercise. *Sport Nutrition for Health and Performance, Human Kinetics, 2000*, 105–128.

38. Zimmermann, M. B., (2003). Vitamin and mineral supplementation and exercise performance. *Schweizerische Zeitschrift fur Sport Medizin und Sport Traumatologie (Swiss Journal for Sports Medicine and Sports Traumatology), 51*(1), 53–57.
39. Market, A. P. R., (2018). *Global Industry Analysis and Forecast 2016–2024.* Online; https://www.persistencemarketresearch.com/market-research/xylanase-market.asp (accessed on 2 November 2023).
40. Nielsen, P., & Nachtigall, D., (1998). Iron supplementation in athletes. *Sports Medicine, 26*(4), 207–216.
41. Pandey, S., & Singh, V., (2013). Nutritional needs of athletes. *Pedagogics, Psychology, Medical-biological Problems of Physical Training and Sports, 4*, 88–92.
42. Parnell, J. A., Wiens, K. P., & Erdman, K. A., (2016). Dietary intakes and supplement use in pre-adolescent and adolescent Canadian athletes. *Nutrients, 8*(9), 526–533.
43. Parnell, L. K., (2013). Sport nutrition for young athletes. *Paediatrics & Child Health, 18*, 200–205.
44. Perrier, R., Pirog, A., Jaffredo, M., Gaitan, J., Catargi, B., Renaud, S., & Lang, J., (2018). Bioelectronic organ-based sensor for microfluidic real-time analysis of the demand in insulin. *Biosensors and Bioelectronics, 117*, 253–259.
45. Pramuková, B., Szabadosová, V., & Šoltésová, A., (2011). Current knowledge about sports nutrition. *The Australasian Medical Journal, 4*(3), 107–111.
46. Yagüe, M., Collado-Yurrita, L., Ciudad-Cabañas, M. J., & Cuadrado-Cenzual, M. A., (2020). Role of vitamin D in athletes and their performance: Current concepts and new trends. *Nutrients, 12*(2), 579–585.
47. Quinlivan, A., Irwin, C., Grant, G. D., Anoop-Kumar, S., Skinner, T., Leveritt, M., & Desbrow, B., (2015). The effects of Red Bull energy drink compared with caffeine on cycling time-trial performance. *International Journal of Sports Physiology and Performance, 10*(7), 897–901.
48. Ratamess, N. A., Hoffman, J. R., Ross, R., Shanklin, M., Faigenbaum, A. D., & Kang, J., (2007). Effects of an amino acid/ creatine energy supplement on the acute hormonal response to resistance exercise. *International Journal of Sport Nutrition and Exercise Metabolism, 17*(6), 608–623.
49. Saat, M., Singh, R., Sirisinghe, R. G., & Nawawi, M., (2002). Rehydration after exercise with fresh young coconut water, carbohydrate-electrolyte beverage, and plain water. *Journal of Physiological Anthropology and Applied Human Science, 21*(2), 93–104.
50. Sahlin, K., (2014). Muscle energetics during explosive activities and potential effects of nutrition and training. *Sports Medicine (Auckland, N.Z.), 44*(Suppl 2), S167–S173.
51. Sharman, I. M., Down, M. G., & Sen, R. N., (1971). The effects of vitamin E and training on physiological function and athletic performance in adolescent swimmers. *British Journal of Nutrition, 26*(2), 265–276.
52. Thomas, D. T., Erdman, K. A., & Burke, L. M., (2016). Nutrition and athletic performance. *Med. Sci. Sports Exerc., 48*, 543–568.
53. Volpe, S. L., (2007). Micronutrient requirements for athletes. *Clin. Sports Med., 26*(1), 119–130.
54. Watson, P., (2008). Nutrition, the brain, and prolonged exercise. *European Journal of Sport Science, 8*(2), 87–96.
55. Williams, M. H., (1984). Vitamin and mineral supplements to athletes: Do they help? *Clinics in Sports Medicine, 3*(3), 623–637.

56. Wyon, M. A., Koutedakis, Y., Wolman, R., Nevill, A. M., & Allen, N., (2014). The influence of winter vitamin D supplementation on muscle function and injury occurrence in elite ballet dancers: A controlled study. *Journal of Science and Medicine in Sports, 17*(1), 8–12.

CHAPTER 3

Functional Foods for Geriatrics

Kirti Joshi, Ranjana Singh, and Manisha Singh

ABSTRACT

The worldwide population is growing older day by day. The increase in life expectancy is the outcome of recent scientific, technological, and social advancements. In terms of health, old age is one of the most fragile and prone stages. The increase in life expectancy, however, is not accompanied by a healthy aging process, as older persons suffer a number of physiological changes that can lead to the development of a variety of diseases or chronic ailments. Nutrition is a critical determinant of geriatric mass, particularly in those over the age of 60. Adequate nutrition is essential for healthy aging. Physical activities are also part of the process of making aging easier. Before beginning any type of physical activity, health concerns such as hypertension, cholesterol, renal failure, joint problems, and cardiac problems should be considered. Malnutrition is common among elderly persons and is a prominent geriatric condition with multiple etiology and serious repercussions for health and quality of life. In recent years, there has been a lot of interest in the creation of functional foods with customized qualities for the aged population, with the goal of improving the bio-accessibility and bioavailability of essential nutrients in their diets. Nutrition is essential for human survival. The health and well-being of older persons can now be considerably enhanced by using revolutionary food processing technology. These advancements, combined with growing knowledge of age-related physiological changes, should be used to develop nutritional methods and nutritive products aimed at preventing diseases and ensuring a healthy global population as it ages.

The Functional Foods: Nutrient and Health Benefits.
Megh R. Goyal, Junaid Ahmad Malik, and Anu Kumari (Eds.)

3.1 INTRODUCTION

Geriatrics focuses on the unique requirements of older people, emphasizing functional independence even in the presence of chronic illness. The treatment of geriatric individuals necessitates a unique approach and is quite difficult. Geriatric drugs attempt to improve health in older people by preventing and treating diseases and disabilities. Because aging is linked to an increased risk of chronic disease, disability, and death, developing effective dietary interventions for promoting healthy aging is an active but complex area of research. Aging of population is a global phenomenon. Low immunity is also contributed to the increased morbidity and mortality caused by infectious pathogens in the elderly. Micronutrient (MN) deficiencies are more common among the elderly due to a range of causes such as social, physical, economic, and emotional barriers to eating. As a result, there is a strong need to refocus our efforts on measures to avoid chronic disorders linked to aging. Individuals must make greater efforts to adopt healthy lifestyle behaviors, such as eating a more nutritious food [35].

With the advancement of medical sector, the proportion of older people is expanding at the fastest rate among all segment of population. The life expectancy at birth will have increase from the current 70–77 years by 2045. By 2050, it is predicted that about 2 billion people will be over the age of 60, with almost 15% of the population in various nations (e.g., Japan, Germany, and South Korea) being over the age of 80. According to these projections, there will be over 400 million people over the age of 80 in the world by 2050 [73]. The need of food and nutrition for very old and weak people requires a special attention [46].

Aging is defined as a biological phenomenon characterized by "temporal continuity, heterogeneity on a cellular, somatic, and molecular level, and the ability to be controlled," according to gerontologists [31]. Aging is a multifaceted process that causes changes on a biological, physiological, pathological, psychological, behavioral, and environmental level, and as a result, various basic human functions might be weakened or reduced during this process. These changes that occur as people age may be unharmful in some situations or may lead to more complicated disorders that make daily tasks more difficult, cause impairments, and raise disease susceptibility. As a result, aging is regarded as one of the key risk factors for a variety of human chronic diseases [72].

The loss of physiological functions in older adults is one of the most significant age-related changes since it has a direct impact on their metabolism, reducing their ability to digest and absorb food nutrients, leading to nutritional deficiencies and sometimes, malnutrition. Nutrition is rapidly being connected to the most common age-related chronic disorders, such as cancer, dementia, atherosclerosis, type II diabetes and heart disease. This impairs the quality of life resulting in physical function and cognitive decline. Bone density loss is prevalent among the elderly, raising the risk of osteoporosis [51].

Other age-related issues include sarcopenia, muscle loss, and malnutrition, all of which result in a loss of strength, endurance, disease resistance, and functional decline [3, 104]. Naturally, older adults require a greater and more specific nutrient intake; nonetheless, it is estimated that up to 35% of population who are aged 65 years or above have nutrient deficiencies in protein, calories, minerals, and vitamins [3]. To overcome or mitigate such ailments, older persons are routinely forced to take a mixture of multiple medications, an increasingly prevalent scenario that is both costly and ineffective. The creation and implementation of more effective and healthier ways to improve the nutrition of older persons is a pressing topic that is receiving more attention.

This chapter focuses on the relevance and development of appropriate food products for the geriatrics to improve their nutrition, health, and well-being.

3.2 WORLDWIDE ELDERLY POPULATION

The term "elderly population" is used to refer to those above the age of 65. The proportion of people over the age of 65 is steadily rising as the world population ages. Life expectancy has increased, but worldwide fertility has decreased, resulting in this pattern [3]. United Nations data shows that there were more people 65 and older than children under the age of five, making them the fastest-growing age group, during 2018. Over the last decade, the number of people aged 65 and older has increased from 6% to 9% of the world population, according to estimates from the World Health Organization (WHO) [73]. The proportion of people over the age of 65 is expected to more than quadruple by 2050, from the current 16% to the projected 1.5 billion people, constituting a population that would outnumber those aged 15 to 24 (United Division). In 2019, the world's population at birth had a life expectancy of 72.6 years, which is expected to rise to 77.1 years by 2050.

As a result of these trends, social and economic changes are required to assure long-term growth and active aging. As a result, it is crucial to develop methods for enhancing health and well-being of older people, as well as encouraging them to stay active as they age.

3.3 AGING-ASSOCIATED PHYSIOLOGICAL CHANGES AND HEALTH-RELATED ISSUES

Molecular reactions and functions in the human body decline or stop as a consequence of the accumulation of cellular degeneration over time. Due to the complexity of the aging process, which includes natural aging, chronic conditions or somatic illnesses, cognitive, and psychological, social/environmental and functional needs, psychological, and mental, practical, and social/environmental changes [47]. It is because of the normal aging process that older people have difficulty in interacting with others and their surroundings. The most typical alterations observed by older persons owing to normal aging include decreased visual acuity, hearing loss, impaired olfaction, and gustation [13].

Sarcopenia, a disorder characterized by the loss of skeletal muscle strength and total mass, is also a common side effect of normal aging. Increased fat mass generally follows, which has an impact on quality of life and on people's ability to take medication [21]. The immune system's dysregulated reactions, such as inadequate vaccination responses, increased tissue oxidation and inflammation, autoimmunity, and decreased generation of autoantibodies, cause Immune senescence, which is a typical aging change [77]. Hence, there is a view that immuno-degeneration enhances the susceptibility and risk of occurrence of numerous somatic illnesses and chronic disorders.

Aside from cancer, cardiovascular disease (CVD; such as coronary artery and peripheral arterial disease, heart attacks) and malignant neoplasms (including colon, breast, lung, and prostate cancer) are the top causes of mortality in the EU among older persons [24]. Aging brings about a number of issues, such as atherosclerosis that causes stiffness and remodeling of the vascular system (i.e., plaque accumulation in the blood vessels). Hypertension, the most frequent chronic disorder among elderlies, is defined by elevation of blood pressures which inevitably raises the possibility of atherosclerosis and coronary heart disease (CHD) [74]. In this demographic, *diabetes mellitus* is becoming more common place, which raises the threat of cardiovascular and microvascular disease [56].

Osteoporosis and osteoarthritis are two more diseases/conditions that predominantly affect elderly persons. Osteoporosis is a condition in which the bone density decreases with age, increasing the risk of fractures. However, a break in articular cartilage leads to hypertrophic development of the bone around it, resulting in the creation of osteophytes, which causes inflammation and damage to soft tissue structures. Osteoarthritis is a degenerative disorder [25, 29, 61].

Aging may bring about a variety of psychological and cognitive changes. There are typical aging results such as difficulties locating words, increased processing times, and even minor memory loss that is short-term [47]. Dementia and other neurodegenerative illnesses are more frequent in the aged population because of the functional and structural changes in the adult brain that occur with aging. In reality, the global mortality from Alzheimer's disease is on the rise [53].

Normal aging and the aforementioned somatic diseases may be blamed for the functional and social/environmental alterations. Aging, mobility, weakness, dependence, and sadness are the most noticeable bodily reactions to the underlying changes that are more difficult to detect.

Modifications in the gut microbiota and the gut-brain axis are both linked to the physiological changes that come with advancing age. Microorganisms that live in digestive system of human are recognized as the human gut microbiome. Our health and well-being are constantly linked to our gut microbiota, which changes and adapts throughout our lives. We don't yet know whether gut microbiota changes are caused by or a result of aging. However, it was shown that the gut microbiota profiles of older persons are consistently different from those of younger ones. Older persons are more likely to have low amounts of short-chain fatty acids (SCFAs), which might explain these results [6]. The disbalance of the gut microbiome in aging adults and the breakdown of the gut barrier may contribute to the elevated levels of lipopolysaccharides, SCFAs, immunoglobulin A (IgA), serotonin, and histamines along with other things. This could disrupt normal gut-brain communication, leading to cognitive impairment and dementia [71].

Although the physiological changes that may occur are interwoven in nature, yet each person will experience them in their own unique manner as they get older. This process association may also cause or enhance the risk factor for other adjacent illnesses to be initiated or increased. Because of these reasons, a variety of nutritional and pharmacological approaches are being researched in an effort to reduce or postpone aging.

3.3.1 AGING-ASSOCIATED PSYCHOSOCIAL CHANGES

Age-related sadness was strongly linked to physical sickness and poverty, as has been seen in research studies. Sad people in low-income neighborhoods are more likely to believe in psychosocial and interpersonal explanations for their depression than in biological ones. Tolerance for depressive symptoms and functional impairment among the elderly may be increased by families' low expectations [51].

Poverty and social exclusion have a complicated link with physical and mental health. About 15% of people over the age of 65 experience hopelessness, the most prevalent reason of unexplained weight loss, and this number rises dramatically among those who reside in long-lasting care facilities.

Dementia, Alzheimer's disease, and other neurological illnesses are causing an alarming surge in memory loss. The decline in income, health issues, and the death of a spouse or close friend all contribute to social isolation among the elderly. As a consequence, a person's nutritional state may suffer [93].

3.3.2 AGE-RELATED CHANGES IN DIETARY REQUIREMENTS

The body's need for calories diminishes with age, although this varies widely across people, depending on their level of activity and diet. Around 1,800 calories are the recommended daily caloric intake for a person of average health. Adults of all ages, regardless of age, have the same protein requirements. Protein has to be consumed at the rate of 0.8–1.0 g/kg of body weight. Vitamins and minerals are found in green leafy vegetables, other vegetables, and fresh fruits; and these help to prevent illness. Preventable diseases such as heart disease may be caused by an excess of plant and animal-based fat in our diets.

ALA acids are found in legumes, green leafy vegetables, fenugreek, and mustard seeds, among other foods. Calcium excretion is increased when salt consumption is increased, which may lead to a decrease in bone density. Current research shows that increased salt consumption may harm the cardiovascular system, bone density, and the gastrointestinal tract (GIT). The average American's sodium consumption far exceeds the recommended daily limit. A daily intake of more than 6 grams is harmful. Due to its widespread usage in the kitchen, salt serves as a food fortification vehicle. For their ability to boost immunity, antioxidant vitamins including vitamin E and carotenoids continue to get attention. For riboflavin, vitamin B_6, and vitamin B_{12}, the demand for vitamin A reduces and zinc has to be increased.

A well-balanced diet that includes items from many different food categories is the best way to get all the nutrients your body needs. Some chronic illnesses, such as CVD, cataracts, and cancer, may be prevented by taking these supplements. Fruit liquids lack the nutritional value of whole fruits. Type 2 diabetes, fatty liver disease, gallstones, high cholesterol and triglycerides, osteoarthritis, hypertension, and other cardiovascular illnesses, as well as certain malignancies and psychological difficulties, are all more common in obese and overweight people.

A well-balanced diet necessitates consuming a wide range of foods. Preservatives, salt, sugar, and fat-laden processed meals may constitute a health hazard if eaten on a regular basis. In order to maintain good health, we need to drink a lot of water. As a general rule of thumb, most people need to drink at least 25–30 ml/kg of body weight in order to maintain a healthy level of water consumption. The basis of good health is based on healthy and pleasant food ideas and culinary methods. To maintain a healthy and active lifestyle for the elderly, they need extra vitamins and minerals. The elderly should make an effort to eat a wide range of meals that are high in nutrients. They must strike a healthy equilibrium between diet and exercise. Eat a lot of little meals throughout the day. Avoid foods with a lot of oil and spices.

Walking, yoga, and other forms of regular exercise all need the elderly to move their bodies. Chronic illnesses such as Type 2 diabetes, high blood pressure, heart disease, osteoporosis, arthritis, and certain kinds of cancer may be reduced by maintaining a healthy weight and body composition. This can be achieved by a variety of diet and exercise plans. The benefits of regular exercise go well beyond physical health; they also include mental health benefits including preventing depression and enhancing one's feeling of well-being and self-esteem. Men and women over the age of 40 should get the advice of a doctor or health care practitioner before embarking on a severe exercise regimen, as well.

3.3.3 DIETARY CONSIDERATIONS

It has become more challenging for health and government services to cope with the increasing lifespan of the world's population. Increasing life expectancy throughout the globe and ensuring good aging and extending healthy active years are crucial. Nutrition has a significant influence on life expectancy and quality of life, mostly through ensuring enough calorie intake and a healthy weight. Furthermore, the risk of "nutritional frailty"—the loss of either weight, muscle mass and strength or key physiologic reserves—is significantly related with the likelihood of older persons' inability to meet basic nutritional demands [86].

For older adults, achieving these conditions can be more difficult due to a variety of reasons including dental and chewing problems; changes in smell and taste perception, reduced mobility, which makes fresh food more difficult to obtain; loss of appetite, alterations in the digestive (GI) tract; reduction in metabolic rate, among others [9].

Since age-related changes are experienced differently in each person, the dietary demands of older individuals have yet to be fully characterized. For this reason, a more nutrient-dense, but less energy-dense diet (i.e., foods rich in nutrients but low in calories) has been recommended for a healthy aging diet [103].

Older individuals are increasingly consuming processed carbohydrates, saturated fats, processed meats, whole-fat dairy products and goods that lead to overnutrition and obesity. This method considers this tendency. Because of this, kids don't get enough nutrients like protein, dietary fibers, omega-3 fatty acids, vitamins (B_6 and D, E, and K, etc.), and certain minerals like calcium [67]. Due to a deficiency in the consumption of fruits, vegetables, and seafood as well as gastrointestinal (GI) abnormalities, malabsorption might occur [86].

3.3.4 MALNUTRITION IN GERIATRIC PATIENTS

In spite of their late age, geriatric patients are still considered elderly because of their extreme fragility and propensity for chronic illness. During aging, several physiological systems degrade, resulting in a decrease in functional reserves and an increase in sensitivity to stress. To retain or regain independence in daily life as far as feasible, comprehensive therapy is needed to address all aspects of a person's physical, mental, and social well-being. Malnutrition (i.e., protein-energy-malnutrition or undernutrition) deserves attention among other common geriatric disorders, such as dementia, delirium, falls, or incontinence, which have a high prevalence, complex etiology, and poor outcomes [100].

3.3.5 HEALTH CARE FOR THE UNDERNOURISHED OLD-AGED PEOPLE

Nutritional treatments for the elderly may have a variety of goals, including preventing disease, enhancing quality of life, and promoting independence.

Muscle mass and function are primarily linked to nutritional status, which may replenish protein and energy storage to meet metabolic stress demands. Muscle function and capacity for rehabilitation are also closely linked. Health-related quality of life is more important than reducing mortality in older adults, and restoring food intake may play a direct role as: (i) an essential mediator of pleasure and well-being; Morbidity reduction, including an improved outcome of underlying chronic diseases; (ii) reduction of mortality as a result; and (iii) reduction of the hospital length of stay, the need for sub-acute care stays, reduction of nursing home admissions, the number of medical examinations and prescriptions. Consequently, the goals and techniques to malnutrition therapy in older patients are similar to those in younger patients, but the preservation of function and quality of life take precedence over mortality.

There are a wide variety of nutritional treatments for older persons that may help ensure optimal intake and go well beyond just giving enough calories and minerals. It is essential to identify and manage several reasons in order to provide enough nourishment to elderly patients. Furthermore, adequate intake may be supported by various strategies—first of all, but not only by help with eating if required. When it comes to implementing direct nutritional measures, oral techniques are always the preferred method. Modifications to common meals and the provision of dietary supplements orally are just a few examples. Also, for elderly patients, enteral and parenteral nutrition are significant alternatives despite the fact that they are seldom used.

In old age, various lifestyle and physiological changes take place in the body. Therefore, the diets of this age group should be altered accordingly. As food is required for human sustenance and help to maintain the body function. As per the data, the global population predictions shows that the old population will continue to expand and overtaking younger populations.

Old age population require more medical care, pharmaceutical therapy and other major preventions to get away from the diseases, which can be accomplished by proper nutrition. That's why food industries and food nutritionists are working together to develop the novel foods for the old-age population. Novel foods include various research fields, including development of functional foods and manufacturing of food by 3D (three dimensional) printing technology. These foods delivered all the important nutrients along with desired organoleptic and rheological properties and also provide satiety to the old age consumers [25].

3.4 FUNCTIONAL FOODS

The term “Functional foods” was coined in mid-1980’s by D.J. Nicholas in Japan. Other terms for functional foods include “nutraceuticals,” “vitafoods,” “medical foods,” “probiotics,” and “pharmafoods,” “health promoting foods,” “designer foods,” and “fortified foods” [80]. The awareness regarding diet as a part of a healthy lifestyle can significantly reduce illness risk and increase health has spawned a profitable industry for a variety of new products such as “functional foods,” “nutraceuticals,” and “novel foods” [5].

Functional foods resemble conventional foods in terms of appearance. Functional foods are part of our regular diet. These types of foods have various physiological benefits and can have the ability to lower down the risk of chronic disease [10]. These foods not only provide basic nutrients such as carbohydrates, proteins, vitamins, fats, etc., in appropriate quantity required for healthy survival of human being but also helps in maintaining gut health as well [27]. These products are designed to improve the metabolic function [25].

In the 1990s, consumers began to perceive food from a whole new perspective. This ‘evolving face’ of food has given rise to the intriguing field of functional foods in the food and nutrition sciences. Functional foods are complete, fortified, enriched, or improved foods that have a potentially beneficial effect on health when consumed on a regular basis at effective amounts as part of a varied diet [37].

The goal of developing functional meals for the elderly is to provide bioactive compounds that may minimize the risk of diet-related illnesses and chronic disorders.

3.4.1 DEFINITION OF A FUNCTIONAL FOOD

According to the “European Commission Concerted Action” on Functional Food Science in Europe (FUFOSE) – “*Functional foods are something that has been shown positive influence on one or more specific functions in the body, in a way that is significant to either better health and well-being and/or reduced illness risk, in addition to adequate nutritional effects. A functional food must be a food and shows its benefits in levels that may be expected to be ingested in the diet: it is not a pill or a capsule, but part of the normal diet*” [63].

According to the Institute of Food Technologists (IFT), functional foods are the foods and components of food that deliver health benefits to the consumer beyond the basic nutrition (for the targeted population), such as, fortified, dietary supplements, enriched or enhanced foods and conventional foods. They also supply essential nutrients in quantities that are typically more than those required for normal maintenance, growth, and development, as well as additional bio-active compounds that have health or physiological impacts on the human body [63].

As per the American Dietetic Association (ADA), foods to provide health advantages in addition to the nutrients they contain are known as functional foods [38].

According to the International Food Information Council (IFIC), any food or dietary component that has the potential to give health benefits in addition to basic nutrition are known as functional foods [45].

Health Canada defines functional foods as they may resemble to a conventional food and has been shown to offer physiological advantages and/or reduce the risk of chronic disease beyond basic nutritional functions when ingested as part of a regular diet [40].

According to the US Food and Nutrition Board of the National Academy of Science, functional food is as any modified food or food ingredient that may provide a health advantage in addition to the regular nutrients it provides is considered potentially healthy [92]

3.4.2 FEATURES OF FUNCTIONAL FOODS

- Any existing or future health claims are scientifically established and are compliant with legislation [80].
- Bioactive components are responsible for their physiological functionality.
- Functional foods are intended for certain consumer groups because they meet specific dietary demands or health issues.
- It is essential to document their safety and effectiveness.
- They are eaten as a part of regular diet.

3.4.3 TYPES OF FUNCTIONAL FOODS

1. **Natural Functional Foods:** These are the foodstuffs that contain naturally occurring components, which are beneficial for health or prevent diseases along with their basic function of supplying nutrients

to the human being. Natural functional foods include blueberries, broccoli, and salmon.

2. **Manipulated Functional Foods:** Standard food can be converted into functional food by simply modifying their production system or by manipulating the animal feed. For example, Omega-3 fatty acids-rich eggs are produced by feeding chickens with tailored foods. Scientists produced a genetically modified strain of tomato which is rich in lycopene and have the ability to lower the risk of cancerous growth and CVD.
3. **Processed Functional Foods:** The features of these foods are usually modified during the factory processing phase, which may include adding or removing chemicals and microorganisms. Artificially enriched functional goods dominate the dairy industry. In recent years, the yogurt industry has effectively expanded into the functional food market by inventing a variety of health-promoting bacteria that may be consumed to enhance physiological functions like digestion and circulation [80].

   ➢ **Common Functional Foods Are [80]:**
   - Omega-3 fatty acids enriched products, such as margarine, bread, and vegan milk.
   - Pre and probiotics foods such as yogurts and dairy drinks.
   - Products containing lower cholesterol content, including yogurts, margarine, salad dressing, milk, etc.
   - Products enriched with vitamin D and calcium, including milk, margarine, yogurt, and breakfast cereals like muesli, cornflakes, etc.

3.4.4 IMPLICATIONS FOR DEVELOPMENT OF FUNCTIONAL FOODS

Due to biological aging, consumer's nutritional needs fluctuate with age, therefore older persons require more of certain nutrients (e.g., calcium, vitamin D, and protein) [97]. The food industry can help to address these demands by creating enticing new foods that account for biological aging as well as human variances in cognition, motivation, and identity. Functional foods seem to have a special interest in the development of innovative foods for elderly persons by the food industry. Functional foods are ordinary foods that have been engineered to offer a physiological benefit [15].

Functional foods are commonly accessible in supermarkets and have been introduced into a variety of categories, including dairy, confectionery, soft-drink, bakery, and baby food [58, 66]. Calcium-incorporated juice, omega-3 incorporated spread, and fiber added cereals are all examples of useful foods [98].

A functional food consists of a base product (carrier) and a functional component (ingredient) that is transmitted to customers through a message (i.e., claim). Each of these elements are responsible to influence the desire to try and repurchase of functional foods among older customers.

3.4.1.1 CARRIERS

Functional ingredients can be amalgamated into variety of carriers such as, bread, dairy products, beverages which will open various doors for the functional food industry to create variety of products by using these functional ingredients and introduce into the market. However, not all ingredient and carrier formulations appeal to customers equally, and certain formulations may not be accepted by consumers due to variety of reasons [57]. In a study on protein enrichment of carriers, 'supplementation of meat is just trifle for one consumer' and for another consumer 'there is no requirement to get protein by eating soup' [96]. It is really very difficult to forecast that which category of functional food would attract the consumers because they reject them for such intuitive reasons. Therefore, it is advised to the researchers that during designing of functional food, each type of carrier should be studied separately [4].

A study was conducted on protein-enriched food products for the old age people which results, old age people prefer healthy food, conventional foods augmented with protein, including milk and bread [95]. Only around 50% of the groups meet the criteria when the sample was divided into six subgroups based on their preferences. Unhealthy carriers were chosen by few groups, while other groups liked innovative carriers and protein-rich snacks. The fact that older people prefer functional foods produced from goods, they already eat could explain this discrepancy in carrier preferences [96]. Only healthiness and familiarity are not sufficient to motivate the older population to consume the functional food. As a result, creating multiple formats of the same sort of functional food is the most effective strategy to suit the demands and desires of the elderly.

3.4.1.2 INGREDIENTS

The functional element in a functional food (e.g., omega-3 fatty acid, fiber, calcium) influences acceptance to a lower extent [59]. Consumers will be more ready to buy a product if it can provide them with the benefits they want, as anticipated by the consumer decision-making framework [33]. Old age people are more likely to prefer the ingredients, which supply individually relevant functionality to them. Functional foods are useful to cure target specific health concerns, especially for old, aged population, since most of the chronic diseases are target specific whose possibility increases with the old age including CVDs, GI function, and bone health [37].

This is the fact that no consumer is ready to change their dietary habits until and unless they are facing any health issues [28]. Old age population are less keen to buy functional foods with unfamiliar ingredients such as, bioactive peptides than those familiar ones.

Old age consumers are keen to buy functional foods with familiar ingredients such as, omega 3 fatty acids than those uncommon ingredients like bioactive peptides [32]. The sensory appeal of food is very important parameter to the consumer to buy. Therefore, these are the important aspects to be considered during the development of functional food for old-aged consumers [61].

3.5 HEALTH CLAIMS TO PURCHASE FUNCTIONAL FOODS

Old-age consumers are less likely to purchase functional foods even if they are specially customized according to their individual needs or provide highly relevant benefits to them [95]. Some customers are more aware about "short" health claims that just report medical benefits (e.g., "reduces artery blockage"), while others are more aware about "long" health claims that include active components and physiological activities (e.g., "contains omega-3, which promotes cardiovascular health") [32]. Overall, old age customers preferred short claims, but demographic characteristics (such as: gender, nationality, and education) predicted significantly less diversity in preference than age. There is likely to be lot of variation among older people when it comes to the types of health claims that will best assist them to infer the intended benefits from functional meals [94].

3.6 EFFECTS OF FUNCTIONAL FOODS ON VARIOUS DISEASES: EXAMPLES

3.6.1 CARDIOVASCULAR DISEASE (CVD)

The 51% of mortality in the world is only due to CVD. Plaque development in arteries is initiated by inflammation of the arterial lining. The main risk factor for CVD suggested by biochemical, cell culture and epidemiological research is insufficient consumption of antioxidants and MNs (such as: vitamins E and C, beta-carotene, zinc, copper, and selenium) [82]. Antioxidant-rich fruits and vegetables should be consumed in greater quantities to protect against CVD [22]. Furthermore, Fish consumption for least twice a week reduces the risk of heart attack and stroke [39]. Omega-3 fatty acids play a crucial physiological role in the prevention of CVD [42]. Eicosapentaenoic acid (EPA) and Docosahexaenoic acid (DHA) in fish oil are essential fatty acids, but they can also be synthesized in the body from alpha-linolenic acid (ALA). Various foods are rich in ALA including, spinach, flaxseeds, and marine organisms.

3.6.2 CANCER

Due to change in environmental factors and dietary habits in developed countries, cancer is the second biggest cause of death [67]. Consumption of nutritious diet can considerably reduce the chance of acquiring cancer, according to epidemiological and clinical observations. Consumption of these phytochemicals and other compounds such as Vitamins A, B complex, C, and D, fibers, flavonoids present in fruits and vegetables, organo-sulfur compounds present in *Allium* plants (onion and garlic), glucosinolatesin cruciferous vegetables including broccoli reduces the risk of cancer [48]. These numerous essential and non-essential nutrients appear to interfere with carcinogenesis by altering the carcinogenic process at certain places. Although nutrition is not the sole factor that promotes cancer development, it is a simple and potentially effective strategy to lower your risk.

3.6.3 OBESITY

Obesity is a major concern among all age groups in developed and developing countries. Obesity increases due to:

- People are switching from traditional foods (balanced diet) to ready-to-eat food products or fast foods that are high in trans and saturated fatty acids, sugar, and other unhealthy ingredients.
- Modern lifestyle increases the stress in daily life of human being which leads to busy people to consume ready to eat convenient food products. Also leads to binge eating of the consumer.
- Due to digitalization of work, causes excessive use of computer in day to-day life.
- Lack of exercise and sedentary lifestyle with sitting idle in front of TV and munching of snacks.

Obesity raises the risk of a variety of diseases including cancer, diabetes, and heart disease. If the body mass index exceeds 30, then the person is considered obese. Dietary elements such as macronutrients (carbohydrate, protein, and fat), MNs (vitamins and minerals, for example), and non-nutrients (phytoestrogens, glucosinolates, capsaicin, and so on) are all important for fat distribution and energy balance in humans. As a result, there is a significant opportunity to produce functional meals to address the obesity epidemic in modern culture [101].

3.6.4 ALZHEIMER'S DISEASE

The risk of Alzheimer's disease can be reduced by 60% by consuming the fish rich in omega-3 fatty acid DHA twice a week. As per the research the DHA had strongest preventive effect against this condition as compared to other omega-3 fatty acids (a-linolenic acid, prevents only individuals having specific genotype). Whereas EPA shows no preventive effect against this disease [70].

3.7 APPLICATIONS OF FUNCTIONAL FOODS

Several research on the effects of functional foods on the elderly population have been conducted. Geriatric animal model (24 to 30 months geriatric Wistar albino rats) was fed on grain-based functional food formulations results in hypoglycemic effect and reduction in serum cholesterol in the animal body [91].

Various bioactive compounds are added in double emulsion form (vitamin A and D in the lipophilic phase; berry polyphenols in the hydrophilic phase;

vitamins C, B_9, and B_{12} in the hydrophilic phase) incorporated into high protein yogurt. This product was described by geriatric patients as "comfortable," with good swallowing, humidity/soft, and textural features, as well as increased bio-accessibility [54]. The abundance of functional foods specifically intended for old age people is a significant boost to their nutrition. However, as people become older, their metabolic pathways for these foods and the components they contain, change. Nutraceuticals' delivery systems could improve the efficiency and functioning of these food products in this aspect [25].

3.7.1 DELIVERY SYSTEMS FOR NUTRACEUTICALS

Older people malnutrition can be overcome by accepting balanced diet which may contain the customized functional foods supplemented with nutraceuticals, such as antioxidants, antimicrobials, and vitamins. Though, nutraceuticals in free form or in a food matrix can be unstable and fragile, especially when they exposed to environmental stimuli such as light, oxygen, temperature, and incompatible compounds contained in food during processing or storage. In the realm of food science, nutraceutical encapsulation has attracted a lot of interest as it allows for the preservation of bioactive compounds and also increasing the efficiency of the delivery process. As a result, it's important to use an encapsulating agent to protect and release nutraceuticals in foods [30].

3.7.2 3D FOOD PRINTING FOR OLD-AGED PEOPLE

Three-dimensional (3D) printing is a new technique that is derived from the additive manufacturing (AM) and rapid prototyping (RP). 3D printing is based on computerized fabrication of three dimensions by layer-by-layer deposition [20]. Selective laser sintering, inkjet printing, stereolithography, and fused deposition modeling are the available techniques of AM [83]. 3D food printing is a technology used in the food business to create novel food textures, like pureed foods for the elderly with dysphagia and other swallowing or chewing problems. Furthermore, it has the potential to increase children's consumption of vegetables and fruit, as well as contribute to the prevention of various diseases linked to lifestyle and food waste [78]. This technique benefited the consumer by providing the customized and acceptable food with enhanced organoleptic properties, nutrient content and increased functionality. As the age of person increases, the chances of many disorders (like dysphagia and malnutrition) increases. Hence, these diseases affect the well-being and lifestyle of the older adults.

Foods with modified textures and customized nutritional content is required for the treatment of medical issues. Customized foods that can be thickened (such as drinks, soft bite-sized moist food and smooth pureed fruit/vegetable meals) are also available in the traditional market but they have less appealing sensory qualities (appearance, taste, and texture) and are not consumed frequently [78]. 3D printed foods have the potential to fill this need and help those with dysphagia and other swallowing and chewing problems. Dysphagia, or difficulty in chewing and swallowing, affects approximately one out of every five older persons. To resolve this problem, the EU-funded project "PERFORMANCE" (personalized food using rapid manufacturing for the nutrition of elderly consumers) created customized food for the patients using this technology. In one study, 3D printing was utilized to successfully make edible food products from a mixture of fruits and vegetables that had appealing sensory and nutritional properties for various customer groups (children, elderly people, pregnant women, etc.) [84]. Additionally, the capacity to manufacture cereal snacks created in 3D with varying texture profiles was investigated in recent research. This study looks into the prospect of 3D printing foods with specific texture requirements, including for older people suffering from dysphagia or other swallowing issues [19].

3.8 DEVELOPMENT OF FOOD PRODUCTS FOR GERIATRICS

As per Functional Food Center (FFC) defines functional foods as the – "*Natural or processed foods that contain biologically-active compounds, which, in defined, effective, non-toxic amounts, provide a clinically proven and documented health benefit utilizing specific biomarkers, to promote optimal health and reduce the risk of chronic/viral diseases and manage their symptoms*" [60].

3.8.1 NATURAL FUNCTIONAL FOODS

Dairy, bakery, beverages, confectionary, and infant foods are the most popular functional food products [51, 66]. The most well-known form of functional food items including plant-based ingredients have been introduced to the market with the goal of lowering high blood pressure, blood sugar, cholesterol, and osteoporosis [11, 52, 58]. The products are based on various natural sources which are as follows:

1. **Cereals:** These are rich in dietary insoluble fibers and water-soluble fibers. Dietary insoluble fibers are natural functional ingredient and act as roughage. Water-soluble fibers preferentially encourage the growth of *lactobacilli* and *Bifidobacteria* in the colon, working as prebiotics and encouraging various favorable effects on physiology [52]. Extracted plant elements have been effectively incorporated into a variety of novel food products, including chews, breakfast cereals, nutrition bars, baked goods, pasta, drinks, yogurt, ice cream, and food supplements. Apart from the extraction of specific food components from cereals, processed whole cereals are high in bioactives that may act as antioxidants, hypocholesterolemia, antiproliferative, enzyme, and hormone stimulants, carcinogen detoxifiers, anti-inflammatory, antibacterial, and antiviral, potential inhibitors of various actions, and ligand to cell walls, influencing the initiation and progression of a variety of pathogenic processes [51]. Resistant starch, which can be found in heated starch and starch-containing cereal food products, is one of the functional fibers that helps bacteria in the colon process carbohydrates, lowering the risk of bowel disease [23].
2. **Tea:** It is primarily composed of polyphenols, especially catechins, which account for 25 to 30% of the weight recovered after drying fresh tea leaves [88]. Green tea polyphenols have anticarcinogenic, antioxidant, antiviral, antifungal, thermogenic, antimutagenic, deodorizing, kidney-protecting, and blood lipid and blood glucose-controlling properties [52]. Consumption of Green tea boosts basal metabolism, resulting in weight loss. Catechins in black tea oxidize to generate aflavins and arubigins, which have antibacterial, antioxidant, anticancer, and anti-inflammatory properties [52].
3. **Vegetables and Fruits-Based:** Due to amount and type of phytochemicals present in fruits and vegetables, they possess numerous health benefits [85]. Berries are rich in polyphenols. Among the polyphenols present in berries, flavonoids (such as: cathecins), flavanols, and anthocyanins, phenolic acids including hydroxycinnamic acids and tannins such as ellagitannins and proanthoscyanidins are the bioactive compounds present in berries. Berries are used as sweetener, colorant, and essence to enhance the aroma of the functional breakfast cereal products [52]. Berry oils, berry seed oils,

powders of berry and berry extracts are the innovative functional food ingredients. It protects against CVD, cancer, diabetes, and premature aging [105]. Lycopene is a carotenoid antioxidant present in red fruits and vegetables, particularly tomatoes and tomato-based products, that has been shown to help reduce hypertension, oxidative stress, diabetes, and atherosclerosis [23, 99]. Intake of tomato and tomato-based products reduces the risk of lung, stomach, and prostate cancer [52, 81]. The total carotene content of carrot pomace was removed using enzymes and put into a functional drink [52, 87].

4. **Herbs:** These contain a variety of functional food elements that are beneficial when used in food, whether whole or extracted [52]. Turmeric is rich in curcumin which is a polyphenolic pigment and yellow in color shows therapeutic effects against Parkinson's and Alzheimer's disease, epilepsy, cerebral injury, multiple sclerosis, CVD, allergy, asthma, cancer, bronchitis, rheumatoid arthritis, colitis, renal ischemia, diabetes, obesity, psoriasis, depression, and fatigue [2, 52]. Curcumin antioxidant has been granted a 'generally recognized as safe' (GRAS) status, allowing it to be used in milk products, baked goods, jams, jellies, oils, fats, gelatins, soups, puddings, candy, snacks, and frozen dairy, as well as seasonings and flavors and imitation non-alcoholic and dairy beverages [52].

3.8.2 Processed Functional Foods

1. **Yogurt:** It is a popular fermented milk product that contains a variety of nutrients in substantial proportions, including calcium in accessible form, protein, phosphorus, riboflavin, thiamine, vitamin B_{12}, niacin, magnesium, zinc, and a great source of foliate. Yogurt has a number of health benefits, including lactose tolerance, digestive health, and a potential function in body weight and weight loss. The bacteria *Lactobacillus delbrueckii*, *L. bulgaricus*, and *Streptococcus thermophiles* found in yogurt inhibited the fermentation of putrefactive bacteria by the GI flora, enhancing intestinal health [65].
2. **Fermented Milk:** Yeasts are used to make fermented milks like kefir, which contain 1–3% alcohol. Fermented milk has more than 8% SNF (solids-not-fat), which is the same as milk. More than 1×10^7 organisms/ml of lactic acid bacteria (LAB) and yeasts can be found in fermented milk. Prevention and treatment of diarrhea, prevention, and treatment of systemic infections, management of inflammatory bowel disease (IBD), immunomodulation, prevention, and treatment

of allergies, anticancer effects, therapy of cholesterolmia, and reduction of lactose intolerance are some of the health effects of fermented milks that have been documented based on appropriate clinical or animal research [68].

3.8.3 FRUITS AND VEGETABLES RESIDUES FLOUR FOR BAKERY PRODUCTS

The residues of fruits and vegetables left from beverage industry have been used to prepare flour and due to its functional properties, it was incorporated into the biscuits and cereal bars in different compositions (20 to 35%). They use the solid residues left from the processing of beverage made from the combination of orange, passion fruit and watermelon in fruits and lettuce, courgetti, carrot, spinach, mint, taro, and cucumber in vegetables. The fruits and vegetables residue were dried and then ground to make flour. This flour was incorporated in the biscuits and cereal bars. The incorporation of this flour in bakery products enhanced the dietary fibers and other functional properties [26].

3.8.4 FUNCTIONAL FOOD POWDER

A functional food powder has been prepared, which is sugar free, trans fat free and lactose free and also contain carbohydrates (oligosaccharide), dietary fiber that acts as a prebiotic. Consumption of this product lowers the risk of diabetes mellitus. This powder can be incorporated into the hot tea or coffee and serve as a beverage and also mix with the premix or ingredients of the pancakes and served as food [34].

3.8.5 FORTIFIED FOODS: BREADS

Phytochemicals and other bioactive compounds have a significant impact on human diet and health [44]. Plants and fruits, such as flaxseed and soybean, naturally contain phytochemicals. Soy flour makes a healthy baked treat, suitable for diabetics, gluten-free, dairy-free, and cholesterol-free. Conforti and Davis tested the use of soy flour (5% and 10%) and flaxseed (15%) in yeast bread to examine how they affected the bread's quality. Even though soybean and flaxseed have high nutritional content, combining them with bread does not improve the outcomes. According to the findings, the bread had a lesser volume, a harder texture, and a dark color in the crumb and crust [12].

Purple sweet potato flour can also be used to make bread as a partial substitute for flour. The antioxidant activity of the resulting bread will be increased. A maximum substitution of 20% sweet potato flour resulted in the best bread fortification [36]. Otherwise, bread quality attributes such as particular volume, hardness, and the pleasing purple color of the bread will be reduced.

Polyphenols, which are natural antioxidants, are found in green tea extract. Green tea is now used as an active ingredient in a wide range of products, particularly in the culinary, nutraceutical, and cosmeceutical industries. The effect of green tea extract as a functional component on bread quality was investigated [102]. They noticed that green tea extract had a substantial impact on the bread's flavor. It also contributed to the bread's hardness, sweetness, stickiness, and astringency. Fortification with bread will offer a functional food product with added health advantages, as green tea is a good source of antioxidants.

Coriander leaves are high in polyphenolic compounds and include vitamins A, B2, and C. Coriander leaf powder added to bread will increase moisture content, delay the pace of aging, increase antioxidant activity, and improve the texture, color, and flavor of the bread [16]. Bread fortified with powder coriander leaf in wheat flour will benefit in many ways, including increased antioxidant content, increased moisture retention capacity, higher baking efficiency, and improved sensory characteristics in terms of color, aroma, and flavor [17].

Banana pseudo-stem flour bread had higher total phenolics and antioxidant activity than regular bread. The study compared control bread (regular bread), commercial bread, bread with 10% banana pseudo-stem flour, xanthan gum, and sodium carboxymethylcellulose, and bread with 10% banana pseudo-stem flour, xanthan gum, and sodium carboxymethylcellulose. The results revealed that using banana pseudo-stem flour affected the end product's physical quality. Sodium carboxymethylcellulose, on the other hand, can improve loaf height, volume, and density [41].

The addition of banana fiber to supplemental vital gluten could maintain the elasticity of dough. Due to the high sugar level, adding more banana powder will make the bread firmer. Due to the additional sugar in the banana powder, samples with higher banana powder concentration will be darker. As the amount of banana powder in the formulation rose, the protein capacity extracted from dry mixes or dough dropped [69].

Fennel (*Foeniculum vulgare var. dulce*) is an edible herb that is commonly used to treat skin conditions such as acne, conjunctivitis, and blepharitis. Fennel seed extracts are strong in vital fatty acids, proteins, minerals, fibers, and antioxidant and antibacterial properties. Bread crumbs with up to 7% fennel seed powder have a high moisture level, higher antioxidant content, and acceptable customer appeal. The more the fennel seed powder content in the bread, the firmer the crumb [18].

3.8.5.1 BRAN FORTIFIED FOOD PRODUCTS

Cereal brans, which are by-products of grain milling, have gained prominence in the development of functional foods. They include non-starch carbohydrates (arabinoxylan, beta-glucan), phenolic acids (ferulic acid), flavonoids (anthocyanin), oil (-oryzanol), vitamins (carotenoids, tocols), oligosaccharides, folates, and sterols, among other things. They are suitable for food fortification due to their physicochemical characteristics. Rice, wheat, oat, barley, sorghum, millet, rye, and maize brans have been found to contain a variety of health-promoting elements. They've been shown to have antilipemic, antiatherogenic, antihypertensive, and hypoglycemic effects. They have been shown to reduce oxidative stress, reduce insulin resistance, prevent obesity by increasing satiety, and reduce cardiovascular problems [76].

Oat bran had more soluble dietary fiber and unsaturated fatty acids, while wheat bran had more total dietary fiber, insoluble dietary fiber, and saturated fatty acids. The nutritional benefits of bran-fortified food are being reported in an increasing number of studies. Wheat and oat bran were used to create dietary fiber-rich chicken meat patties. The addition of bran enhanced the water holding capacity and emulsion stability substantially. Wheat bran showed superior emulsion stability and water retention capacity than oat bran. When bran was added, cooking yield, firmness, total dietary fiber, and unsaturated fatty acids increased significantly, while sensory characteristics, moisture, protein, fat, and cholesterol content decreased significantly. It was determined that oat and wheat bran can be used to make baked and steamed chicken patties in amounts of up to 10% and 15%, respectively [89].

Wheat bran is currently a common ingredient in dairy foods, sauces, dressings, condiments, chilled foods, infant foods, dry foods, pasta, noodles, pizza, soups, spreads, and sweets [79]. The addition of wheat bran arabinoxylan to chitosan-based edible films increased their functionality by adding prebiotic value [14]. Sorghum bran, when added into Turkey patties, displayed antioxidant potential comparable to butylated hydroxyanisole (BHA) and butylated hydroxytoluene (BHT) without sacrificing sensory values [62].

3.8.5.2 FORTIFIED JUICES

When probiotic strains and a prebiotic source were added to fig juice, the functional characteristics of the juice changed dramatically. The nutritional and physiological benefits of probiotic and prebiotic microorganisms were integrated in this fruit juice. In terms of probiotic properties and antioxidants, symbiotic juice incorporating *L. delbrueckii* was found to be superior. The viable count of symbiotic juices was roughly one log cycle greater than that of probiotic juices at the conclusion of the storage period, and it remained over the minimum criterion for probiotic goods (6 log CFU ml^{-1}). Although sensory scores in the treatments were lower than in the control group, concealing medicinal flavor and organoleptic features with other ingredients such as tropical fruits could be a viable option. As a result, fig juice could be a good vehicle for probiotic and prebiotic supplements. Overall, fermentation can increase polyphenol bioavailability, and fermented fig juices with high antioxidant content can be a promising bioprocessing approach for developing novel functional foods [55].

3.8.5.3 Fortified Dairy Products

1. **Ice Cream:** It is a safe, nutritious, and frozen dairy product that is extensively consumed around the world, and it has a large following because it brings joy to people [64]. Ice cream is an excellent source of vital amino acids from milk proteins, vitamins, and minerals, and its components are easily digested and absorbed by the body [1]. Ice cream can be made either naturally or through fermentation. In order to make fermentative ice cream, probiotic bacteria are added to the ice cream recipe. Because of its neutral pH (potential of hydrogen), probiotic ice cream can survive in the human digestive system. The sole issue is ice cream aeration during the freezing process, which can compromise probiotic life because oxygen is present. This issue can be solved by using the encapsulation method (micro-covering) in ice cream [43]. According to Abqari's research, after adding

Lactobacillus acidophilus to a type of fermentative ice cream, the survival rate of this bacteria was evaluated, and it was discovered that the required amount (108 (cfu/ml)) of bacterial mass survived, which is consistent with Probiotics product definition and can be considered functional fermentative ice cream [1].

2. **Whey-based Beverages:** Whey is a byproduct of the cheese-making process. Whey's composition and features are determined by the manufacturing process, the final product, and the quality of milk utilized. Liquid whey is around 93% water and comprises over half of the total solids found in milk, with lactose being the most important component [50]. Lactoferrin, lactoperoxidase, immunoglobulins, growth factors, and other highly desired nutraceutical components can all found in whey [49]. Whey-based products can be classified into the following categories [49]: Thirst-quenching beverages; Dairy-type beverages; and Fruit-flavored beverages. The beverage industry prefers sweet whey over acid cheese whey [8]. For whey-based drinkable yogurt, whey inclusion into the formula or fermentation of liquid UF whey protein retentates appear to be viable options. Whey permeates, which contain no proteins, are typically used to make thirst-quenching whey-based beverages. Rivella, a whey beverage from Switzerland, is a good example of this type of product in the marketplace. Unless any functional food components are added to thirst-quenching whey drinks, they should not be considered functional foods. Rivella has made similar attempts, and Rivella-green was created with green tea herbal extracts [7, 49]. In fruit juice whey beverages, mineral, and vitamin fortification is prevalent. Apart from externally added vitamins and minerals, some nutraceutical components derived from the fruit, such as lycopene in tomato juice added to whey drinks, improve the functional status of the finished product [75].

3.9 SUMMARY

With the world's population aging at an increasing rate, malnutrition issues are becoming more severe, and they may be linked to the incidence of other age-related diseases and chronic ailments. Significant GI changes can also be detected in this population as a result of the multimodal aging process. In this regard, research into biological processes associated with aging should be expanded and detailed, allowing for a better understanding of which preventive interventions could potentially improve the health and well-being of older persons. The use of such foods can have a significant impact not only on disease prevention, but also on maintaining a healthy aging process. Nonetheless, because the literature on this area is sparse, better tools and processes to understand foods' digestive fate in older persons are required.

KEYWORDS

- **Alzheimer's disease**
- **chronic diseases**
- **fortified foods**
- **functional foods**
- **geriatric foods**
- **malnutrition**
- **nutraceuticals**
- **nutrition**

REFERENCES

1. Abghari, A., Zeinoddin, M. S., Soleimanian, S., & Dokhani, S., (2008). *Survival of Lactobacillus Acidophilus in a Non-fermented Ice Cream.* Paper Presentation at 18th National Congress on Food Technology, Mashhad, Iran.
2. Aggarwal, B. B., & Harikumar, K. B., (2009). Potential therapeutic effects of curcumin, the anti-inflammatory agent, against neurodegenerative, cardiovascular, pulmonary, metabolic, autoimmune, and neoplastic diseases. *The International Journal of Biochemistry and Cell Biology, 41*(1), 40–59.
3. Amarya, S., Singh, K., & Sabharwal, M., (2015). Changes during aging and their association with malnutrition. *International Journal of Clinical Gerontology and Geriatrics, 6*(3), 78–84.
4. Ares, G., Gimenez, A., & Gambaro, A., (2008). Influence of nutritional knowledge on perceived healthiness and willingness to try functional foods. *Appetite, 51*(3), 663–668.
5. Arvanitoyannis, I. S., & Houwelingen-Koukaliaroglou, M. V., (2005). Functional foods: A survey of health claims, pros and cons, and current legislation. *Critical Reviews in Food Science and Nutrition, 45*(5), 385–404.
6. Badal, V. D., Vaccariello, E. D., Murray, E. R., Yu, K. E., Knight, R., Jeste, D. V., & Nguyen, T. T., (2020). The gut microbiome, aging, and longevity: A systematic review. *Nutrients, 12*(12), E-article ID 3759.

7. Barth, R., (2003). From the original idea of a product to the introduction into the market and finally to the creation of a real brand name. In: Jelen, P., (ed.), *The Importance of Whey and Whey Components in Food and Nutrition, Proceedings of the 3rd International Whey Conference* (Vol. 13, pp. 147–156). Hamburg – Germany: Elsevier Ltd.
8. Britten, M., & Giroux, H. J., (2001). Acid-induced gelation of whey protein polymers: Effect of pH and calcium concentration during polymerization. *Food Hydrocolloids, 15*(4–6), 609–617.
9. Brownie, S., (2006). Why are elderly individuals at risk of nutritional deficiency. *International Journal of Nursing Practice, 12*(2), 110–118.
10. Cencic, A., & Chingwaru, W., (2010). Role of functional foods, nutraceuticals, and food supplements in intestinal health. *Nutrients, 2*(6), 611–625.
11. Charalampopoulos, D., Wang, R., Pandiella, S. S., & Webb, C., (2002). Application of cereals and cereal components in functional foods: A review. *International Journal of Food Microbiology, 79*(1, 2), 131–141.
12. Conforti, F. D., & Davis, S. F., (2006). The effect of soya flour and flaxseed as a partial replacement for bread flour in yeast bread. *International Journal of Food Science and Technology, 41*(2), 95–101.
13. Correia, C., Lopez, K. J., Wroblewski, K. E., Huisingh-Scheetz, M., Kern, D. W., Chen, R. C., Schumm, L. P., et al., (2016). Global sensory impairment in older adults in the United States. *Journal of the American Geriatrics Society, 64*(2), 306–313.
14. Costa, M. J., Cerqueira, M. A., Ruiz, H. A., Fougnies, C., Richel, A., Vicente, A. A., Teixeiraa, J. A., & Aguedo, M., (2015). Use of wheat bran arabinoxylans in chitosan-based films: Effect on physicochemical properties. *Industrial Crops and Products, 66*, 305–311.
15. Crowe, K. M., & Francis, C., (2013). Position of the academy of nutrition and dietetics: Functional foods. *Journal of the Academy of Nutrition and Dietetics, 113*(8), 1096–1103.
16. Das, L., Raychaudhuri, U., & Chakraborty, R., (2012). Effect of baking conditions on the physical properties of herbal bread using RSM. *International Journal of Food, Agriculture and Veterinary Sciences, 2*(2), 106–114.
17. Das, L., Raychaudhuri, U., & Chakraborty, R., (2012). Supplementation of common white bread by coriander leaf powder. *Food Science and Biotechnology, 21*(2), 425–433.
18. Das, L., Raychaudhuri, U., & Chakraborty, R., (2013). Herbal fortification of bread with fennel seeds. *Food Technology and Biotechnology, 51*(3), 434–440.
19. Derossi, A., Caporizzi, R., Paolillo, M., & Severini, C., (2021). Programmable texture properties of cereal-based snack-mediated by 3D printing technology. *Journal of Food Engineering, 289*, E-article ID 110160.
20. Derossi, A., Caporizzi, R., Ricci, I., & Severini, C., (2019). Critical variables in 3D food printing; Chapter 3. In: Godoi, F. C., Bhandari, B. R., Prakash, S., & Zhang, M., (eds.); *Fundamentals of 3D Food Printing and Applications* (pp. 41–92). London – UK: Elsevier Ltd. – Academic Press.
21. Dionyssiotis, Y., (2019). Sarcopenia in the elderly. *European Endocrinology, 15*(1), 13–19.
22. Duthie, G. G., & Brown, K. M., (1994). Reducing the risk of cardiovascular disease; Chapter 2. In: Goldberg, I., (ed.), *Functional Foods: Designer Foods, Pharmafoods, Nutraceuticals* (pp. 19–38). London – UK: Springer Science + Business Media.
23. Eksi, G., Kurbanoglu, S., & Ozkan, S. A., (2019). Fortification of functional and medicinal beverages with botanical products and their analysis. In: Grumezescu, A.

M., & Holban, A. M., (eds.), *Engineering Tools in the Beverage Industry* (Vol. 3, pp. 351–404). London – UK: Elsevier Ltd.
24. Eurostat, (2019). *Causes of Death Statistics-People Over 65 Statistics Explained.* https://ec.europa.eu/eurostat/statistics-explained/index.php?title=Causes_of_death_statistics_-_people_over_65&oldid=220614 (accessed on 14 November 2023).
25. Fernandes, J. M., Araujo, J. F., Vieira, J. M., Pinheiro, A. C., & Vicente, A. A., (2021). Tackling older adults' malnutrition through the development of tailored food products. *Trends in Food Science & Technology, 115*, 55–73.
26. Ferreira, M. S. L., Santos, M. C. P., Moro, T. M. A., Basto, G. J., Andrade, R. M. S., & Gonçalves, E. C. B. A., (2015). Formulation and characterization of functional foods based on fruit and vegetable residue flour. *Journal of Food Science and Technology, 52*(2), 822–830.
27. Food Quality and Standards Service (AGNS), (2007). Food and Agriculture Organization of the United Nations (FAO). *Report on Functional Foods* (p. 27). Rome-Italy: FAO.
28. Frewer, L., Scholderer, J., & Lambert, N., (2003). Consumer acceptance of functional foods: Issues for the future. *British Food Journal, 105*(10), 714–731.
29. Glowacki, J., & Vokes, T., (2016). Osteoporosis and mechanisms of skeletal aging. In: Sierra, F., & Kohanski, R., (eds.), *Advances in Geroscience* (pp. 277–308). Cham – Switzerland: Springer International Publishing.
30. Goncalves, R. F. S., Martins, J. T., Duarte, C. M. M., Vicente, A. A., & Pinheiro, A. C., (2018). Advances in nutraceutical delivery systems: From formulation design for bioavailability enhancement to efficacy and safety evaluation. *Trends in Food Science & Technology, 78*, 270–291.
31. Grassi, M., Petraccia, L., Mennuni, G., Fontana, M., Scarno, A., Sabetta, S., & Fraioli, A., (2011). Changes, functional disorders, and diseases in the gastrointestinal tract of elderly. *Nutricionhospitalaria, 26*(4), 659–668.
32. Grunert, K. G., & Van, T. H. C. M., (2014). Consumer-oriented new product development. In: Van, A. N. K., (ed.), *Encyclopedia of Agriculture and Food Systems* (Vol. 1, pp. 375–386). San Diego – USA: Academic Press.
33. Grunert, K. G., Lahteenmaki, L., Boztug, Y., Martinsdottir, E., Ueland, O., Astrom, A., & Lampila, P., (2009). Perception of health claims among Nordic consumers. *Journal of Consumer Policy, 32*(3), 269–287.
34. Gunawan, H., (2022). *U.S. Patent No. 11,297,867*. Washington, DC: U.S. Patent and Trademark Office, US Government.
35. Gupta, C., & Prakash, D., (2015). Nutraceuticals for geriatrics. *Journal of Traditional and Complementary Medicine, 5*(1), 5–14.
36. Hardoko, Hendarto, L., & Siregar, T. M., (2010). Utilization of purple sweet potato (*Ipomoea batatas* L. Poir) as a partial substitute for wheat flour and a source of antioxidants in white bread. *Journal of Food Technology and Industry, 21*(1), 25–32.
37. Hasler, C. M., (2000). The changing face of functional foods. *Journal of the American College of Nutrition, 19*(sup 5), 499S–506S.
38. Hasler, C. M., Brown, A. C., & American Dietetic Association, (2009). Position of the American dietetic association: Functional foods. *Journal of the American Dietetic Association, 109*(4), 735–746.
39. He, K., Song, Y., Daviglus, M. L., Liu, K., Van, H. L., Dyer, A. R., & Greenland, P., (2004). Accumulated evidence on fish consumption and coronary heart disease mortality: A meta-analysis of cohort studies. *Circulation, 109*(22), 2705–2711.

40. Health Canada, (1998). *Policy Paper: Nutraceuticals/ Functional Foods and Health Claims on Foods*. Ottawa, Ontario, Canada: Health Canada, Health Protection Branch, Therapeutic Products Program and Food Directorate. https://www.canada.ca/en/health-canada/services/food-nutrition/food-labelling/health-claims/nutraceuticals-functional-foods-health-claims-foods-policy-paper.html (accessed on 2 November 2023).
41. Ho, L. H., Aziz, N. A. A., & Azahari, B., (2013). Physico-chemical characteristics and sensory evaluation of wheat bread partially substituted with banana (*Musa acuminata* x *balbisiana* cv. awak) pseudo-stem flour. *Food Chemistry, 139*(1–4), 532–539.
42. Holub, B. J., (2002). Clinical nutrition: Omega-3 fatty acids in cardiovascular care. *Canadian Medical Association Journal, 166*(5), 608–615.
43. Homayouni, A., Azizi, A., Ehsani, M. R., Yarmand, M. S., & Razavi, S. H., (2008). Effect of microencapsulation and resistant starch on the probiotic survival and sensory properties of synbiotic ice cream. *Food Chemistry, 111*(1), 50–55.
44. Ibrahim, U. K., Salleh, R. M., & Maqsood-ul-Haque, S. N. S., (2015). Bread towards functional food: An overview. *International Journal of Food Engineering, 1*(1), 39–43.
45. International Food Information Council (IFIC) Foundation, (2002). *The Consumer View on Functional Foods: Yesterday and Today* (p. 5). Food Insight.
46. IUFoST., (2014). Meeting the food needs of the aging population: Implications for food science and technology. *Over 65-Year-Olds – An Overlooked Segment of the Population.* http://iufost.org/iufostftp/IUF.SIB.Meeting%20the%20Food%20Needs%20of%20the%20Ageing%20Population.pdf (accessed on 2 November 2023).
47. Jaul, E., & Barron, J., (2017). Age-related diseases and clinical and public health implications for the 85 years old and over population. *Frontiers in Public Health, 5*, E-article ID 335.
48. Jeffrey, E. H., & Jarrell, V., (2000). Cruciferous vegetable and cancer prevention. In: Wildman, R. E. C., (ed.), *Handbook of Nutraceuticals and Functional Foods* (pp. 169–192). Boca Raton – FL: CRC Press.
49. Jelen, P., (2009). Whey-based functional beverages. In: Paquin, P., (ed.), *Functional and Specialty Beverage Technology* (pp. 259–296). Boca Raton – FL: CRC Press.
50. Jelicic, I., Bozanic, R., & Tratnik, L., (2008). Whey based beverages – a new generation of dairy products. *Mljekarstvo, 58*(3), 257–274.
51. Kaur, D., Rasane, P., Singh, J., Kaur, S., Kumar, V., Mahato, D. K., Dey, A., et al., (2019). Nutritional interventions for elderly and considerations for the development of geriatric foods. *Current Aging Science, 12*(1), 15–27.
52. Kaur, S., & Das, M., (2011). Functional foods: An overview. *Food Science and Biotechnology, 20*(4), 861–875.
53. Kerchner, G. A., & Wyss-Coray, T., (2016). The role of aging in Alzheimer's disease. In: Sierra, F., & Kohanski, R., (eds.), *Advances in Geroscience* (pp. 197–227). Cham – Switzerland: Springer International Publishing.
54. Kersiene, M., Jasutiene, I., Eisinaite, V., Pukalskiene, M., Venskutonis, P. R., Damuleviciene, G., & Leskauskaite, D., (2020). Development of a high-protein yoghurt-type product enriched with bioactive compounds for the elderly. *LWT, 131*, E-article ID 109820.
55. Khezri, S., Mahmoudi, R., & Dehghan, P., (2018). Fig juice fortified with inulin and *Lactobacillus delbrueckii*: A promising functional food. *Applied Food Biotechnology, 5*(2), 97–106.

56. Kirkman, M. S., Briscoe, V. J., Clark, N., Florez, H., Haas, L. B., Halter, J. B., Huang, E. S., et al., (2012). Diabetes in older adults: Diabetes care. *American Diabetes Association, 35*(12), 2650–2664.
57. Kleef, E. V., Van, T. H. C. M., & Luning, P., (2005). Functional foods: Health claim-food product compatibility and the impact of health claim framing on consumer evaluation. *Appetite, 44*(3), 299–308.
58. Kotilainen, L., Rajalahti, R., Ragasa, C., & Eija, P., (2006). *Health Enhancing Foods: Opportunities for Strengthening the Sector in Developing Countries (Agriculture and Rural Development Discussion Paper 30)* (Vol. 37067, No. 1). World Bank Group.
59. Krutulyte, R., Grunert, K. G., Scholderer, J., Lähteenmäki, L., Hagemann, K. S., Elgaard, P., Neilsen, B., & Graverholt, J. P., (2011). Perceived fit of different combinations of carriers and functional ingredients and its effect on purchase intention. *Food Quality and Preference, 22*(1), 11–16.
60. Leem, C., & Martirosyan, D., (2019). The bioactive compounds of probiotic foods/supplements and their application in managing mental disorders. *Bioactive Compounds in Health and Disease, 2*(10), 206–220.
61. Loeser, R. F., & Lotz, M., (2016). Osteoarthritis in the Elderly. In: Sierra, F., & Kohanski, R., (eds.), *Advances in Geroscience* (pp. 309–353). Cham – Switzerland: Springer International Publishing.
62. Luckemeyer, T. J., Luckemeyer, T., Hesteande, B., Miller, R., Kerth, C., Awika, J., & Rooney, L., (2015). Sorghum bran addition in bratwurst, pre-cooked pork patties and pre-cooked Turkey patties. *Meat Science, 101*, 118–122.
63. Macaulay, J. (2003). Increasing funding for food and agriculture research. Food technology. *Food technology Magazine, 57*(11), 35–43. https://www.ift.org/news-and-publications/food-technology-magazine/issues/2003/november/columns/news-and-analysis_science-communications-and-government-relations (accessed on 2 November 2023)
64. Mahmoudi, R., Fakhri, O., Farhoodi, A., Kaboudari, A., Rahimi, P. M. S. F., Tahapour, K., & Chegini, R., (2015). A review on probiotic dairy products as functional foods reported from Iran. *International Journal of Food Nutrition Safety, 6*(1), 1–12.
65. Mckinley, M. C., (2005). The nutrition and health benefits of yoghurt. *International Journal of Dairy Technology, 58*(1), 1–12.
66. Menrad, K., (2003). Market and marketing of functional food in Europe. *Journal of Food Engineering, 56*(2, 3), 181–188.
67. Milner, J. A., (1994). Reducing the risk of cancer; Chapter 3. In: Goldberg, I., (ed.), *Functional Foods: Designer Foods, Pharmafoods, Neutraceuticals* (pp. 39–70). London – UK: Springer Science + Business Media.
68. Mitsuoka, T., (2011). History and evolution of probiotics. *Japanese Journal of Lactic Acid Bacteria, 22*(1), 26–37.
69. Mohamed, A., Xu, J., & Singh, M., (2010). Yeast leavened banana-bread: Formulation, processing, color, and texture analysis. *Food Chemistry, 118*(3), 620–626.
70. Morris, M. C., Evans, D. A., Bienias, J. L., Tangney, C. C., Bennett, D. A., Wilson, R. S., Agarwal, N., & Schneider, J., (2003). Consumption of fish and n-3 fatty acids and risk of incidence of Alzheimer disease. *Archives of Neurology, 60*(7), 940–946.
71. Nagpal, R., Mainali, R., Ahmadi, S., Wang, S., Singh, R., Kavanagh, K., Kitzman, D. W., et al., (2018). Gut microbiome and aging: Physiological and mechanistic insights. *Nutrition and Healthy Aging, 4*(4), 267–285.

72. National Institute on Aging, (2020). *Understanding the Dynamics of the Aging Process.* National Institute on Aging. U.S. Department of Health and Human Services. https://www.nia.nih.gov/about/aging-strategic-directions-research/understanding-dynamics-aging (accessed on 2 November 2023).
73. Nations, U., (2015). World population prospects: 2015 revision. *United Nations Economics and Social Affairs, 33*(2), 1–66.
74. Oliveros, E., Patel, H., Kyung, S., Fugar, S., Goldberg, A., Madan, N., & Williams, K. A., (2020). Hypertension in older adults: Assessment, management, and challenges. *Clinical Cardiology, 43*(2), 99–107.
75. Özer, B. H., & Kirmaci, H. A., (2010). Functional milks and dairy beverages. *International Journal of Dairy Technology, 63*(1), 1–15.
76. Patel, S., (2015). Cereal bran fortified-functional foods for obesity and diabetes management: Triumphs, hurdles, and possibilities. *Journal of Functional Foods, 14*, 255–269.
77. Pawelec, G., (2018). Age and immunity: What is immune senescence? *Experimental Gerontology, 105*, 4–9.
78. Prakash, S., Bhandari, B. R., Godoi, F. C., & Zhang, M., (2019). Future outlook of 3D food printing; Chapter 13. In: Godoi, F. C., Bhandari, B. R., Prakash, S., & Zhang, M., (eds.), *Fundamentals of 3D Food Printing and Applications* (pp. 373–381). London – UK: Elsevier Ltd.- Academic Press.
79. Prückler, M., Siebenhandl-Ehn, S., Apprich, S., Höltinger, S., Haas, C., Schmid, E., & Kneifel, W., (2014). Wheat bran-based biorefinery 1: Composition of wheat bran and strategies of functionalization. *LWT – Food Science and Technology, 56*(2), 211–221.
80. Ramesh, C. K., Jamuna, K. S., & Raghavendra, (2012). Concepts and trends of functional foods: A review. *International Journal of Pharmaceutical Research and Development (IJPRD)- UPRD, 4*(6), 273–290.
81. Rao, L. G., Guns, E., & Rao, A. V., (2003). Lycopene: Its role in human health and disease. *Agro. Food Industry Hi-Tech., 1*, 25–30.
82. Schwingshackl, L., Boeing, H., Stelmach-Mardas, M., Gottschald, M., Dietrich, S., Hoffmann, G., & Chaimani, A., (2017). Dietary supplements and risk of cause-specific death, cardiovascular disease, and cancer: A systematic review and meta-analysis of primary prevention trials. *Advances in Nutrition, 8*(1), 27–39.
83. Severini, C., & Derossi, A., (2016). Could the 3D printing technology be a useful strategy to obtain customized nutrition? *Journal of Clinical Gastroenterology, 50*(Suppl 2), S175–S178.
84. Severini, C., Derossi, A., Ricci, I., Caporizzi, R., & Fiore, A., (2018). Printing a blend of fruit and vegetables: New advances on critical variables and shelf life of 3D edible objects. *Journal of Food Engineering, 220*, 89–100.
85. Shahidi, F., (2009). Nutraceuticals and functional foods: Whole versus processed foods. *Trends in Food Science & Technology, 20*(9), 376–387.
86. Shlisky, J., Bloom, D. E., Beaudreault, A. R., Tucker, K. L., Keller, H. H., Freund-Levi, Y., Fielding, R. A., et al., (2017). Nutritional considerations for healthy aging and reduction in age-related chronic disease. *Advances in Nutrition: An International Review Journal, 8*(1), 17–26.
87. Stoll, T., Schweiggert, U., Schieber, A., & Carle, R., (2003). Process for the recovery of a carotene-rich functional food ingredient from carrot pomace by enzymatic liquefaction. *Innovative Food Science & Emerging Technologies, 4*(4), 415–423.

88. Sun-Waterhouse, D., (2011). The development of fruit-based functional foods targeting the health and wellness market: A review. *International Journal of Food Science & Technology, 46*(5), 899–920.
89. Talukder, S., & Sharma, D. P., (2010). Development of dietary fiber rich chicken meat patties using wheat and oat bran. *Journal of Food Science and Technology, 47*(2), 224–229.
90. Taylor, R., (2011). Older people and functional foods: A role to play, but no silver bullet. *Nutrition Bulletin, 36*(3), 403–407.
91. Teradal, D., Joshi, N., & Aladakatti, R. H., (2017). Therapeutic evaluation of grain based functional food formulation in a geriatric animal model. *Journal of Food Science & Technology, 54*(9), 2789–2796.
92. Thomas, P. R., & Earl, R., (1994). *Opportunities in the Nutrition and Food Sciences* (p. 310). Washington, DC – USA: National Academy Press.
93. Tripathi, K. M., Singh, Y., Dubey, S. K., & Seven, T., (2016). Geriatric nutrition: Need for better aging. *South Asian Journal of Food Technology and Environment, 2*(3, 4), 432–437.
94. Van, D. Z. L. D. T., & Van, T. H. C. M., (2017). Designing new and functional foods for the aging; Chapter 16. In: Raats, M. M., & De Groot, L., (eds.), *Food for the Aging Population* (2nd edn., pp. 323–347). London – UK: Woodhead Publishing.
95. Van, D. Z. L. D. T., Van, K. E., De Wijk, R. A., & Van, T. C. M., (2015). Examining heterogeneity in elderly consumers' acceptance of carriers for protein-enriched food: A segmentation study. *Food Quality Preference, 42*, 130–138.
96. Van, D. Z. L. D. T., Van, K. E., De Wijk, R. A., & Van, T. C. M., (2014). Knowledge, perceptions, and preferences of elderly regarding protein-enriched functional food. *Appetite, 80*, 16–22.
97. Vanm S. W. A., & De Groot, L. P. G. M., (2011). Evidence-based dietary guidance and the role of dairy products for appropriate nutrition in the elderly. *Journal of the American College of Nutrition, 30*(5 Suppl 1), 429S–437S.
98. Verbeke, W., Scholderer, J., & Lähteenmäki, L., (2009). Consumer appeal of nutrition and health claims in three existing product concepts. *Appetite, 52*(3), 684–692.
99. Viuda-Martos, M., Sanchez-Zapata, E., SayasBarberá, E., Sendra, E., Pérez-Álvarez, J. A., & Fernández-López, (2014). Tomato and tomato byproducts. Human health benefits of lycopene and its application to meat products: A review. *Critical Reviews in Food Science and Nutrition, 54*(8), 1032–1049.
100. Volkert, D., Beck, A. M., Cederholm, T., Cruz-Jentoft, A., Goisser, S., Hooper, L., Kiesswetter, E., et al., (2019). ESPEN guideline on clinical nutrition and hydration in geriatrics. *Clinical Nutrition, 38*(1), 10–47.
101. Wahlqvist, M. L., (1994). Functional foods in the control of obesity; Chapter 4. In: Goldberg, I., (ed.), *Functional Foods: Designer Foods, Pharmafoods, Neutraceuticals* (pp. 71–86). London – UK: Springer Science + Business Media.
102. Wang, R., Zhou, W., & Isabelle, M., (2007). Comparison study of the effect of green tea extract (GTE) on the quality of bread by instrumental analysis and sensory evaluation. *Food Research International, 40*(4), 470–479.
103. World Health Organization (WHO), (2015). *Nutrition for Older Persons* (pp. 8, 9). http://www.who.int/nutrition/topics/ageing/en/index1.html (accessed on 2 November 2023).

104. Wyka, J., Biernat, J., Mikołajczak, J., & Piotrowska, E., (2012). Assessment of dietary intake and nutritional status (MNA) in Polish free-living elderly people from rural environments. *Archives of Gerontology Geriatrics, 54*(1), 44–49.
105. Yildiz, Ö., & Eyduran, S. P., (2009). Functional components of berry fruits and their usage in food technologies. *African Journal of Agricultural Research, 4*(5), 422–426.
106. http://documents.worldbank.org/curated/en/314421468324275640/Health-enhancing-foodsopportunities-for-strengthening-developingcountries (accessed on 2 November 2023).

CHAPTER 4

Cultivation of Highly Demanded Medicinal Plants: Case Study in Uttarakhand, India

Nishant Gupta, Ashok Kumar Agarwal, and Poonam Gusain

ABSTRACT

This study assessed the most demanding medicinal plant cultivation awareness, government subsidies, and traditional uses. India represents 12% of medicinal plants in the world with the maximum species being reported in its Himalayan state, Uttarakhand. The cultivation and medicinally important plant data is derived from the National Medicinal Plant Board, Ministry of AYUSH (Ayurveda, Yoga and Naturopathy, Unani, Siddha, and Homeopathy) India and State Medicinal Plant Board (SMPB), Uttarakhand. The data was further augmented by the information provided by 38 informants (8 popular herb dealers and 30 local farmers) in Dehradun. The results showed that 65 medicinal plants belong to 36 families; and 61 genera are being cultivated at various levels (low, moderate, significant) in the Uttarakhand state. An estimated 92% of species are covered under 30 to 75% government subsidies. Most species belong to the families *Fabaceae* and *Apocynaceae*. The most cultivated medicinal plant species are: *Aegle marmelos*, *Nardostachys jatamansi*, *Phyllanthus emblica*, *Swertia chirayita*, *Tinospora cordifolia*, *Withania somnifera*, *Citrus* × *aurantium*, and *Zingiber officinale*. Common medicinal applications include liver diseases, blood disorders, skin problems, digestive disorders, fever, and diabetes. This study suggests that medicinal plant cultivation is low in Uttarakhand. Improved medicinal plant cultivation in Uttarakhand could be helpful for the agro-economy and conserve the valuable medicinal flora and its traditional knowledge.

The Functional Foods: Nutrient and Health Benefits.
Megh R. Goyal, Junaid Ahmad Malik, and Anu Kumari (Eds.)

4.1 INTRODUCTION

Medicinal herbs have been a pivotal part of human lives around the world. Plants have been a vital part of Indian culture, religion, and traditional medicinal systems. Even at present, around 58% of Indian inhabitants rely on plants and agriculture for their basic necessities, including water, oxygen, food, and medicine. Additionally, plants are the main source of their income and livelihood, especially in mountain and low-land areas [10, 16].

There are 52,885 medicinal plant species found around the world. India is among one of the most geographically diverse countries, which represents a significant portion of the world's medicinal flora [8, 15]. There are 7,500 medicinal plant species in India (1,748 in the Indian Himalayas and 1,127 in Uttarakhand) (Figure 4.1). In India, the highest number of medicinal plants has been observed in Uttarakhand. Over 200 medicinal plant species from the Himalayas are consumed as food and fodder. Local inhabitants use many medicinal plant species for nutrition, spices, energy drinks, pickles, jams, and oil. For instance, *Terminalia bellirica* and *Myrica esculenta* are used for medicines, food, and fuel [8].

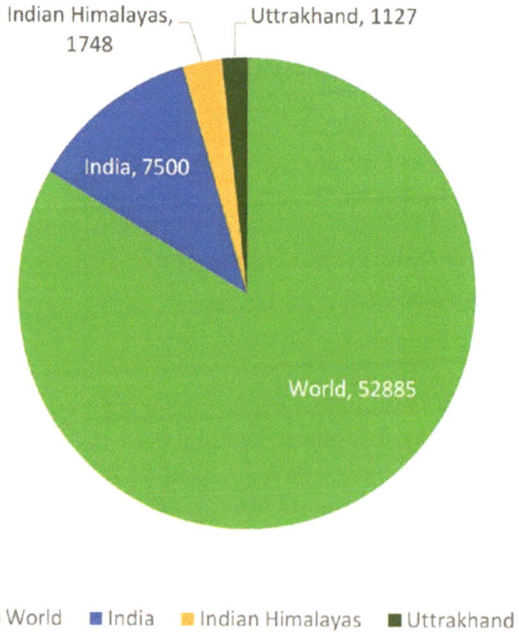

FIGURE 4.1 Worldwide distribution of medicinal plants.

In Uttarakhand, medicinal herb cultivation has attracted government as well as local farmers due to the increased demand. The Ministry of AYUSH (Government of India) started the project on the National Mission on Medicinal Plants (NMMP) in 2008–2009. This flagship program promotes cultivation of medicinal plants by providing subsidies and financial assistance throughout the country [11].

According to Sharma et al. [14], the cultivation of medicinal plants is limited despite the support from the Government of India (GoI); and most plant species are still harvested from their limited wild sources to fulfill the local, national, and international demands. On the other hand, constraints like weather, rain-dependent irrigation, and lack of modern farming techniques remain a major hurdle in medicinal plant cultivation in Uttarakhand.

The objective of the present study is to provide an insight into the most demanded medicinal plant cultivation status, traditional medicinal uses and Indian government subsidies in the state of Uttarakhand.

4.2 MATERIALS AND METHODS

4.2.1 STUDY AREA

The study area consisted of the Himalayan state of Uttarakhand that is spread in 53,485 km^2 area that is divided into two major divisions called Garhwal and Kumaon. Both divisions have 13 districts. About 93% area of Uttarakhand is mountainous and covered by 65% forest. The interim capital is Dehradun. Geographical coordinates are 30.0668°N, 79.0193°E, elevation is 320 to 5,000 m, annual rainfall is 1,075 mm, average temperature ranges from 1 to 35°C. Uttarakhand shares its border with Himachal Pradesh, Uttar Pradesh, Nepal, and China [7]. The most popular Indian rivers, Ganga and Yamuna, also originate from the state (Figure 4.2).

The study comprises some commercially important medicinal plants, their traditional uses, current cultivation status, and government subsidies.

The information about the most popular and demanding medicinal plant species is derived in two main steps:

1. **Data Collection:** The medicinal plants list was derived from several sources, i.e., the SMPB, Uttarakhand, National Medicinal Plant Board, Ministry of AYUSH India [11, 15], and regional floras. Online research and review articles were searched on Google Scholar; and a total of 14,800 results appeared when the term '*Medicinal plant*

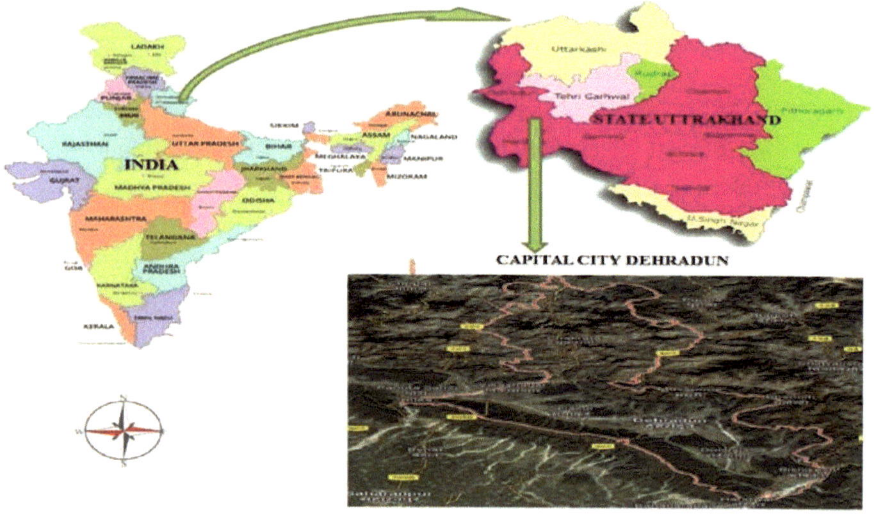

FIGURE 4.2 Study area: Himalayan state Uttarakhand and its capital Dehradun.

cultivation in Uttarakhand' was used, with a customized time period between 2010 and 2022. While 18,400 results appeared when searching for *'Medicinal plant cultivation in India.'* However, only the most cited and highly relevant articles were selected for this study. Further, information was collected from the eight herbal medicine dealers and 30 local farmers belonging to the villages Sahaspur and Shankerpur in Dehradun; and they participated willingly in a brief questionnaire session regarding medicinal plant cultivation status, local names, and traditional uses.

2. **Verification:** Selected medicinal plants taxonomical status was verified by using online plant databases, such as: plant of the world online (POWO) and International Plant Names Index (IPNI) [6, 12]. The data was processed in Microsoft Excel.

4.3 RESULTS AND DISCUSSION

Around 65 most common medicinal plants belong to 36 families; and 61 genera are being cultivated at various levels (low, moderate, significant) in Uttarakhand. Though, most species belong to *Fabaceae* and *Apocynaceae* (Table 4.1). Observed plant families are:

- *Acanthaceae* 2 spp.
- *Apocynaceae* 5 spp.
- *Asparagaceae* 2 spp.
- *Asphodelaceae* 1 sp.
- *Asteraceae* 3 sp.
- *Berberidaceae* 2 spp.
- *Bignoniaceae* 1 sp.
- *Burseraceae* 1 sp.
- *Capparaceae* 1 sp.
- *Caprifoliaceae* 1 sp.
- *Celastraceae* 1 sp.
- *Colchicaceae* 1 sp.
- *Combretaceae* 3 spp.
- *Convolvulaceae* 1 sp.
- *Dioscoreaceae* 1 sp.
- *Elaeagnaceae* 1 sp.
- *Fabaceae* 9 sp.
- *Gentianaceae* 1 sp.
- *Lamiaceae* 3 spp.
- *Lauraceae* 1 sp.
- *Meliaceae* 1 sp.
- *Menispermaceae* 1 sp.
- *Moringaceae* 1 sp.
- *Nyctaginaceae* 1 sp.
- *Phyllanthaceae* 2 spp.
- *Plantaginaceae* 2 spp.
- *Primulaceae* 1 sp.
- *Ranunculaceae* 1 sp.
- *Rubiaceae* 2 spp.
- *Rutaceae* 3 spp.
- *Saxifragaceae* 1 sp.
- *Smilacaceae* 1 sp.
- *Solanaceae* 2 spp.
- *Urticaceae* 1 sp.
- *Violaceae* 1 sp.
- *Zingiberaceae* 1 sp.

Among all reported medicinal plant species, most are suggested for general infection (5 spp.), liver diseases (7 spp.), cancer (5 spp.), blood disorders (7

TABLE 4.1 List of Most Demanded Medicinal Plants: Cultivation Status, Traditional Uses, and Government Subsidies in Uttarakhand [11, 14]

Taxonomy	Vernacular Name	Family	Government Subsidy	Cultivation Status	Traditional Uses
Abrus precatorius L.	Ratti	*Fabaceae*	30%	Low	Tetanus, rabies, cancer
Acacia catechu Willd	Katha	*Fabaceae*	50%	Low	Dysentery, colitis
Aconitum ferox Wall. ex Ser.	Vatsnabh	*Ranunculaceae*	75%	Low	Ear, nasal infection
Aegle marmelos (L.) Correa	Bel	*Rutaceae*	50%	Significant	Stomach disorders
Albizia lebbeck (L.) Benth.	Sherish	*Fabaceae*	50%	Low	Brain tonic
Aloe vera (L.) Burm.f.	Ghrit Kumar	*Asphodelaceae*	30%	Moderate	Skin and blood disorders
Andrographis paniculata (Burm.f.) Nees	Kaal Megh	*Acanthaceae*	30%	Moderate	Cancer, digestive disorders
Artemisia annua L.	Artemisia	*Asteraceae*	30%	Low	Fever, headache
Asparagus racemosus Willd.	Shatavari	*Asparagaceae*	30%	Moderate	Stomach disorders
Atropa bella-donna L.	Atropa	*Solanaceae*	50%	Moderate	Sedative, antispasmodic
Azadirachta indica A. Juss.	Neem	*Meliaceae*	30%	Rare	Blood disorders
Bacopa monnieri (L.) Wettst.	Brahmi	*Plantaginaceae*	30%	Low	Brain tonic
Berberis aristata DC.	Daruhaldi	*Berberidaceae*	75%	Moderate	Eye, skin disorders
Bergenia ciliata (Haw.) Sternb.	Pashnabheda	*Saxifragaceae*	30%	Low	Digestive ailments
Boerhavia diffusa L.	Punarnava	*Nyctaginaceae*	30%	Low	Liver, blood disorders
Cassia angustifolia Vahl.	Senna	*Fabaceae*	30%	Moderate	Laxative
Catharanthus roseus (Linn.) G. Don	Sadabahar	*Apocynaceae*	30%	Low	Diabetes, cancer, wound
Celastrus paniculatus Willd.	Malkangani, Jyothismathi	*Celastraceae*	30%	Low	Anemia
Centella asiatica (L.) Urb.	Mandookparni	*Apiaceae*	30%	Low	Skin disorders
Chlorophytum borivilianum Santapau & R.R.Fern.	Shwet Musali	*Asparagaceae*	30%	Low	Diabetes, cancer

TABLE 4.1 *(Continued)*

Taxonomy	Vernacular Name	Family	Government Subsidy	Cultivation Status	Traditional Uses
Cinnamomum tamala (Buch. Ham.) T. Nees & C.H.Eberm.	Tejpaat,	*Lauraceae*	30%	Moderate	Respiratory infections
Citrus × *aurantium* L.	Malta	*Rutaceae*	Non	Significant	Energetic drinks, antiemetic
Citrus × *limon* (L.) Osbeck	Mosambi, Mosmi	*Rutaceae*	Non	*Rutaceae*	Aid digestion, anti-emetic
Clitoria ternatea L.	Aparajita	*Fabaceae*	30%	Low	anxiolytic
Commiphora wightii (Arn.) Bhandar	Guggal	*Burseraceae*	75%	Moderate	Blood disorders
Convolvulus prostratus Forssk.	Shankhpushpi	*Convolvulaceae*	30%	Moderate	Weakness
Crataeva nurvala Buch–Ham.	Varun	*Capparaceae*	50%	Low	Digestive ailments
Dioscorea bulbifera L.	Rotalu, Gethi	*Dioscoreaceae*	30%	Low	Diabetes, asthma
Eclipta prostrata (L.) L	Ajagara, Bringaraj	*Asteraceae*	30%	Moderate	Hair fall, Skin disorders
Embelia ribes Burm. f.	VaiVidang	*Primulaceae*	30%	Low	Infections
Gloriosa superba L.	Agnishikha, fire lily	*Colchicaceae*	50%	Low	Snake bite, skin disorders
Glycyrrhiza glabra L.	Licorice Roots, Mulethi	*Fabaceae*	50%	Moderate	Liver disorders
Gmelina arborea Roxb. ex Sm.	Gambhari	*Lamiaceae*	50%	Low	Weakness, tonic
Gymnema sylvestre (Retz.) R.Br. ex Sm.	Gudmar	*Apocynaceae*	30%	Rare	Diabetes
Hippophae rhamnoides L.	Seabuckthorn	*Elaeagnaceae*	50%	Low	Blood disorders
Holarrhena pubescens Wall. ex G.Don	Kurchi/Kutaj	*Apocynaceae*	30%	Low	Dysentery, colitis
Inula racemosa Hook.f.	Pushkarmool	*Asteraceae*	50%	Low	Chest pain
Justicia adhatoda L.	Adusa	*Acanthaceae*	30%	Moderate	Cough
Leptadenia reticulata (Retz.) Wight & Arn.	Jivanti	*Apocynaceae*	50%	Low	TB, fever
Moringa oleifera Lam.	Drumstick tree, Sahajan	*Moringaceae*	0%	Moderate	Weakness
Mucuna pruriens (L.) DC.	Konch	*Fabaceae*	30%	Rare	Infertility

TABLE 4.1 *(Continued)*

Taxonomy	Vernacular Name	Family	Government Subsidy	Cultivation Status	Traditional Uses
Nardostachys jatamansi (D.Don) DC.	Jatamansi	*Caprifoliaceae*	75%	Significant	Hair fall, antiepileptic.
Ocimum tenuiflorum L.	Tulsi	*Lamiaceae*	30%	Moderate	Infection, fever
Oroxylum indicum (L.) Kurz	Syonaka	*Bignoniaceae*	75%	Low	Liver and gastric disorders
Phyllanthus amarus Schumach. & Thonn.	Bhumiamlaki	*Phyllanthaceae*	30%	Moderate	Liver, urine, kidney problems
Phyllanthus emblica L.	Amla	*Phyllanthaceae*	30%	Significant	Digestive ailments
Picrorhiza kurroa Royle ex Benth.	Kutki	*Plantaginaceae*	75%	Moderate	Liver, respiratory
Pleurolobus gangeticus (L.) J.St. Hil. ex H.Ohashi & K.Ohashi	Sariwan	*Fabaceae*	50%	Low	Febrifuge
Podophyllum hexandrum Royle	Bankakri, Indian podophyllum	*Berberidaceae*	75%	Low	Purgative
Pueraria tuberosa (Roxb. ex Willd.) DC.	Vidarikand	*Fabaceae*	50%	Low	Tonic
Rauvolfia serpentina (L.) Benth. ex Kurz	Sarpgandha	*Apocynaceae*	50%	Moderate	Hypertension
Rheum webbianum Royle	Archa	*Rubiaceae*	50%	Low	Cancer, kidney problems
Rubia cordifolia L.	Manjishtha	*Rubiaceae*	30%	Moderate	Skin disorders
Smilax china L.	Hrddhatri (Madhu Snuhi), Chob Chini Lokhandi	*Smilacaceae*	50%	Low	Arthritis
Swertia chirayita (Roxb.) H.Karst.	Chirata, Charayatah	*Gentianaceae*	75%	Significant	Liver disorders
Terminalia arjuna (Roxb. ex DC) Wight & Arn	Arjun	*Combretaceae*	30%	Moderate	Heart and blood disorders
Terminalia bellirica (Gaertn.) Roxb.	Behera	*Combretaceae*	30%	Moderate	Liver and gastric disorders
Terminalia chebula Retz	Harad	*Combretaceae*	30%	Moderate	Digestive ailments

TABLE 4.1 *(Continued)*

Taxonomy	Vernacular Name	Family	Government Subsidy	Cultivation Status	Traditional Uses
Tinospora cordifolia (Willd.) Hook.f. & Thomson	Giloy	*Menispermaceae*	30%	Significant	Infections, fever
Urtica dioica L.	Bicchu Ghass, Bichhu Buti	*Urticaceae*	Non	Rare	Edible, nutritive
Viola odorata L.	Banafsha, Vanafsha, Gul Banafsha	*Violaceae*	0%	Moderate	Fever, headache
Vitex negundo L.	Nirgundi	*Lamiaceae*	30%	Low	Skin and blood disorders
Withania somnifera (Linn.) Dunal	Ashwagandha	*Solanaceae*	30%	Significant	Weakness, never disorders
Zanthoxylum alatum Roxb.	Timoor	*Rutaceae*	50%	Rare	Asthma, dental problem
Zingiber officinale Roscoe (Dried)	Sounth	*Zingiberaceae*	0%	Significant	Respiratory and digestive diseases

spp.), skin disorders (7 spp.), digestive disorders (6 spp.), stomach disorders (2 spp.), fever (5 spp.) and diabetes (4 spp.) (Figure 4.3).

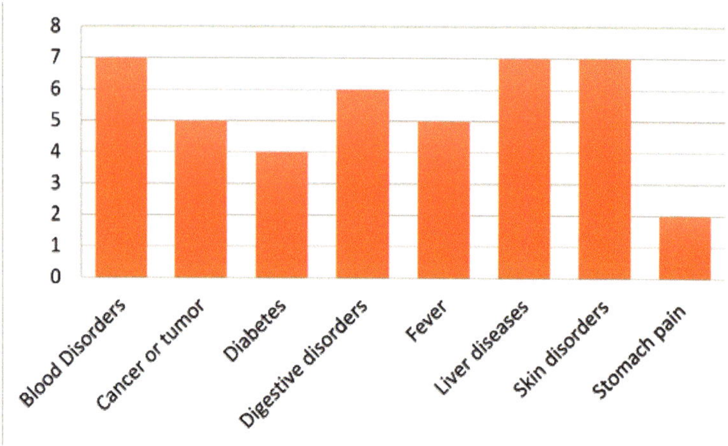

FIGURE 4.3 Number of medicinal plants species used for treating different diseases.

The cultivation status of observed plant species is only based on the questionnaire session outcomes like participants' response to variables like low, moderate, significant or rare. Out of 65 medicinal plant species, 29 spp. less, 22 spp. moderately, 9 spp. significantly while 5 spp. are less cultivated (Figure 4.4). The most cultivated species include *Aegle marmelos*, *Citrus* × *aurantium*, *Nardostachys jatamansi*, *Phyllanthus emblica*, *Swertia chirayita*, *Tinospora cordifolia*, *Zingiber officinale* and *Withania somnifera*.

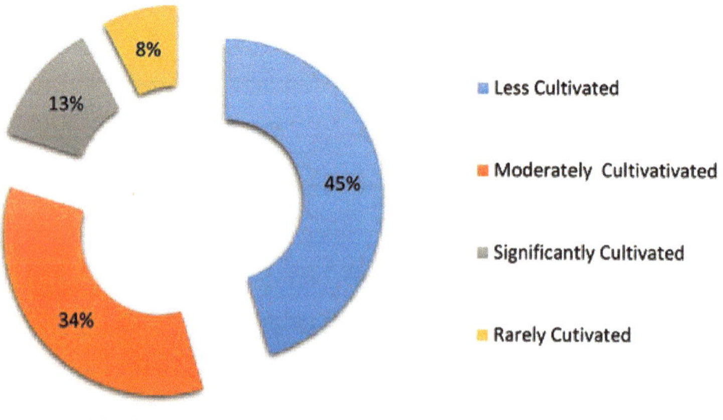

FIGURE 4.4 Cultivation frequency of most demanded medicinal plants in Uttarakhand.

Since the GoI provides subsidies (30 to 75%) for numerous medicinal plants growing throughout India, therefore, the reported 65 medicinal plants fall under various subsidy species: 34 species (30%), 17 species (50%), 8 species (75%), while 5 species are not covered by subsidies (Figure 4.5). Around 92% of species are covered under 30 to 75% central government subsidies, while only 12% of species are being cultivated significantly.

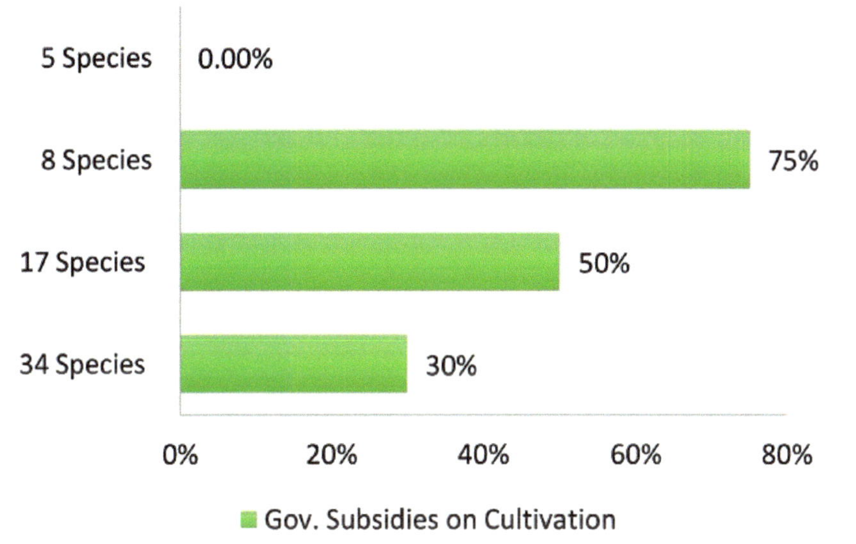

FIGURE 4.5 Percentage of government subsidies on medicinal plant cultivation.

The worldwide demand for herbal medicines and plant-based products has been increasing remarkably in recent years, possibly due to minimal or no side effects. However, despite a growing demand for medicinal plants and significant research on plant products and phytochemicals [13, 17]; agricultural sector, including medicinal plant cultivation in Uttarakhand, still has some natural and social constraints such as limited irrigation, and lack of modern cultivation techniques.

Uttarakhand has sufficient cultivated area (0.8 million hectares) of cultivated area. However, more than 50% of cultivated area is rain-fed [5]. Some people at higher elevations usually depend on traditional water sources. A small pit-like structure called 'Bavadi' (Figure 4.6) is the main water source for their daily needs. Sometimes the whole village depends on such traditional water source.

FIGURE 4.6 Water source 'Bavadi' at high altitude in Uttarakhand.

However, local people are quite aware about the medicinal plants' traditional uses. A previous study revealed that more than 200 plant species are employed to treat liver diseases by indigenous communities (such as: nomadic Gujjars, Tharu, and Bhoxa of Uttarakhand); *Boerhavia diffusa, Tinospora cordifolia* and *Phyllanthus amarus* are the most popular herbs used against liver diseases [3].

Traditional knowledge of medicinal herbs may be a promising indicator for further scientific evaluation as most of the traditional medicine plants are found significantly effective in several *in vivo* and *in vitro* studies. For instance, the most common traditional medicinal herb, *Tinospora cordifolia*, showed its remarkable immune stimulatory, antimicrobial, hepatoprotective, and antioxidant properties in experiments [1].

Similarly, the most commonly used and cultivated species, *Zingiber officinale*, is not only one of the most suggested traditional herbs in Uttarakhand, but also recognized as 'safe' by the Food and Drug Administration (FDA). In several studies, *Z. officinale* has been shown its numerous remarkable therapeutic constituents and properties, including antidiarrheal, antimicrobial, antioxidant, anti-inflammatory, and anticoagulant activities [2, 4].

Uttarakhand is a natural home of substantial medicinal plant species, possibly less explored so far. According to the Uttarakhand Medicinal Plants Database (UMPDB), there are about 1,127 medicinal plant species growing across the 13 districts of Uttarakhand. Albeit, maximum medicinal plants are reported in their natural habitat, only 65 species are being considered for cultivation (Table 4.1), while only 9 medicinal plant species have grown significantly. Near 70% of medicinal herbs from the Himalayas are an already subjected to ruinous harvesting method, which indicates a severe shortage in future. Similarly, knowledge of traditional medicinal plants is disappearing rapidly [9, 14].

Since, most medicinal herbs are collected from their natural habitat and sold to market. The cultivation is rare, especially for those who do not have farmland or facilities to grow the medicinal herbs [18].

This study can provide a preliminary view of the current medicinal plant cultivation conditions and government aid. We regret that the study was limited to the capital Dehradun and its rural area. We further suggest more comprehensive district-wise studies.

4.4 SUMMARY

The Himalayan state of Uttarakhand favors a variety of medicinal plant growth. However, despite knowing the significance of medicinal plant cultivation and government subsidies, only a few species are being cultivated in the state due to the various constraints. Due to the increased demand, most medicinal species are harvested destructively from their wild sources. Improved medicinal plant cultivation in the state could be helpful for agro-economy as well as to conserve valuable medicinal flora and its traditional knowledge.

KEYWORDS

- **agro-economy**
- **cultivation**
- **herbal medicine dealers**
- **medicinal plants**
- **natural habitat**
- **subsidies**
- **traditional water sources**

REFERENCES

1. Alsuhaibani, S., & Khan, M. A., (2017). Immune-stimulatory and therapeutic activity of *Tinospora cordifolia*: Double-edged sword against salmonellosis. *Journal of Immunology Research, 2017*, 1–9.

2. Beristain-Bauza, S. D. C., Hernández-Carranza, P., Cid-Pérez, T. S., Ávila-Sosa, R., Ruiz-López, I. I., & Ochoa-Velasco, C. E., (2019). Antimicrobial activity of ginger (*Zingiber officinale*) and its application in food products. *Food Reviews International, 35*(5), 407–426.
3. Bhat, J. A., Kumar, M., & Bussmann, R. W., (2013). Ecological status and traditional knowledge of medicinal plants in Kedarnath Wildlife Sanctuary of Garhwal Himalaya, India. *Journal of Ethnobiology and Ethnomedicine, 9*(1), 1–18.
4. Daswani, P. G., Brijesh, S., Tatali, P., Antia, N. H., & Birdi, T. J., (2010). Anti-diarrheal activity of *Zingiber officinale*. *Current Science, 2010*, 222–229.
5. ICAR, (2022). https://icar.org.in/sites/default/files/2023-02/Indian-Agriculture-after-Independence.pdf (accessed on 28 November 2023).
6. *International Plant Names Index,* (2022). https://www.ipni.org (accessed on 2 November 2023).
7. *Introduction to Uttarakhand: Geographic Personality of Uttarakhand.* https://www.uou.ac.in/sites/default/files/slm/BTTM-601.pdf (accessed on 2 November 2023).
8. Kala, C. P., Dhyani, P. P., & Sajwan, B. S., (2006). Developing the medicinal plants sector in Northern India: Challenges and opportunities. *Journal of Ethnobiology and Ethnomedicine, 2*(1), 1–15.
9. Kumar, A., Kumar, R., Sharma, M., Kumar, U., Gajula, M. P., & Singh, K. P., (2018). Uttarakhand medicinal plants database (UMPDB): A platform for exploring genomic, chemical, and traditional knowledge. *Data, 3*(1), 7–9.
10. Manida, M., & Nedumaran, G., (2020). Agriculture in India: Information about Indian agriculture and its importance. *Aegaeum Journal, 8*(3), 729–736.
11. *National Medicinal Plant Board* (2022). https://nmpb.nic.in/content/medicinal-plants-cultivation (accessed on 2 November 2023).
12. *Plant of the World Online,* (2022). https://powo.science.kew.org (accessed on 2 November 2023).
13. Rokaya, M. B., Uprety, Y., Poudel, R. C., Timsina, B., Münzbergová, Z., Asselin, H., & Sigdel, S. R., (2014). Traditional uses of medicinal plants in gastrointestinal disorders in Nepal. *Journal of Ethnopharmacology, 158*, 221–229.
14. Sharma, J., Gairola, S., Gaur, R. D., & Painuli, R. M., (2012). The treatment of jaundice with medicinal plants in indigenous communities of the Sub-Himalayan region of Uttarakhand – India. *Journal of Ethnopharmacology, 143*(1), 262–291.
15. *State Medicinal Plant Board Uttarakhand (*2014*).* http://www.smpbuk.org (accessed on 2 November 2023).
16. Tuteja, U., (2013). *Agriculture Profile of Uttarakhand* (p. 12). New Delhi – India: Agricultural Economics Research Centre, University of Delhi.
17. Vasisht, K., Sharma, N., & Karan, M., (2016). Current perspective in the international trade of medicinal plants material: An update. *Current Pharmaceutical Design, 22*(27), 4288–4336.
18. Zenderland, J., Hart, R., Bussmann, R. W., Paniagua-Zambrana, N. Y., Sikharulidze, S., Kikvidze, Z., & Batsatsashvili, K., (2019). The use of use value: Quantifying importance in ethnobotany. *Economic Botany, 73*(3), 293–303.

PART II

ANTIVIRAL ACTIVITIES OF FUNCTIONAL FOODS – COVID-19

CHAPTER 5

Antiviral Activities of Ocimum Species (Tulsi) to Combat RNA Virus-Mediated Pandemic Situations: COVID-19

Hariswami Das, Rajkrishna Biswas, Arif Jaman, and
Anup Kumar Sarkar

ABSTRACT

Over the last few decades, the world has experienced many virus outbreaks caused by RNA viruses. Some of these have directly infected humans and some are harbored in wild mammals or birds when transmitted to humans. Research in microbiology for decades has relied heavily on the use of medicinal herbs and natural compounds as templates for developing new antimicrobial drugs with different modes of action. Many species of genus Ocimum (Tulsi) have been recognized for their ability to inhibit RNA viruses. Different parts of Tulsi plant are edible and frequently consumed as raw, extract, capsules, tinctures, powder, and as herbal tea. The medicinal virtues of various plant-based formulations were also mentioned in ayurvedic literature. Recent studies have documented that methanol, ethanol, aqueous, and chloroform extract of such plants have anti-viral property against several pathogenic strains of RNA viruses. Scientific study conducted in the modern era on Ocimum sp. also indicates numerous psychological and physiological benefits associated with its consumption. This chapter highlights anti-RNA virus efficacy of Ocimum sp. and practical significance of their bioactive compounds to combat epidemic and pandemic situations (such as: COVID-19; SARS-CoV-2).

The Functional Foods: Nutrient and Health Benefits.
Megh R. Goyal, Junaid Ahmad Malik, and Anu Kumari (Eds.)

5.1 INTRODUCTION

Most common human pathogenic RNA viruses are Rhinovirus, Yellow fever virus, Dengue virus (DENV), Polio virus, Measles virus, human immunodeficiency virus (HIV), hepatitis C virus (HCV), respiratory syncytial virus (RSV), Zika virus, Ebola virus, middle east respiratory syndrome (MERS) virus, severe acute respiratory syndrome (SARS) virus [13, 28, 77]. Due to RNA viruses, world have faced number of epidemic and pandemic situations in several times [57].

When preventative vaccinations and antiviral medicines are unavailable, virus outbreaks pose grave threat to public health. Many plant molecules have been reported to inhibit RNA viruses including corona virus disease-19 (COVID-19) [56]. Therefore, humans on earth are completely dependent on variety of plants for their livelihood and medicinal needs [79]. Biologically the antiviral activities include suppression of viral replication, endocytosis, transcription of viral genome and biosynthesis of viral protein component [12].

Plant-specific phytochemicals (such as: lignans, saponins, alkaloids, apigenin, baicalin, catechins, flavonoids, kaempferol, luteolin, polysulfates, and quercetin) have different roles in preventing viral entrance, disrupting the nucleocapsid and genetic material, and suppressing viral replication [22].

Medicinal plants and their byproducts have demonstrated prospective health effects (such as: antiviral activity) and are consequently employed by health healers to protect humans from variety of health conditions [22, 24, 94]. Herbal medications and their active phytochemicals have been demonstrated to be beneficial against different pathogenic viruses, such as, coxsackievirus, HIV, HSV (herpes simplex virus), Rhinovirus, SARS, MERS, along with SARS-CoV-2 (which caused the COVID-19 pandemic) [31].

As a result of the persistent threat posed by earlier RNA virus-mediated epidemics, scientific communities are now significantly better equipped with infrastructure, allowing for rapid and precise research on the major scientific elements of the RNA virus-mediated disease, with consequences for its management and lessons for upcoming epidemics. Since previous epidemic conditions, the development of innovative anti-RNA viral phytomolecules have become the primary objective of medical research in an effort to obtain the advantage over such widespread diseases and avoid their recurrence [2, 17].

Recent scientific publications on SARS-CoV-2 or COVID-19 have revealed that polyphenols, which are conjugated with fused ring structures, are the most promising phytochemicals found to control RNA viruses [56].

RNA viruses are known to affect multiple organ systems, including the respiratory system, cardiovascular system, gastrointestinal (GI) system, renal, and excretory system, central nervous system (CNS), reproductive system, etc., ultimately impairing normal functioning of the entire body [75]. Most viruses can be neutralized in vitro by medications like interferon (IFN) and ribavirin, but this is not always the case [12].

Some species of Ocimum have shown to be useful in preventing the spread of various dangerous viruses including RNA viruses. They have both prophylactic and curative efficiency against such viruses. Many traditional medical literatures have reported that these plants hold strong antiviral activity. In case of pandemic scenario, priorities should be given on the development of novel natural medications and repeated clinical trials of existing drugs and immuno-boosting natural products, as well as the discovery and improvement of vaccinations against these viral strains [76].

During COVID-19 pandemic situation, people were advised by several national and international organizations to take handful of O. tenuiflorum leaves along with boiled water [53]. Because of its multi-modal therapeutic advantages, some experts believe that Ocimum could be effective in the prevention and treatment of COVID-19 and other RNA virus-mediated disorders [27]. Although some current evidence supports the use of Ocimum to treat RNA virus-mediated symptoms, yet there is no standard formulation. Experts claim that strong immune boosting compounds can speed up the pace of healing even when one gets infected by the RNA viruses [95].

O. tenuiflorum is an essential immune-boosting herb that has shown to increase immunity to fight against RNA virus infection by increasing levels of IFN-, IL-4, T-helper cells and natural killer cells [66]. Anti-viral bioactive compounds may bind either to surface proteins, bio-membrane, DNA (deoxyribonucleic acid) or RNA of the viral particle [32, 51].

This chapter has enlightened anti-RNA virus efficacy of Ocimum species and practical significance of their components to combat epidemic and pandemic situations.

5.2 BOTANICAL ASPECTS OF OCIMUM SPECIES

Most of the species of genus Ocimum are widely used as medicinal plants in the past. It contains appreciable quantities of therapeutic ingredients. Ocimum is one of the largest genera in the Lamiaceae family, which includes more than 150 aromatic species and is considered as a source of essential oils (EOs) [96]. These are most significant medicinal plants mentioned in several ancient and recent medical literature for their therapeutic properties in several aspects [29]. Leaves, seeds, and roots of this plant are used for their health benefits. It is commonly used in home remedies to treat injuries, respiratory problems, hepatic syndrome, viral infection, ear-ache, spinal discomfort, hiccup, and conjunctivitis in neonates [19, 73]. Ocimum have antibiotic properties against huge number of microorganisms including RNA viruses. Several studies have shown that methanol, ethanol, aqueous, and chloroform extracts of several species have anti-viral properties against several animal RNA viruses [103].

In many regions, several species of Ocimum are consumed as leafy vegetables and flavoring agents [79]. Regular addition of Ocimum to the diet and/or medicinal formulation can potentially assist in the immune-enhancement and can improve the recovery status of a person.

5.3 GENERAL STRUCTURE AND PATHOGENESIS OF RNA VIRUSES

RNA viruses contain ss-RNA (majorly) or ds-RNA (minorly) as genetic materials. Hundreds of notable diseases in humans are caused by RNA viruses, such as: influenza, common cold, dengue, mumps, polio, rubella, rabies, e-bola virus disease, measles, hepatitis, Zika fever, MERS, SARS, coronavirus-19, etc. [76, 97]. Several other deadly animal diseases like avian flu (bird flu), swine flu is also caused by the RNA viruses.

Following the COVID-19 pandemic in December 2019, approximately 143 million individuals became infected within 16 months, causing over 3.05 million deaths [14]. Studies showed that around 700 million to 1.4 billion cases of swine flu occurred including asymptomatic and mild cases in the outbreak of swine flu in 2009 outbreak. Spanish flu outbreak in 1918–1920 was one of the deadliest virus outbreaks leading the death of 17 to 100 million people [90]. A total of almost 7,11,000 Zika virus disease cases were found in the epidemic expansion in 2015–2016 in several countries of North and South America [104].

The SARS-CoV-2 genome is delivered into host cells as ss-RNA and then transformed into viral polyproteins via the protein synthesizing machinery of the host [81]. Taxonomically SARS-CoV-2 belongs to the family coronaviridae, which also includes some viruses causing mild common cold, viz., hCoV-OC43, HKU, and 229E [72, 80]. In most cases, these viruses impart some common symptoms including cold, fever, weakness, hypertension, cerebral infarction, etc. Several studies confirmed that Ocimum based medical therapy can significantly improve these symptoms.

Unlike other pathogens, viruses are non-autonomous microorganisms, which require living cells to replicate. For multiplication of viral genome, normal cellular metabolic pathways of host cells are utilized and thus it is very challenging to devise a therapy that attacks the virion directly or indirectly without causing harm to the infected cells [100]. Viral replication is completely dependent on the viral enzyme, which also plays a key role in triggering disease [36].

RNA viruses have an enzyme called RNA-dependent RNA polymerase (RdRP) that helps to make copies of the g-RNA and is also involved in transcription of that g-RNA into m-RNA [38, 105]. The viral RdRP is the priority target for most of the potential antiviral drugs [10]. RdRPs are known to be error-prone and thus RNA virus genomes are susceptible to potentially modifications owing to their own error-prone RNA-synthesis machinery [9]. Once the viruses get into human body, it will be able to cause several morphological, histological, physiological, and genetic alterations. Within 24–48 h of viral infection, up to 25% of all RNA molecules present in host cells are pathogenic vRNA, which comprise viral genomic RNA (gRNA), messenger RNA (mRNA) transcribed from gRNA and ds-replication intermediates (dsRNA: double stranded RNA) [57, 70]. Controlling of such viral disorders is quite complex as many of these evolve very rapidly due to short genome length and high mutation rates of RNA [23].

5.4 DIFFERENT SPECIES OF OCIMUM WITH ANTIVIRAL ACTIVITIES

Ocimum has a complex chemical makeup, containing vast number of nutrients as well as bioactive compounds, the levels of which vary greatly between species or variants, and even between same variants growing in the same location [16]. Various preparations of Ocimum have been explored for viral pathogen-related issues, including aqueous extracts and extracts in numerous different organic solvents (such as: benzene, chloroform, diethyl ether, ethyl alcohol, ethyl acetate, methyl alcohol, petroleum ether, etc.) [37]. Even various plant portions possess varied biochemical compositions. Most of the compounds have anti-microbial properties. The EO in case of O. tenuiflorum has abundance of bioactive compounds, such as [85, 96]:

- (EZ)-Famesol;
- 14-hydroxy-α-humulene;
- Benzene;
- Borneol;
- Calamine;
- Camphene;
- Carvacrol;
- Caryophyllene;
- Cis-sesquisainene hydrate;
- Citronellel;
- Dimethylbenzene;
- Elemol;
- Ethyl-2-Methyl butyrate;
- Eugenol;
- Euginal;
- Geraneol;
- Humulene oxide;
- Iso-eugenol;
- Isocaryophyllene;
- Ledol;
- Limatrol;
- Limocene;
- Linalool;
- Methyl carvicol or Estragol;
- Nerolidol;
- Octane;
- Sabinene;
- Selin-11-en-4-α-ol;
- Terpiniolene;
- Tetradecanal;
- Toluene;
- Urosolic acid;
- α-amorphene;
- α-bisbolol;
- α-guaiene;
- α-humulene;
- α-murolene;
- α-terpeneol; and
- β-elemene.

Seeds of this plant are rich in Hexourenic acid, Linodilinolin, Linolenic acid, Oleic acid, Palmitic acid, and Stearic acid [96]. Seed mucilage of O. tenuiflorum contains some level of sugars like xylose and other polysaccharides [87]. EOs are more potent and versatile than the commercial drugs [54]. This plant would also be source of monoterpenes and sesquiterpenes, including campene, cholesterol, neral, and stigmasterol [96].

Vitamins C and Vitamins A are also found in O. tenuiflorum, which promote antibody production up to 20% to provide improved immunity against several diseases [64, 99]. In addition to stearic acid, oleic acid, palmitic acid, linoleic acid, myristic acid, α-linolenic acid, carpic acid, and lauric acid, O. basilicum also contains antioxidant compounds, such as caffeic acid, vanillic acid, rosmarinic acid, quercetin, rutin, apigenin, chlorogenic, and p-hydroxybenzoic acids [99].

Researchers have indicated that O. americanum is also abundant in alkaloids, phenolics, tannins, lignin, saponins, flavonoids, terpenoids, α-pinene, camphene, sabinene, β-pinene, β-mircene, and anthraquinones [4, 21].

It has been pointed out the presence of huge bioactive compounds, such as: carbohydrates, alkaloids, flavonoids, tannins, steroids, and triterpenoids in O. gratissimum [65, 71]. Eugenol, one of the key phytoconstituents of Ocimum sp., has shown promising activity against E-bola virus [48]. Some of the compounds are acclaimed to inhibit one of the following events (Table 5.1):

- Assemblage of virus particles;
- Entry of virus;
- Formation of structural and enzymatic proteins like protease, helicase, methyl transferase, polymerase, etc.;
- Organization of capsid;
- Viral replication, etc.

During the COVID-19 pandemic, Patanjali Research Institute (Hardwar–India) and the National Institute of Medical Science (Jaipur–India) jointly developed an Ocimum-based drug coronil. According to Patanjali's researchers, 69% of patients with COVID-19 proved negative after only three days during clinical trials of this drug. And after seven days, they were entirely healed [95]. Researchers have also demonstrated that eugenol from O. tenuiflorum reduced pulmonary inflammation, decreased temperature and suppressed arrhythmias, and enhanced locomotor activity in animal model of COVID-19 [66]. The O. basilicum extracts and purified components showed wide range of anti-viral activity against both DNA virus and RNA viruses [85].

TABLE 5.1 Ocimum Derived Anti-RNA Viral Components and Their Mode of Action

Plant Species	Chemical Components	Mode of Action	Type of Virus	References
Ocimum americanum	Rosmarinic	Inhibition of penetration of virus.	HIV-I	[5, 18, 67, 96]
	Oleanolic acid			
	Dichloromethane extract of leaves.		EV71	
	Methyl extract of leaves.			
Ocimum basilicum	Essential oils and monoterpenes.	Viral particle inhibitor.	Bovine viral diarrhea virus.	[43]
	Aqueous extract.	Weak activity against reverse transcriptase.	HIV	[102]
	Aqueous extract.	Antiviral activity	HSV-1	[15]
	Ethanolic extract.		HSV-2	
	Eugenol	Inhibition of virus replication.	HIV-1 (RNA non-enveloped virus).	[11]
	Eugenol epoxide			
	Apigenine	Inhibition of virus replication.	EV (RNA non-enveloped virus).	[15]
			CVB1 (RNA non-enveloped virus).	
		Inhibition of chymotrypsin-like protease.	SARS-CoV2	[58]
	Linalool	Immuno-enhancement against viral infection.	ADV-3,8,11, HBV	[15]

TABLE 5.1 *(Continued)*

Plant Species	Chemical Components	Mode of Action	Type of Virus	References
	Ursolic acid	Infection and replication inhibitor.	Coxsackie virus Enterovirus 71	[15]
		Main protease	SARS-CoV-2	[45]
		Immuno-enhancement against viral infection.	ADV-8, -11; HIV, HSV, HBV, EV, CVB1	[6, 15]
		Inhibition of chymotrypsin-like protease.	SARS-CoV2	[58]
	Rosmarinic acid	Main protease	SARS-CoV-2	[45]
	Oleanolic acid	Main protease	SARS-CoV-2	[45]
		Inhibition of chymotrypsin-like protease.	SARS-CoV-2	[58]
	Methyl eugenol	Main protease	SARS-CoV-2	[45]
	Crude extract and terpenoid.	Inhibition of infection.	H9N2 virus	[25]
Ocimum gratissimum	Eugenol Thymol	Inhibition of infection.	HSV I HSV II	[46]
	Aqueous extract	Activity against reverse transcriptase and proviral DNA copying.	HIV	[7]

TABLE 5.1 *(Continued)*

Plant Species	Chemical Components	Mode of Action	Type of Virus	References
Ocimum forsskaolii	Ursolic acid	Prophylactic immunity.	HCV	[84]
Ocimum kilimands-charicum	Aqueous extract	Inhibition of reverse transcriptase.	HIV	[93]
	Methanolic extract			
Ocimum labiatum	Leaf extract Pheophytin-a	Inhibition of protease and reverse transcriptase.	HIV-I	[40]
Ocimum menthiifolium	Butanolic extract	Proteolytic activity	SARS-CoV-2	[3]
Ocimum tenuiflorum	Leaf extract	Antiviral activity	COVID-19	[25, 63, 88]
	Hydro-alcoholic extract	Inhibit intracellular multiplication.	H9N2	[1, 39]
	Crude extract	Promising antiviral	H9N2 virus	[25]
	Terpenoid			
	Orientin	Attached to amino acid residues of both structural and non-structural protein.	SARS-CoV-2	[10]
	Vicenin-1		SARS-CoV-2	[10]
	Methanolic extract of aerial parts.	Protease inhibitors	HIV	[74]
	Methanolic extract	Inhibition of CPE formation as well as viral replication.	DENV-1	[52]

5.5 BIOCOMPOUNDS OF OCIMUM WITH SPECIFIC MECHANISM OF INHIBITION OF RNA VIRUSES

The Ocimum species have gained significant attention as anti-RNA viral formulation, because it can balance homeostasis of the body and can increase life span. The antiviral properties of Ocimum and its phytoconstituents have been studied in numerous clinical studies, including cell line, animal model, and human trials. Although some natural substances do not engage in the direct inhibitory response to RNA viruses, yet adequate immunity is critical in the prevention of such viruses. Remarkable number of research studies have confirmed that most of the species of the genus have immune-modulating efficacy [34].

The innate immune system is critical for recognizing RNA viruses and has a significant impact on antiviral responses and the protection from RNA virus-related disorders [77]. Innate immunity is the primary line of defense against any infective pathogens including RNA viruses. Molecular pathogen-recognition receptors have been implicated in the pathogenesis of diseases caused by RNA viruses, and remarkable number of investigations have validated this fact. Viral nucleic acids are among the most prominent pathogen-associated molecular patterns (PAMPs), eliciting specific immunological response and are typically recognized by certain pattern recognition receptors (PRRs) distributed throughout the cell, where they detect the existence of viral particles [57].

In some cases, these plant extracts can contribute to antiviral efficacy to some level, despite their inability to demonstrate considerable activity against certain RNA viruses [33]. O. gratissimum, O. suave and O. tenuiflorum have been reported as traditional herbal treatments for HIV [30, 42, 47].

O. tenuiflorum is being utilized to treat many disorders associated with RNA viral pathogenesis. Many researchers have confirmed that some biocompounds of Tulsi can improve typical symptoms of COVID-19, which include discomfort, diarrhea, coughing, and fever [50, 55, 62, 69]. The discovery of the antiviral mechanisms inside these naturally occurring substances has paved the way for additional study focusing on the interactions between viruses and their hosts under various environmental conditions. There have been numerous reports of Ocimum-based natural substances that are capable of efficiently preventing the early stage of viral infection, which includes adhesion and invasion of the virus.

Anti-viral activity is usually dependent on the feature of solvent in which the plant part was crushed. Patil et al. [68] reported that water extract of O.

tenuiflorum did not affect paramyxovirus but suppressed orthomyxovirus significantly, whereas ethanol extract suppressed paramyxovirus somewhat but had no effect on orthomyxoviruses [68]. It was also found that methanol extract of the plant had little effect on paramyxovirus but considerable effect on orthomyxovirus. Several viruses are capable of adsorbing particularly to host cells when anti-microbials are administered to host cell for shorter durations, but their ability to infect host depends on the viability of the viral genome within host [26].

Among these natural bioactive compounds, EOs in different parts of O. tenuiflorum have exhibited significant anti-viral activity. Different investigations demonstrated that various kinds of O. tenuiflorum extracts containing EO are effective remedies against many pathogenic viruses. Significant number of Ocimum-based therapy for viral illnesses have been effective and safe for humans. In a controlled randomized study, an improved immunoresponse against viral infection was observed when the individuals were given diet frames supplemented with one g of ethanolic leaf extract of O. tenuiflorum for two weeks [68].

Leaf extract of O. tenuiflorum can significantly improve VO2 max, decreased fatigue, reduced creatine kinase, and enhanced immunity to viral infection, and it simultaneously down regulated the load of human herpesvirus-6 [98]. Methanol extract of O. tenuiflorum had no inhibitory action on paramyxovirus while it has significant inhibitory action on orthomyxovirus. The chloroform extract of O. tenuiflorum has no inhibitory action on paramyxovirus but it significantly inhibits orthomyxovirus [68].

In VERO cells, ethanolic extract of O. tenuiflorum leaves (@ of 22.5 mg/ml) suppressed poliovirus-III multiplication [8]. Wide range of antiviral activities have been reported for Ursolic acid, Linalool, and Apigenin that are present in various species of Ocimum [44, 78, 89]. Molecular docking analyzes suggested that the COVID-19 protease (6LU7) can be significantly inhibited by the extracts O. tenuiflorum [91]. Eugenol, a vital constituent of O. tenuiflorum leaves, was shown to be crucial in preventing the interaction among spike S1 and ACE2 (angiotensin-converting enzyme 2) [66].

According to an in-silico investigation, luteolin-7-O-glucuronide and chlorogenic acid in O. tenuiflorum could covalently attach to Cys145 of SARS-CoV-2 MPRO and could block viral enzymes [63].

It was determined that O. tenuiflorum (syn. O. sanctum) seed oil (OSSO) at a concentration of 3 ml/kg significantly increased the production of anti-sheep red blood cells antibody titer and decreased the percentage of leucocyte migration inhibition, which is intricately associated with cell-mediated immune responses [59]. It also reduced the percentage of histamine released from peritoneal mast cells in sensitized rats, which is intimately linked to humoral immune responses [59, 69]. Adding mild concentration of diazepam to OSSO enhanced ability of OSSO to counteract stress-induced alterations in both cell-mediated and humoral immune responses in resistant stress [59]. OSSO has an effect on both humoral and cell-mediated immunological response, likely through a GABAergic route [49].

Studies showed that Tulsinol A, B, C, D, E, F, G, and Dihydrodieuginol-B from O. tenuiflorum could be employed as possible inhibitors for papain-like protease (PLpro) and main protease (Mpro), which are essential for the pathogenesis of the majority of viruses in the Coronaviridae family [60, 61, 101]. In a molecular docking study using AutoDock Vina-based YASARA software, it was discovered that Ocimum derived natural compounds (such as: vicenin, sorientin 4′-O-glucoside 2″-O-phydroxy-benzoagte, and ursolic acid) inhibited the Mpro of SARS-CoV-2 [83], due to high binding affinity of these molecules for Mpro. There was wide range of antiviral activity observed in the aqueous extract of O. basilicum and its separated bioconstituents, like apigenin, linalool, and ursolic acid [15].

Molecular prospects for the treatment and prevention of COVID-19 include O. basilicum nutraceuticals (such as: apigenin, geranyl acetate, nerolidol, nevadenin, nonyl acetate, β-sesquiphellandren and ursolic acid), as these compounds appear to serve as polymerase inhibitors of SARS-CoV2 [44]. Desai & Pillai [20] reported that nanoparticles from O. tenuiflorum significantly enhanced immunity against COVID-19 situation. In addition, it was discovered that O. tenuiflorum may be able to exert its anti-influenza effect via mechanism unrelated to virus suppression, notably the action of neuraminidase [33].

The genetic material of RNA viruses is either a single- or double-stranded RNA molecule (Figure 5.1). Few of them may use RdRPs to replicate their genome, whereas in other cases, single-stranded RNA (ss-RNA) is catalyzed by reverse transcriptase and generates viral DNA that can be integrated into the host DNA via integrase activity. Most of the species of Ocimum show their antiviral activity and some of these are found effective against RNA viruses. Some Ocimum based natural products can act directly on gRNA, where as some can inhibit the viral enzymes [40, 45, 58, 93].

Recent in-silico studies also highlighted the importance of EOs from the Ocimum spp. in the inhibition of several RNA viruses [63, 66]. Some of the EOs have high binding affinities with viral proteins [63]. O. tenuifloram and O. basilicum have been widely used as edible leafy vegetables and ethnomedicine. Many ethnopharmacological studies and pre-clinical studies have confirmed the anti-viral efficacy of these plants.

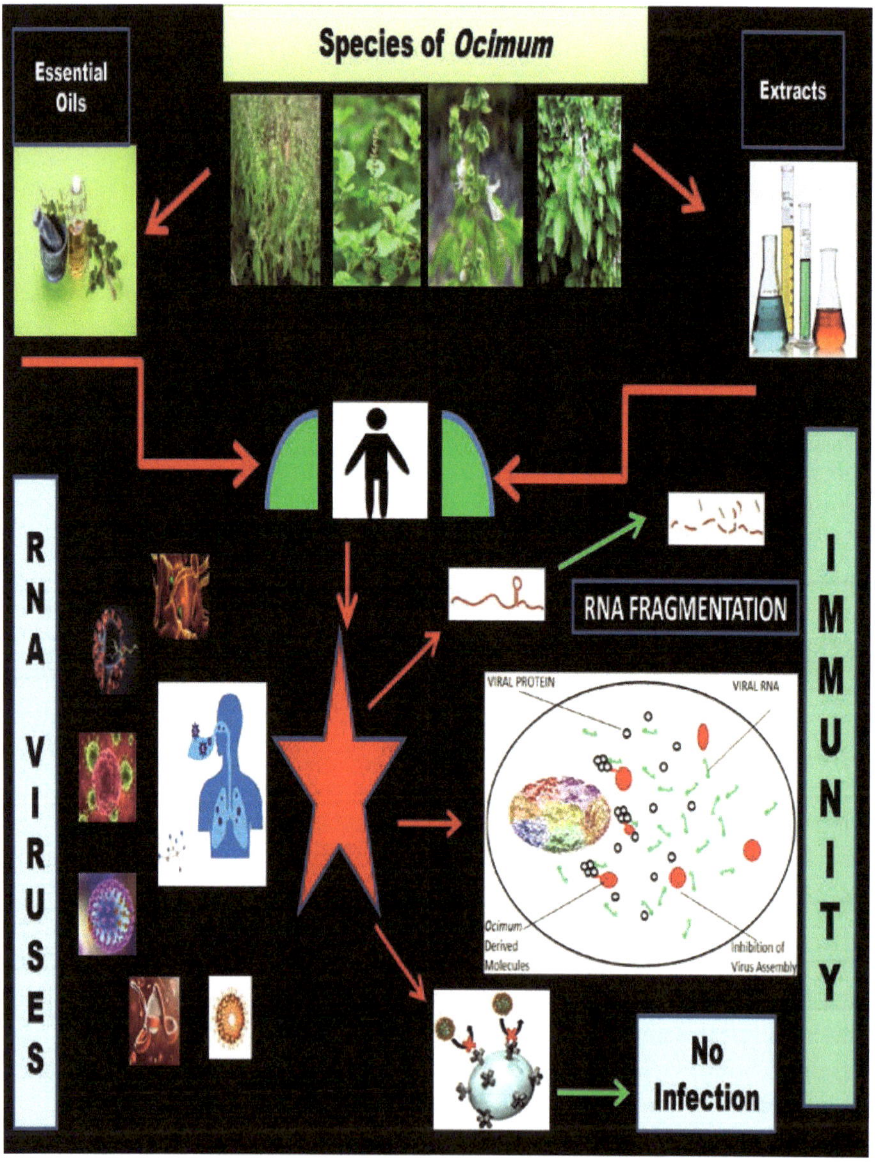

FIGURE 5.1 Antiviral activity of Ocimum sp.

Recent studies have further confirmed the ethnopharmacological data; and in some cases, the antiviral effects were experimentalized on animal models. Most importantly, the antiviral components of Ocimum spp. exhibit few or no negative effects, regardless of concentration, formulation, age, or sex of a patient. However, the knowledge regarding the pathophysiology of RNA viruses has recently increased, necessitating more detailed research on the prophylactic and curative measures of RNA virus-mediated disorders. In addition, it is also necessary to conduct large-scale clinical experiments by using different extracts of under-utilized species of the genus.

5.6 SUMMARY

The globe has witnessed the emergence and reemergence of RNA viral outbreaks during the past few decades. Some of them are presently posing a threat to the world's population. The infectious nature of viral pathogens, their capacity to evade host-virus interactions, and the lack of cost-effective antiviral compounds have further complicated the therapeutic management of viral infections in human. The scientific community has proposed a list to the FDA approved drugs that can be used for the management of respective patients affected by RNA virus based on the characteristics of genome and protease structure of them; but unfortunately, either they have side effects, or their efficacy is still under debate. Treatment for RNA virus-mediated disease is a very tough job. The present treatment options mostly consist of anti-retroviral drugs, such as, remdesivir, plasma therapy, and inhibitors of host cell proteins (such as: ACE-2 and protein synthesis machinery), with only minor success rates [10]. Contrary many recent works advocated for anti-RNA viral efficiency of several species of Ocimum. The participation of certain chemicals identified in EOs and extracts of O. basilicum and O. tenuiflorum in the inhibition of RNA virus glycoprotein attachment on the host surface, as well as the prevention of viral growth, boosts the immune system have also been confirmed. However, there are few clinical studies assessing the benefits of phytoconstituents from underused Ocimum species (i.e., species other than O. basilicum and O. tenuiflorum). Furthermore, a thorough knowledge of the molecular processes underlying host and RNA virus interactions might enable for the development of antiviral medicines derived from Ocimum species. Eradicating viral illnesses looks tough since many viruses lack preventative vaccinations and efficient antiviral medicines. Despite this, ingredients from Ocimum spp. are promising sources for finding novel anti-viral, creating new structure–activity correlations, and developing effective protective as well as therapeutic strategies against viral infections, specifically RNA-viral infections. Since identifying medications and treatment alternatives for corona viruses, medicinal plants and their derived phytoconstituents might give solid foundation for creating and developing unique alternatives and complementary treatment via phytotherapy not only for corona virus but also for other RNA viruses.

KEYWORDS

- **anti-viral activity**
- **immunity**
- **Ocimum**
- **pandemic situation**
- **phytoconstituents**
- **reverse transcriptase**
- **RNA virus**
- **Tulsi**

REFERENCES

1. Ahmad, S., Zahiruddin, S., Parveen, B., Basist, P., Parveen, A., Parveen, R., & Ahmad, M., (2021). Indian medicinal plants and formulations and their potential against COVID-19 preclinical and clinical research. Frontiers in Pharmacology, 11, E-article ID 578970.
2. Akram, M., Tahir, I. M., Shah, S. M. A., Mahmood, Z., Altaf, A., Ahmad, K., Munir, N., et al., (2018). Antiviral potential of medicinal plants against HIV, HSV, influenza, hepatitis, and coxsackie virus – review. Phytotherapy Research, 32(5), 811–822.
3. Albohy, A., Zahran, E. M., Abdelmohsen, U. R., Salem, M. A., Al-Warhi, T., Al-Sanea, M. M., Abelyan, N., et al., (2022). Multitarget in-silico studies of Ocimum menthii folium against SARS-CoV-2 supported by molecular dynamics simulation. Journal of Biomolecular Structure & Dynamics, 40(9), 4062–4072.
4. Alcileia, N. Y., Renata, G., Aureo, T. Y., Expedito, L. S., Saulo, E. S., Marcio, J. D., Marcia, M. O. D., et al., (2013). Anti-inflammatory activity of Ocimum Americanum essential oil in experimental model of zymosan-induced arthritis. The American Journal of Chinese Medicine, 41(4), 913–926.
5. Aluko, B., Oloyede, O., & Afolayan, A., (2012). Phytochemical and nutrient compositions of the leaves of Ocimum Canum Sims. African Journal of Biotechnology, 11(63), 12697–12701.

6. Amber, R., Adnan, M., Tariq, A., & Mussarat, S., (2017). A review on antiviral activity of the Himalayan medicinal plants traditionally used to treat bronchitis and related symptoms. The Journal of Pharma and Pharmacology, 69(2), 109–122.
7. Ayisi, N. K., & Nyadedzor, C., (2003). Comparative in vitro effects of AZT and extracts of Ocimum gratissimum, Ficus polita, Clausena anisata, Alchornea cordifolia, and Elaeophorbia drupifera against HIV-1 and HIV-2 infections. Antiviral Research, 58(1), 25–33.
8. Bano, N., Ahmed, A., Tanveer, M., Khan, G. M., & Ansari, M. T., (2017). Pharmacological evaluation of Ocimum sanctum. Journal of Bioequivalence & Bioavailability, 9(3), 387–392.
9. Barr, J. N., & Fearns, R., (2010). How RNA viruses maintain their genome integrity. The Journal of General Virology, 91(Pt 6), 1373–1387.
10. Baruah, T. J., Dutta, S. P., & Patar, A. K., (2021). A preliminary in-silico analysis of Ocimum sanctum flavonoids, orientin and vicenin-1, as potential drugs against Sars-Cov-2 infection. International Journal of Pharmaceutical Science & Research., 12(5), 2950–2956.
11. Behbahani, M., Mohabatkar, H., & Soltani, M., (2013). Anti-HIV-1 activities of aerial parts of Ocimum basilicum and its parasite Cuscuta campestris. Journal of Antivirals and Antiretrovirals, 5(3), 57–61.
12. Ben-Shabat, S., Yarmolinsky, L., & Porat, D., (2020). Antiviral effect of phytochemicals from medicinal plants: Applications and drug delivery strategies. Drug Delivery and Translational Research, 10(2), 354–367.
13. Bloom, D. E., & Cadarette, D., (2019). Infectious disease threats in the twenty-first century: Strengthening the global response. Frontiers in Immunology, 10, 549–555.
14. Bluhm, R., & Pinkovskiy, M., (2021). Spread of COVID-19 and the BCG vaccine: A natural experiment in reunified Germany. The Econometrics Journal, 24(3), 353–376.
15. Chiang, L. C., Ng, L. T., Cheng, P. W., Chiang, W., & Lin, C. C., (2005). Antiviral activities of extracts and selected pure constituents of Ocimum basilicum. Clinical and Experimental Pharmacology and Physiology, 32(10), 811–816.
16. Chowdhury, T., Mandal, A., Roy, S. C., & De Sarker, D., (2017). Diversity of the genus Ocimum through morpho-molecular (RAPD) and chemical (GC-MS) analysis. Journal of Genetic Engineering & Biotechnology, 15(1), 275–286.
17. Chung, T., Kim, J., Kim, M., Choi, S., Kim, S., Chung, J., Lee, I., Kim, S. H., Hahn, K., & Lee, I., (1995). Investigation of Korean plant extracts for potential phytotherapeutic agents against B-virus hepatitis. Phytotherapy Research, 9(6), 429–434.
18. Chung, Y. C., Hsieh, F. C., Lin, Y. J., Wu, T. Y., Lin, C. W., & Lin, C. T., (2015). Magnesium lithospermate B and rosmarinic acid, two compounds present in Salvia Miltiorrhiza, have potent antiviral activity against Enterovirus 71 infections. European Journal of Pharmacology, 755, 127–133.
19. Das, S. K., & Vasudevan, D. M., (2006). Tulsi: Indian holy power plant. Natural Product Radiance, 5(4), 279–283.
20. Desai, D., & Pillai, T. G., (2021). Nanoparticles from Ocimum sanctum in immune enhancement against COVID-19 situation. Determinations Nanomed Nanotechnology, 2(3), E-article ID 000538.
21. Dhale, D. A., Birari, A. R., & Dhulgande, G. S., (2010). Preliminary screening of antibacterial and phytochemical studies of Ocimum Americanum. Journal of Ecobiotechnology, 2(8), 11–13.

22. Dhama, K., Karthik, K., Khandia, R., Munjal, A., Tiwari, R., Rana, R., Khurana, S. K., et al., (2018). Medicinal and therapeutic potential of herbs and plant metabolites/extracts countering viral pathogens: Current knowledge and future prospects. Current Drug Metabolism, 19(3), 236–263.
23. Duffy, S., Shackelton, L. A., & Holmes, E. C., (2008). Rates of evolutionary change in viruses: Patterns and determinants. Nature Reviews Genetics, 9(4), 267–276.
24. Ganjhu, R. K., Mudgal, P. P., Maity, H., Dowarha, D., Devadiga, S., Nag, S., & Arunkumar, G., (2015). Herbal plants and plant preparations as remedial approach for viral diseases. Virusdisease, 26(4), 225–236.
25. Ghoke, S. S., Sood, R., Kumar, N., Pateriya, A. K., Bhatia, S., Mishra, A., Dixit, R., et al., (2018). Evaluation of antiviral activity of Ocimum sanctum and Acacia arabica leaves extracts against H9N2 virus using embryonated chicken egg model. BMC Complementary and Alternative Medicine, 18(1), 174–180.
26. Gilling, D. H., Kitajima, M., Torrey, J. R., & Bright, K. R., (2014). Mechanisms of antiviral action of plant antimicrobials against murine norovirus. Applied and Environmental Microbiology, 80(16), 4898–4910.
27. Goothy, S. S. K., Goothy, S., Choudhary, A., Potey, G. G., Chakraborty, H., Kumar, A. H. S., & Mahadik, V. K., (2020). Ayurveda's holistic lifestyle approach for the management of coronavirus disease (COVID-19): Possible role of Tulsi. International Journal of Research and Pharmaceutical Science, 11(Suppl. 1), 16–18.
28. Grubaugh, N. D., Lander, J. T., Lemey, P., Pybus, O. G., Rambaut, A., Holmes, E. C., & Andersen, K. G., (2019). Tracking virus outbreaks in twenty-first century. Nature Microbiology, 4(1), 10–19.
29. Gurav, T. P., Dholakia, B. B., & Giri, A. P., (2022). A glance at the chemo-diversity of Ocimum species: Trends, implications, and strategies for the quality and yield improvement of essential oil. Phytochemistry Reviews: Proceedings of the Phytochemical Society of Europe, 21(3), 879–913.
30. Gurrapu, S., & Mamidala, E., (2016). Medicinal plants used by traditional medicine practitioners in the management of HIV/AIDS-related diseases in tribal areas of Adilabad District in Telangana region. The American Journal of Science and Medical Research, 2(1), 239–245.
31. Gyawali, R., Paudel, P., Basyal, D., Setzer, W. N., Lamichhane, S., Paudel, M., Gyawali, S., & Khanal, P., (2020). A review on ayurvedic medicinal herbs as remedial prospective for COVID-19. Journal of Karnali Academy of Health Sciences, 3(COVID-19 special issue), 1–21.
32. Herrmann, F., Romero, M. R., Blazquez, A. G., Kaufmann, D., Ashour, M. L., Kahl, S., Marin, J. J. G., et al., (2011). Diversity of pharmacological properties in Chinese and European medicinal plants: Cytotoxicity, antiviral and anti-trypanosomal screening of 82 herbal drugs. Diversity, 3(4), 547–580.
33. Jadhav, P., Lal, H., & Kshirsagar, N., (2014). Assessment of potency of PC-complexed Ocimum sanctum methanol extract in embryonated eggs against influenza virus (H1N1). Pharmacognosy Magazine, 10(37), 86–90.
34. Jagadale, S. R., (2020). Formulation of herbal medicine on COVID-19 from Azadirachta indica, Ocimum sanctum and Citrus limon. International Journal of Innovative Science and Research Technology, 5(5), 673–675.
35. Jain, N. K., Agarawal, A., Kulkarni, G. T., & Tailang, M., (2021). Molecular docking study on phytoconstituents of traditional ayurvedic drug Tulsi against COVID-19 MPRO

Enzyme: An in-silico study. International Journal of Pharmacy and Pharmaceutical Sciences, 14(4), 44–50.
36. Jassim, S. A., & Naji, M. A., (2003). Novel antiviral agents: A medicinal plant perspective. Journal of Applied Microbiology, 95(3), 412–427.
37. Jayati, A. K., Kumar, A., Goel, A., Gupta, S., & Rahal, A., (2013). In vitro antiviral potential of Ocimum sanctum leaves extract against new castle disease virus of poultry. International Journal of Microbiology and Immunology Research, 2(7), 051–055.
38. Jiang, Y., Yin, W., & Xu, H. E., (2021). RNA-dependent RNA polymerase: Structure, mechanism, and drug discovery for COVID-19. Biochemical and Biophysical Research Communication, 538, 47–53.
39. Joshi, G., Sharma, S., Acharya, J., & Parida, M., (2014). Assessment of in vitro antiviral activity of Ocimum sanctum (Tulsi) against pandemic swine flu H1N1 virus infection. World Research Journal of Antimicrobial Agents, 3(1), 62–67.
40. Kapewangolo, P., Kandawa-Schulz, M., & Meyer, D., (2017). Anti-HIV activity of Ocimum labiatum extract and isolated pheophytin-A. Molecules, 22(11), 1763–1777.
41. Kikkert, M., (2020). Innate immune evasion by human respiratory RNA viruses. Journal of Innate Immunity, 12(1), 4–20.
42. Kisangau, D. P., Lyaruu, H. V. M., Hosea, K. M., & Joseph, C. C., (2007). Use of traditional medicines in the management of HIV/ AIDS opportunistic infections in Tanzania: A case in the Bukoba rural district. Journal of Ethnobiology and Ethnomedicine, 3(1), 29–34.
43. Kubiça, T. F., Alves, S. H., Weiblen, R., & Lovato, T. L., (2014). In vitro inhibition of the bovine viral diarrhea virus by the essential oil of basil and monoterpenes. Brazilian Journal of Microbiology, 45(1), 209–214.
44. Kumar, A., Rahal, A., Chakraborty, S., Tiwari, R., Latheef, S. K., & Dhama, K., (2013). Ocimum sanctum: A miracle herb and boon to medical science–A review. International Journal of Agronomy and Plant Production, 4(7), 1580–1589.
45. Kumar, A. H. S., (2020). Molecular docking of natural compounds from Tulsi and neem against SARS-CoV-2 protein targets. BEMS Reports, 6(1), 11–13.
46. Lai, W. L., Chuang, H. S., Lee, M. H., Wei, C. L., Lin, C. F., & Tsai, Y. C., (2012). Inhibition of herpes simplex virus type 1 by thymol-related monoterpenoids. Planta Medica, 78(15), 1636–1638.
47. Lamorde, M., Tabuti, J. R. S., Obua, C., Kukunda-Byobona, C., Lanyero, H., Byakika, P., Bbosa, G. S., et al., (2010). Medicinal plants used by traditional medicine practitioners for the treatment of HIV/ AIDS and related conditions in Uganda. Journal of Ethnopharmacology, 130(1), 43–53.
48. Lane, T., Anantpadma, M., Freundlich, J. S., Davey, R. A., Madrid, P. B., & Ekins, S., (2019). The natural product eugenol is an inhibitor of the E-bola virus in vitro. Pharmaceutical Research, 36(7), 104–110.
49. Lata, M., (2020). Role of bioactive compounds in Ocimum sanctum and Zingiber officinale conferring immunity against COVID-19. IOSR Journal of Environmental Science, Toxicology and Food Technology, 14(8), 10–18.
50. Ledford, H., (2020). How does COVID-19 kill? Uncertainty is hampering doctors' ability to choose treatments. Nature, 580, 311–312.
51. Lim, S. Y. M., Chieng, J. Y., & Pan, Y., (2021). Recent insights on anti-dengue virus (DENV) medicinal plants: Review on in vitro, in vivo and in-silico discoveries. All Life, 14(1), 1–33.

52. Ling, A. P. K., Khoo, B. F., Seah, C. H., Foo, K. Y., Cheah, R. K., Chye, S. M., & Koh, R. Y., (2014). Inhibitory activities of methanol extracts of Andrographis paniculata and Ocimum sanctum against dengue virus. In: International Conference on Biological, Environment and Food Engineering (BEFE) (pp. 47–52). Bali – Indonesia.
53. Logeswari, J., Shankar, S., Guha, B. P., & Muninathan, N., (2020). Role of medicinal plants in the prevention of COVID-19 pandemic. Medico-legal Update, 20(4), 2303–2306.
54. Ma, L., & Yao, L., (2020). Antiviral effects of plant-derived essential oils and their components: An updated review. Molecules, 25(11), 2627–2633.
55. Machhi, J., Herskovitz, J., Senan, A. M., Dutta, D., Nath, B., Oleynikov, M. D., Blomberg, W. R., et al., (2020). The natural history, pathobiology, and clinical manifestations of SARS-CoV-2 infections. Journal of Neuroimmune and Pharmacology, 15(3), 359–386.
56. Mani, J. S., Johnson, J. B., Steel, J. C., Broszczak, D. A., Neilsen, P. M., Walsh, K. B., & Naiker, M., (2020). Natural product-derived phytochemicals as potential agents against coronaviruses: A review. Virus Research, 284, E-article ID 197989.
57. Markiewicz, L., Drazkowska, K., & Sikorski, P. J., (2021). Tricks and threats of RNA viruses – towards understanding the fate of viral RNA. RNA Biology, 18(5), 669–687.
58. Matondo, A., Kilembe, J. T., Ngoyi, E. M., Kabengele, C. N., Kasiama, G. N., Lengbiye, E. M., Mbadiko, C. M., et al., (2021). Oleanolic acid, ursolic acid and apigenin from Ocimum basilicum as potential inhibitors of the SARS-CoV-2 main protease: A molecular docking study. International Journal of Pathogen Research, 6(2), 1–16.
59. Mediratta, P. K., Sharma, K. K., & Singh, S., (2002). Evaluation of immuno-modulatory potential of Ocimum sanctum seed oil and its possible mechanism of action. Journal of Ethnopharmacology, 80(1), 15–20.
60. Mengist, H. M., Fan, X., & Jin, T., (2020). Designing of improved drugs for COVID-19: Crystal structure of SARS-CoV-2 main protease MPRO. Signal Transduction and Targeted Therapy, 5(1), 1–2.
61. Mittal, L., Kumari, A., Srivastava, M., Singh, M., & Asthana, S., (2021). Identification of potential molecules against COVID-19 main protease through structure-guided virtual screening approach. Journal of Biomolecular Structure and Dynamics, 39(10), 3662–3680.
62. Mohan, L., Amberkar, M. V., & Kumari, M., (2011). Ocimum sanctum Linn. (Tulsi) – An overview. International Journal of Pharmaceutical Sciences Review and Research, 7(1), 51–53.
63. Mohapatra, P. K., Chopdar, K. S., Dash, G. C., Mohanty, A. K., & Raval, M. K., (2021). In-silico screening and covalent binding of phytochemicals of Ocimum sanctum against SARS-CoV-2 (COVID 19) main protease. Journal of Biomolecular Structure & Dynamics, 1–10, E-article ID 2007170.
64. Narwal, S., Rana, S. C., Tiwari, V., Gangwani, S., & Sharma, R., (2011). Review on chemical constituents and pharmacological action of Ocimum charicum. Indo Global Journal of Pharmaceutical Sciences, 1(4), 287–293.
65. Offiah, V. N., & Chikwendu, U. A., (1999). Anti-diarrheal effects of Ocimum gratissimum leaf extract in experimental animals. Journal of Ethnopharmacology, 68(1–3), 327–330.
66. Paidi, R. K., Jana, M., Raha, S., McKay, M., Sheinin, M., Mishra, R. K., & Pahan, K., (2021). Eugenol, a component of holy basil (Tulsi) and common spice clove, inhibits the interaction between SARS-CoV-2 Spike S1 and ACE2 to induce therapeutic responses. Journal of Neuroimmune Pharmacology., 16(4), 743–755.

67. Pandey, S., Singh, S. K., Kumar, N., & Manjhi, R., (2017). Antiviral, antiprotozoal, antimalarial, and insecticidal activities of Ocimum gratissimum L. Asian Journal of Pharmaceutical Research and Development, 5(5), 1–9.
68. Patil, U., (2018). Studies on antiviral activity of Ocimum sanctum crude extracts on selected viruses on veterinary importance. International Journal of Ayurveda and Pharma Research, 6(4), 17–21.
69. Pattanayak, P., Behera, P., Das, D., & Panda, S. K., (2010). Ocimum sanctum reservoir plant for therapeutic applications: An overview. Pharmacognosy Reviews, 4(7), 95–105.
70. Pichlmair, A., Lassnig, C., Eberle, C. A., Gorna, M. W., Baumann, C. L., Burkard, T. R., Burckstummer, T., et al., (2011). IFIT1 is an antiviral protein that recognizes 5′-triphosphate RNA. Nature Immunology, 12(7), 624–630.
71. Prabhu, K. S., Lobo, R., Shirwaikar, A. A., & Shirwaikar, A., (2009). Ocimum gratissimum: A review of its chemical, pharmacological and ethnomedicinal properties. The Open Complementary Medicine Journal, 1(1), 1–15.
72. Prajapat, M., Sarma, P., Shekhar, N., Avti, P., Sinha, S., Kaur, H., Kumar, S., et al., (2020). Drug targets for corona virus: A systematic review. Indian Journal of Pharmacology, 52(1), 56–65.
73. Prajapati, N. D., Purohit, S. S., Sharma, A. K., & Kumar, T., 2003). Handbook of Medicinal Plants: A Complete Source Book (p. 367). Jodhpur – India: Agrobios.
74. Rege, A., & Chowdhary, A. S., (2014). Evaluation of Ocimum sanctum and Tinospora cordifolia as probable HIV protease inhibitors. International Journal of Pharmaceutical Sciences Review and Research, 25(1), 315–318.
75. Renu, K., Prasanna, P. L., & Gopalakrishnan, A. V., (2020). Coronaviruses pathogenesis, comorbidities, and multiple-organ damage: A review. Life Science, 255, E-article ID 117839.
76. Roychoudhury, S., Das, A., Sengupta, P., Dutta, S., Roychoudhury, S., Choudhury, A. P., Ahmed, A., et al., (2020). Viral pandemics of the last four decades: Pathophysiology, health impacts and perspectives. International Journal of Environmental Research and Public Health, 17(24), E-article ID 9411.
77. Said, E. A., Diaz-Griffero, F., Bonte, D., Lamarre, D., & Al-Jabri, A. A., (2018). Immune responses to RNA viruses. Journal of Immunology Research, 2018, E-article ID 5473678.
78. Sangeetha, P., & Poornamathy, J. J., (2015). In vitro assessment of anti-inflammatory activity of Ocimum sanctum leaves. International Journal of Pharma and Bio Sciences, 6(2), B1387–B1391.
79. Sarkar, A. K., Dey, M., & Mazumder, M., (2017). Ecological status of medicinal plants of Chalsa forest range under Jalpaiguri division in West Bengal, India. International Journal of Herbal Medicine, 5(5), 196–215.
80. Sarma, P., Sekhar, N., Prajapat, M., Avti, P., Kaur, H., Kumar, S., Singh, S., et al., (2021). In-silico homology assisted identification of inhibitor of RNA binding against 2019-nCoV N-protein (N terminal domain). Journal of Biomolecular Structure and Dynamics, 39(8), 2724–2732.
81. Sen, P. S., Biswal, S., Panda, S. K., Ray, A. K., & Rana, M. K., (2022). Binding mechanism and structural insights into the identified protein target of COVID-19 and importin-α with in-vitro effective drug ivermectin. Journal of Biomolecular Structure and Dynamics, 40(5), 2217–2226.

82. Shahrajabian, M. H., Sun, W., & Cheng, Q., (2020). Chemical components and pharmacological benefits of Ocimum basilicum: A review. International Journal of Food Properties, 23(1), 1961–1970.
83. Shree, P., Mishra, P., Selvaraj, C., Singh, S. K., Chaube, R., Garg, N., & Tripathi, B. Y., (2022). Targeting COVID-19 (SARS-CoV-2) main protease through active phytochemicals of ayurvedic medicinal plants: Ashwagandha, Giloy and Tulsi. Journal of Biomolecular Structure and Dynamics, 40(1), 190–203.
84. Silva, M. G., Vieira, I. G., Mendes, F. N., Albuquerque, I. L., Dos-Santos, R. N., & Silva, F. O., (2008). Variation of uronic acid content in eight Ocimum species from Northeastern Brazil. Molecules, 13(10), 2482–2487.
85. Singh, N. A., Kumar, P., & Kumar, N., (2021). Spices and herbs: Potential antiviral preventives and immunity boosters during COVID-19. Phytother. Research, 35(5), 2745–2757.
86. Singh, P., Chakraborty, P., He, D. H., & Mergia, A., (2019). Extract prepared from the leaves of Ocimum basilicum inhibits the entry of Zika virus. Acta Virologica, 63(3), 316–321.
87. Singh, E., Dwivedi, J., & Sharma, S., (2012). Diversified potentials of Ocimum sanctum: An exhaustive survey. Journal of Natural Product and Plant Resource, 2(1), 39–48.
88. Soni, K., Lawal, T., Wicks, S., Patel, U., & Mahady, G., (2015). Boswellia serrata and Ocimum sanctum extracts reduce inflammation in an ova-induced asthma model of BALB/c mice. Planta Medica, 81(11), B4-B10.
89. Sood, R., Bhatia, S., Bhatnagar, H., Gupta, V., & Kumar, M., (2013). Phytochemical analysis and in vitro screening of selected Indian medicinal plants for antiviral activity against highly pathogenic avian influenza virus. Spatula DD, 3(3), 81–88.
90. Spreeuwenberg, P., Kroneman, M., & Paget, J., (2018). Reassessing the global mortality burden of the 1918 influenza pandemic. American Journal of Epidemiology, 187(12), 2561–2567.
91. Srivastava, A. K., Kumar, A., & Misra, N., (2020). On the Inhibition of COVID-19 Protease by Indian Herbal Plants: An In-Silico Investigation. arXiv Preprint, E-article ID 03411.
92. Suganya, J., Kumar, G. R., Radha, M., & Dhananya, S., (2021). In-silico investigation of natural compounds identified from Ocimum species as dengue NS3 and NS5 protein inhibitors. Research Journal of Pharmacy and Technology, 14(12), 6621–6626.
93. Thayil, S. M., & Thyagarajan, S., (2016). Methanol and aqueous extracts of Ocimum kilimands charicum (Karpura Tulsi) inhibits HIV-1 reverse transcriptase in vitro. International Journal of Pharmacognosy and Phytochemical Research, 8(7), 1099–1103.
94. Tiwari, R., Latheef, S. K., Ahmed, I., Iqbal, H., Bule, M. H., Dharma, K., Samad, H. A., et al., (2018). Herbal immunomodulators – A remedial panacea for designing and developing effective drugs and medicines: Current scenario and future prospects. Current Drug Metabolism, 19(3), 264–301.
95. Tiwari, S., & Talreja, S., (2020). An overview on coronil drug. Journal of Global Trends in Pharmaceutical Sciences, 11(3), 8242–8247.
96. Tshilanda, D. D., Ngoyi, E. M., Kabengele, C. N., Matondo, A., Bongo, G. N., Inkoto, C. L., Mbadiko, C. M., & Gbolo, B. Z., (2020). Ocimum species as potential bioresources against COVID-19: A review of their phytochemistry and antiviral activity. International Journal of Pathogen Research, 5(4), 42–54.

97. Van, D. H. R., (2021). The epidemiology of emerging infectious diseases and pandemics. Medicine, 49(10), 659–662.
98. Venu-Prasad, M. P., (2014). Antifatigue and Neuroprotective Properties of Selected Species of Ocimum L. (p. 231). Doctor of Philosophy Thesis, University of Mysore, Mysore, India.
99. Vidhani, S. I., Vyas, V. G., Parmar, H. J., Bhalani, V. M., & Hassan, M. M., (2016). Evaluation of some chemical composition, minerals fatty acid profiles, antioxidant, and antimicrobial activities of Tulsi (Ocimum sanctum) from India. American Journal of Food Science and Technology, 4(2), 52–57.
100. Wagner, E. K., & Hewlett, M. J., (2004). Basic Virology (2nd edn., p. 454), London – UK: Blackwell Publishing.
101. Wrapp, D., Wang, N., Corbett, K. S., Goldsmith, J. A., Hsieh, C. L., Abiona, O., Graham, B. S., & McLellan, J. S., (2020). Cryo-EM structure of the 2019-nCoV spike in the prefusion conformation. Science, 367(6483), 1260–1263.
102. Yamasaki, K., Nakano, M., Kawahata, T., Mori, H., Otake, T., Ueba, N., Oishi, I., et al., (1998). Anti-HIV-1 activity of herbs in Labiatae. Biological & Pharmaceutical Bulletin, 21(8), 829–833.
103. Yucharoen, R., Anuchapreeda, S., & Tragoolpua, Y., (2011). Anti-herpes simplex virus activity of extracts from the culinary herbs Ocimum sanctum L., Ocimum basilicum L. and Ocimum Americanum L. African Journal of Biotechnology, 10(5), 860–866.
104. Zhang, Q., Sun, K., Chinazzi, M., Pastore, Y., Piontti, A., Dean, N. E., Rojas, D. P., et al., (2017). Spread of Zika virus in the Americas. Proceedings of the National Academy of Sciences of the USA, 114(22), E4334–E4343.
105. Zhu, W., Chen, C. Z., Gorshkov, K., Xu, M., Lo, D. C., & Zheng, W., (2020). RNA-dependent RNA polymerase as target for COVID-19 drug discovery. SLAS Discovery: Advancing Life Sciences R & D, 25(10), 1141–1151.

CHAPTER 6

Functional Foods: Formulation, Fortification, and Combating COVID-19

Meghali Nag, Fatima Siddiqua, Nabanita Baruah, Sumi Das, Beauty Basumatary, Manab Deka, Kandarpa Kumar Saikia, and Pinki Malick

ABSTRACT

Food plays an indispensable role for our survival. Through the centuries, food has been used to maintain healthy lifestyle, overall growth, and development. Due to increased diseases because of micronutrient (MN) deficiency, fortification, and supplementation of food has become an effective strategy to increase the nutritional demands required by the targeted group. Nowadays, people are also focusing on customized foods to live a healthy lifestyle by reducing the impact of some life-threatening diseases like obesity, heart disease, neurological disease, etc. This chapter deals with the development of fortified foods based on MNs, dietary fiber, probiotics, and others for diabetic, geriatric, and infant groups. Also, it emphasizes on the formulation of various foods related to sports person and COVID patients.

6.1 INTRODUCTION

Escalating global concern about wellness and wellbeing is provoking larger part of people to gauge better ways of life and practice good eating habits. The extending food handling market demonstrates that end-clients are looking for negligibly handled food with extra healthful advantages and organoleptic esteem. Considering the global scenario of increasing diseased population, it is crucial to develop food products with high nutraceutical

The Functional Foods: Nutrient and Health Benefits.
Megh R. Goyal, Junaid Ahmad Malik, and Anu Kumari (Eds.)

value along with providing an alternate way to encourage functional food intake by means of food fortification and supplementation by using functional ingredients.

Fortification of foods adds variety to the commonly consumed foods by incorporating nutrients to increase the nutritional value. Food supplements have specific nutritional and physiological effects and marketed as different capsules and powder forms. The importance of fortified foods and food supplements is increasing more with the gloom of unprecedented pandemic and the consciousness of the people to remain free from disease. Food supplements with vitamins and minerals are meant to suffice the deficiency in individuals. Hence, fortified food should be developed in respect with public health and dietary nutritional guideline to meet the global dietary requirements.

This chapter highlights on functional ingredients using various dietary approaches, nutritional requirements, management of food habits and lowering risk of diseases.

6.2 DEVELOPMENT OF FOOD PRODUCTS FOR THE DIABETIC POPULATION

In global prospect, often people are more conscious about healthy diet to maintain a hassle-free daily lifestyle. Growing population and increasing disease related issues; development of food product with higher nutritional and functional value is an utmost important. Food fortification in that relation is a broad aspect of working on functional food may provide value and supplemented with nutrients, marks an alternative way to develop sustainable growth and provide an opportunity to encourage the global need of functional food intake.

Based on pharmacological aspects, it was majorly classified in two groups as non-insulin dependent diabetes mellitus and insulin dependent diabetes mellitus, non-insulin dependent diabetes mellitus. Sub-classified as gestational diabetes and diabetics with malnutrition. Cases of diabetes mellitus growing globally as its prevalence in adults was found 4% in 1995 and it is estimated that it will reach 5.4% by 2045 with 42% increase in developed countries and 70% increases in developing countries with most affected countries being India, China, and United states by 2030 [35].

6.2.1 GLOBAL SCENARIO OF MANAGEMENT OF DIABETICS BASED ON LIFESTYLE

Diabetes is one of the most targeted risk factors of premature death and one of the most non-communicable diseases (NCDs) that can be prevented by approaches with affordable, cost-effective, and based on population in a multidimensional intervention. Environmental and genetic factors are the important roles in the development of diabetes.

Promote physical activity, weight management and healthier nutritional therapy includes walking, gardening, running, and other work that has significant controlled in glycemic level, effect on insulin level, reduces cardiovascular risk factors and in metabolic abnormalities associated with diabetic patients. Replacement of sugars with aspartame, saccharine, and saturated fats, cholesterol, polyunsaturated fats is recommended.

6.2.2 DIETARY GUIDELINES AND FOOD HABITS

WHO and Food and Agriculture Organization (FAO) [67] provides dietary guidelines for prevention of type 2 diabetes that limits intake of less than 10% saturated fatty acid of total energy intake (which consists of minimum 20 g of dietary fiber intake daily through regular consumption of legumes, wholegrain cereals, fruits, and vegetables). Nutrient recommendations for diabetics and profile of dyslipidemia are listed in Table 6.1.

TABLE 6.1 Nutrient Recommendations for Diabetics and Profile of Dyslipidemia

Source	Contents
Carbohydrates	45%
Cholesterol	<200 mg/day
Fatty acids (mono-unsaturated)	20%
Fatty acids (poly-unsaturated)	<8%
Fatty acids (saturated and trans-)	<7%
Protein	15%–20%
Total Fats	35%

6.3 FORTIFIED FOODS AND DIETARY APPROACHES FOR PRODUCT DEVELOPMENT

Fortification of food is an integrated approach that provides support and is applied at various levels to reduce malnutrition and other group of diseases.

A range of different combinations are available for food fortification (such as: food fortified with micronutrient (MN; such as: zinc, iron, selenium) and vitamins, probiotic, dietary fiber, etc.).

6.3.1 MICRONUTRIENT (MN)-BASED

Development of fortified food for diabetic patients to ensure free radical activation of lipids, balance glucose utilization and insulin secretion to stabilizes the water-soluble vitamin contents. Thiamin deficiency linked with type 2 *diabetes mellitus* and breaks down the glycolysis pathway and joins with tricarboxylic acid cycle, inhibiting glucose utilization and tolerance of carbohydrate [52]. Vitamin C, D, and carotenoids deficiency exhibit insulin resistance and atherosclerotic vascular changes and also associated with diabetic nephropathy [62]. Chromium deficiency associated with impaired glucose and lipid metabolism, which reduces number of insulin receptors. Selenium as trace mineral acts as a cofactor for the enzyme glutathione peroxidase (GP) and provides an antioxidant protection.

6.3.2 MACRONUTRIENT-BASED

Macronutrients are major nutrients to maintain blood glucose level and controlled glycemic index (GI). Major nutrients carbohydrates, protein, and fats should take as daily consumption to receive the quality of individual macronutrients.

6.3.2.1 CARBOHYDRATES

Carbohydrates play major role in diabetes management depending on severity of hyperglycemia, weight, and other parameter for balancing insulin level in the body. To ensure a healthy insulin to carbohydrate ratio (ICR), carbohydrate counting is a method to regulate blood glucose in postprandial and fasting conditions, which influence glycemic responses. Nutritive sweetener sucrose composed of one glucose and fructose unit where fructose is digested slowly in the intestine but to prevent adverse metabolic complications. Sugar substitute intensively used in formulated food for diabetic population. Recently used new sweetener neo-aspartame shows higher thermal stability that does not affect the post-dermal glycemia of diabetic patients [25].

Intake of protein should be based on eating pattern particularly 20–30% of protein may contribute to increase diabetic kidney disease. To stabiles the kidney function for diabetic people, total protein recommendation is 0.8 g protein per kilogram desirable by body weight that reduces albuminuria [25]. "Fat Facts" have been developed by the American Heart Association to encourage replacement of trans-fatty acids with unsaturated fats.

6.3.3 DIETARY FIBER-BASED

Food product regulations by FDA claim fiber as low-fat diet consisting of fruits and vegetables, grains that may reduce the risk of some disease like cancer. Dietary fiber is the portion for metabolic maintenance of soluble and insoluble fiber used widely as deficiency of dietary fiber associated with risk of colon cancer, cholelithiasis, type 2 *diabetes mellitus*, coronary heart disease (CHD), atherosclerosis. In general population or DM patients, a person should consume 20–35 g of fiber per day from fresh fruits and vegetables.

6.3.4 PROBIOTICS-BASED

Probiotics are the beneficial living microorganisms which is found in fermented foods, kimchi, kefir, yogurt, that can use as probiotic supplements and found in human gut microbiota. Gut microbiota alternation can be an environmental factor that results in rise of metabolic disturbance, hypertension, diabetes, stroke. To improve gut flora, probiotics are used for treatments of diarrhea, irritable bowel syndrome (IBS), genito-urinary infections and constipation. In a recent review, it was reported that *Lactobacillus* spp. and *Bifidobacterium* species were most commonly used in the research studies and these show moderate improvement in glycemic and lipid parameters.

6.4 DEVELOPMENT OF GERIATRIC-FOODS FOR OLD AGED POPULATION

Various physical, physiological, and intellectual changes occur in old-aged population [65]. In order to age without much health complication, proper nutrition is a must, without which elderly people might develop several chronic diseases and other health issues related to geriatrics [37].

According to the the United Nations, old aged population is growing faster in the world [65]. Courtesy to the current scenario of global increase in geriatric population, the causes behind this rise are supposed to be progressions in medical field and socio-economic development leading to triumph over diseases and early demises. Old-aged populations have weaker immune system and are more prone to diseases, malnutrition, and issues associated with old age. Hence to live a normal life during old age, proper nutritional need of geriatric population must be met as their body weakens with age and it is nearly difficult for them to cope up with proper dietary habit [37].

Escalating global concern about wellness and wellbeing is provoking larger part of people to gage better ways of life, practice more, and practice good eating habits. The extending food handling market demonstrates that end clients are looking for negligibly handled food with extra healthful advantages and organoleptic esteem. Considering the global scenario of increasing old aged population associated with disease, it is to a great extent crucial to develop food products with high nutraceutical value along with providing an alternate way to encourage functional food intake by means of food fortification. While developing fortified functional food products for old aged population, certain food safety concerns are to be paid attention. Ignorance to such concerns may cost a huge price causing more trouble for elderly people [32].

6.4.1 POPULATION AGEING IN THE WORLD: CURRENT SCENARIO

In accordance with the data provided by United Nations, in 2019, existence of 0.703 billion 65 years or more than 65 years was recorded in the entire globe. It has been estimated that the number of geriatric persons likely to go higher by around 1,500 million by the end of 2050. The figure of geriatric population is rising faster in Latin America, Caribbean, and Eastern and South-Eastern Asia. Around 11% of old aged population has been recorded in Eastern and South-Eastern Asia compared with 9% subsequently in Latin America and Caribbean in 2019. During the period of 2023 to 2050, the figure of old aged population might be doubled in Latin America, Central, Western, and Southern counterpart of Asia, Caribbean. Globally, the existence after the age 65 is progressing. Worldwide, it has been anticipated that 65 years old person during the period of 2015 to 2020 may live further 17 years more. This figure might rise to 19 years during the years 2045 to 2050 [65].

6.4.2 CONSIDERATIONS FOR DEVELOPMENT OF GERIATRIC FOODS

If the diet assigned for old aged population fails to provide adequate nutrients required to sustain optimal health, it might lead to nutrient deficiencies plus development of degenerative diseases. Old aged people possess higher risk of evolving nutritional deficiencies due to a little dietary consumption or digestive issues. Energy needs might decline with age, but for normal functioning of the body, protein, and other nutrients are much needed, deficiency of which may result in cognitive dysfunction in geriatric population. Hence while developing food for such population; following considerations are to be kept in mind: There are certain recommendations of nutrients for older people, which are described in subsections [37].

6.4.2.1 PROTEIN

Poor regular diet among old aged population causes insufficiency of protein intake resulting in sarcopenia and affects bone health triggering functional loss and fragility. Hence it has been recommended that geriatric people should consume equal proportion of protein in a day. It has been estimated that on average, an old aged person should include protein in his diet at the range 0.9 g/kg body weight per day. Previously, amount of protein in diet to be included for a human population irrespective of age and sex counseled for was 0.8 g/kg body weight per day. Nevertheless, this amount of protein intake is inadequate for maintaining health of old aged population. Geriatric population suffering from ailments requires more amount of protein, ranging from 1.2 g/kg body weight per day to 1.5 g/kg body weight per day due to fault in anabolism of protein. Better preservation of muscle mass is associated with ingestion of protein from animal-based food rather than plant-based food [37].

6.4.2.2 CALCIUM AND VITAMIN D_3

Calcium as well as vitamin D_3 both plays an essential part in maintaining optimal bone health. Reduction of bone mineral density is very usual in geriatric population leading to critical osteoporotic fractures. In comparison to men, elderly women have higher risk of bone loss mainly because of lack of estrogen after menopause and as a result, decreased calcium absorption in intestine is highly witnessed, kidney re-absorb calcium poorly and parathyroid hormone is secreted in higher amount than usual along with bone resorption. Vitamin D_3 deficiency causes increased bone turnover in geriatric population which affects calcium homeostasis. The decreased exposure to sun and inability to synthesize vitamin D_3 by the skin of elderly people most of the time contributes to vitamin D_3 deficiency. Amplified weakness of muscles and declined physical activity among geriatric population are linked with serum 25 (OH) D-level below 50 nmol/l. For the maintenance of optimum bone health, the amount of calcium for old aged population varies from 1,000 to 1,200 mg per day. High sodium diets can also cause more calcium loss in the urine. Hence, levels of both calcium and vitamin D_3 should be managed in elderly population that can be accomplished through dietary sources and supplementation [37].

6.4.2.3 ZINC

The elderly people have low amount of serum zinc resulting in poor immunity leading to surges the risk of morbidity. Zinc deficiency in geriatric population can cause T-cell-mediated dysfunction. Zinc insufficiency is mainly due to inappropriate dietary zinc intake joined with additional intrinsic and extrinsic reasons. Inadequate food chewing, oral problems restraining diet, numerous medications are some of the major factors of insufficiency of zinc in old aged population. Hence, in order to normalize serum zinc concentration, zinc intake can be aided both dietary sources and supplementation [36].

6.4.2.4 IRON

Old aged populations are more prone to iron deficiency which confer anemia to them. The lessened food intake, recurrent medications, gastrointestinal (GI) issues are the usual reasons behind deficiency of iron in older age. Increased hepcidin levels also can cause age related anemia. Hence, iron supplementation via fortification of food can be a good substitute to treat anemia caused by deficiency of iron [37].

6.4.2.5 VITAMIN-B COMPLEX

Vitamin-B complex maintains the function of cells and emaciation of brain. Lack of vitamin B6 and B12, and folate affects cognitive functions among geriatric population by increasing serum homocysteine levels, which in turn elevates the chances of dementia and Alzheimer's diseases. As animal-based foods are chief source of vitamin B12, thus vitamin B12 fortified foods can be a substitute for people, who are vegan or vegetarian to normalize their serum vitamin B12 levels during older age [32].

6.4.3 FOOD SAFETY CONCERNS FOR OLD-AGED POPULATION

While developing food products to meet the dietary needs for old aged population, one important reason that should always be given utmost importance is food safety concern because in comparison to young adults, elderly people are more susceptible to food borne diseases and food contaminants due to age related weakened immune system. Hence, a broad study on the identification of pathogens which can potentially cause infectious diseases in old aged population along with the study of their susceptibility to its infection and severity of illness, study of adverse effects and collection of safety data of often beneficiary dietary supplements designed for old aged population on their health, a detailed study of food processing and technologies for food product formulation in order to defend against microbes and impurities, development of irradiated food products aimed to market directly to older people susceptible to risk of infections from food borne pathogens are studied in details when it comes to food safety concerns for old aged population [32].

In relation to the study of pathogens that cause food borne disease to older population, *Salmonella*, *Listeria monocytogenes*, and *E. coli* O157:H7 are greater risk aimed at geriatric people than the young adults. Therefore, while developing food products for geriatric population, it is necessary to take measures to prevent chances of contamination by such pathogens [32].

The adverse effects and safety data of beneficial dietary supplements aimed to fortify with food products designed for old aged population are necessary to study because with age, body metabolism changes and it increases the adverse effects of some beneficiary dietary supplements which in fact have positive effect in young adults [32].

Foodborne ailments are mostly related with ingestion of raw and uncooked foods in old aged people. The food industry around the globe has industrialized numerous different interventions and remains to grow novel ideas to guarantee food safety. Such interventions can be physical (such as: heat, high pressure), UV light, ionizing radiation, or gas that can be modified atmospheres or biological or chemical which may be oxidizing agents, organic acids, bacteriocins [32].

Recent inventions in food packaging, accentuates that food packaging can offer protection against microbiological contamination which can be key factor for consideration while developing food products for geriatric population. For example, a new advancement called aseptic packaging includes sterilization of food product and packaging separately and after that bring them together in a sterile atmosphere [32].

6.4.4 FORTIFICATION-AIDED DIETARY APPROACH FOR DEVELOPMENT OF GERIATRIC-FOODS

Due to weakened immune system, digestive malfunction and other health problems, old aged people suffer from malnutrition. Hence to meet proper dietary needs in order to maintain normal functioning of the body, it is important to develop food products fortified with nutrients. Fortification of food can be MN-based, vitamin based, or probiotic based [61]. In MN-based fortification of food, MNs such as iron, iodine, zinc, calcium, etc., are fortified with different food products which are supposed to meet the need of old aged people with MN deficiency [61]. Vitamin based fortification of food developed for geriatric population includes the addition of vitamin A, vitamin B complex, vitamin E, which old aged people essentially need [61].

Not all bacteria are pathogenic. Some bacteria provide health benefits too. Probiotics have many health benefits and help old aged population fight with many health issues. Food products such as yogurt, cottage cheese contain beneficial probiotic cell culture. Hence fortification of probiotics with food product aimed to be developed for geriatric population can meet the nutritional need [61].

6.5 FOODS FOR CUSTOMIZED NUTRITION

As we can see in today's world everything can be customized and people prefer customized things and online shows. There is also a trend of customized food which is more prevalent in Western countries and also it is spreading on other parts of the world too. In recent times when people are visiting doctors for their health issues, they are being made familiar with the connection between their food choices and health issues. It is now possible to access someone's genetic and metabolic history and information, which is connected with their food eating habits and lifestyles. As people are becoming familiar with internet, they are becoming more conscious about health and fitness are part of their everyday lifestyle.

Some easily accessible gadgets of today's worlds like social media, magazine articles are helping shaping people's minds and changing their perception about health and fitness and to improve and personalized their dietary intakes. But all magazine articles, social media posts were not written on proper scientific knowledge and therefore not everyone is getting benefits from customized food. It is discovered recently that common diseases like cancer, diabetes, and heart diseases are connected with dietary intakes of person effected [46]. Therefore, there is a need to know what is personalized nutrition and what type of strategy and food one should choose to stay fit and healthy to avoid health ailments.

6.5.1 WHAT IS A CUSTOMIZED NUTRITION?

This personalized nutrition can be defined as a strategy and planning which deals with metabolic and genetics of an individual and formulation of dietary packages to make that person stay healthy and fit. This personalized nutrition can also be called tailored nutrition. Nutri-genomics is a field which shows that metabolism of a person which resulted from his genetic makeup is highly influenced by type of food consumed [51]. Genetic and phenotypic make up of a person is very important for researchers and dieticians to formulate the customized nutrition [45].

The tailored nutrition helps to reduce the risk of various health ailments like diabetes, obesity, cholesterol by controlling our food choices and eating habits and also fitness regimes. Some companies working in the field of tailored nutrition examines people's food intake history, food related health issues to formulate the personalized diet chart. Although this examination is done by online questionnaires which is less costly, yet now-a-days some devices are being developed which can be worn by people regularly; and this device can help predict and tract their health problems, daily body and heart functioning, etc., which can help researcher in a better way to formulate personalized nutrition [20].

6.5.2 TYPES OF FOODS FOR CUSTOMIZED NUTRITION

As we said earlier the choice of food for customized nutrition will vary from person to person due to their genetic makeup. The market value and demand of personalized nutrition is increasing day by day despite some difficulties, companies which are working in this field are also increasing day by day. Consumers of this field are also increasing each day, which mostly include young professionals along with middle aged people. There is a constant relationship between chronic disease and food intake because these diseases become worse in the long run due to intake of wrong food products, which can worsen the health condition. Therefore, it is necessary to track people's food intake to determine any underline health issues and to reduce some chronic illness from becoming worse and to control their spread [11].

6.5.3 BENEFITS OF NUTRIENT-RICH FOODS

The field of tailored nutrition has progressed in a better way in recent era, and it is spreading knowledge worldwide by making a revolution in food and fitness industry. To achieve the world changing aim to make food a medicine does not poison for people, food and personal nutrition companies are constantly working with health professionals and researchers to make a reliable field for person of every age, culture, and community. Table 6.2 lists nutrient rich foods that are in use for nutrient deficiency diseases.

TABLE 6.2 List of Nutrient Deficiency Diseases versus Nutrient-Rich Foods

Disease	Nutrients Involved	Nutrient-Rich Food
Goiter	Iodine	Iodine salt, marine fish
Iron deficiency	Iron	Marine food, meat product, etc.
Osteo-Malacia	Vitamin D	Sun rays, milk products, fish product, etc.
Osteoporosis	Calcium	vegetables, fermented milk products, etc.
Pellagra	Vitamin B3	Various grains, meat products, etc.

However, in customized nutrition and food selection sector, there exist drawbacks, such as: the genetic relations and metabolic outcomes of various food habits and lifestyle changes; and these fields require more attention and research to make the nutrition industry more compatible and beneficial for future generations. There are still many steps to go to decide a proper diet for complex health related issues [68].

6.5.4 FUTURE ASPECTS OF CUSTOMIZED FOODS

As we have seen that although the customized food industry is gaining popularity in our younger as well as older generations there are still some drawbacks and shortness of knowledge and manpower, more research is the need of the hour for this industry. More research should be done to decipher the connection between genes and metabolism and food items and lifestyle habits. The pathways which are get influenced by food intake should be studies carefully and the information's should be stored and generate data to work in various labs to make proper diet planning for people of various age and community. There should be some trial-and-error methods to help cure and reduce impact of various chronic diseases by using customized nutrition. Awareness regarding food habit and disease related issues should be done through popular communication gadgets like social media post, blogs, and magazine articles so that this kind of information's spreads in every field and section of the world and people start customizing their food from an early age to avoid development of various chronic diseases [11].

6.6 FOOD FORMULATIONS TO COMBAT COVID-19 PANDEMIC

The current supra infection COVID-19 has stormed the public health, causing havoc to millions of people around the globe through a contributory agent called severe acute respiratory syndrome (SARS) corona virus (SARS CoV-2). To combat the current divulged pandemic, worldwide efforts have been taken to develop vaccines and immunity boosting formulated products. Researchers, Clinicians, Food manufacturers have strived to find the active compounds in the food that can inhibit the entry of virus and encapsulation to foster fast recovery in COVID-19 patients. Formulated foods or functional foods are meant to suffice the lack of availability of newly discovered medicines, underdeveloped vaccines, etc. The choice of foods that leads to an effective functioning and sound immune system has become clearer in recent decades, as foods were one of the options despite suggested antiviral medicines chosen by the populace, when a sudden turmoil took place in Dec. 2019 [33].

Foods ranging from fruits to vegetables, pulses, drinks and beverages, and macro-MNs loaded products are sources that can modulate the body to inhibit the COVID-19 disease. Consumers in the world around also rely on legume-based products, protein rich and energy dense food to endure good health during the pandemics. Formulated food has also become crucial to post COVID-19 patients to alleviate the chances of re-infection so far.

6.6.1 BRIEF HISTORY OF COVID-19

The existence of SARS COVID-19 unfolded within the city of Wuhan, Hubei territory, China at end most of the year 2019. The grievous severe acute respiratory syndrome-coronavirus-2 (SARS-CoV-2) was documented to be the first pandemic of the century in March 2020 [33]. And fifth pandemic after 1918 flu pandemic [12]. The obscure general origin of its zoonotic properties can be tracked down back to the ecosystem alteration that have declined biodiversity, largely enabling the contacts to link humans and the animal pool that carry pathogens, counting SARS-CoV-2. These pools are the bats, or we can say that the virus is closely related to bat coronavirus (RaTG13) [69]. However, as of documentation done, this virus and its infectious effects existed in Chinese region much prior to Wuhan, China explosion. It also manifested outside China, for example in Europe [69]. Though it persisted, yet it remained dormant or did not encounter anything that could lead to outbreaks, and if managed to cause disease it remained limited within space and time. Different variants of SARS CoV2 in different geographic locations are reported, such as, SARS-CoV2 belongs to Coronaviridae subfamily of RNA viruses, the Coronavirinae family. Sub-families such as alpha, beta, gamma, and delta are categorized on the basis of genomic sequence variants of coronavirus [69].

6.6.2 RECENT TRENDS WITH COVID-19 AND DIABETICS

Outbreak of corona virus pandemic marks identified risk factors with other complications including death. Such risk factors include *diabetes mellitus*, CVDs, obesity, respiratory, CHD [39]. Furthermore, diabetic patients show effect on liver and pancreatic dysfunction, endocrine organ damage that leads to metabolic abnormalities that may develop [36]. Other complications requited ICU treatment, also suffer from cognitive impairment with post-traumatic stress depression, rhabdomyolysis that generate long term abnormality. After the pandemic it was found that new onset of diabetes rises 1% in general population after discharging from hospital and required new treatments for glycemic control [47]. COVID-19 affects glucose homeostasis, where insulin intravenous infusion is used to maintain glycemic control and extreme fluctuation of glucose that has the ability to destroy islets through receptors [16].

Food exerts a major influence on human health system. With sudden onset of COVID-19 global pandemic and high mortality rates with limited knowledge of medications and food access, the food formulation to tackle COVID-19 has started to take the momentum. Various pharmaceutical and food industries and research institutes along with the government agencies have set efforts to formulate required foods to fight COVID-19. Since no 100% effective medicines and vaccines are still available to cure COVID-19, therefore it must be relied on food products that can speed-up prevention of the viral entry. One can mostly rely on the immunity boosting, anti-inflammatory food products of both natural and synthesized. It can hold to adjust the body with COVID-19 via several mechanisms including the reduction in delivering pro-inflammatory cytokines and reduce the diction of ACE2 receptors in cells and hinder crucial enzymes in SARS-CoV-2 [40].

Various plants and herbs act as source for formulation of therapeutic food products namely Japonica alnus Tunstead's product Hirsuten on ethanol extract of 0–600 µl for 60 mints showed positive actions on *invitro* SARSCoVPLpro *Angelica keiskei* leaves; and bioactive product Xanthoangelo-IE also produces inhibition of SARS CoV 3CL up to 50 µl of doses. Whole *Aglaia perverdis* Hiern inhibits SARs-CoV helicase by affecting the ATPase activity. Natural plant products with bioactive compounds (such as: phenolic and flavonoids) in *Malva Silvestri* are gaining importance for COVID-19 [33] prevention. Consumers throughout the pandemic have been using energy-dense foods and protein-rich source food butters, nuts, seeds, milk eggs, legumes chickpeas, lentils, hummus, tofu, protein shakes enriched with vitamins, minerals.

6.7 VARIOUS TYPES OF FOOD FORMULATIONS

6.7.1 PLANT-BASED

Plants have been acting as a big source with many superior benefits to human health. Plant based products are rich in proteins, vitamin, carbohydrates, fats, and minerals. Plant derivatives are processed to impart better properties and uses.

Some plant product formulations that gained more claims during pandemics are oral nano-curcumin of *Curcuma longa* (turmeric) with bioactive polyphenolic compound. Compounds are often extracted from turmeric rhizome's dried powder. Curcumin delivers various pharmacological activity including removal of free radicles, anti-angiogenic, antimicrobials such as bacteria and viruses. It also acts as antidiabetic and anti-inflammatory [31]. Curcumin and piperin formulated tablets allow recovery from symptoms like fever, cough sore throat, and breathlessness. Foods from animal sources includes dairy products and meat.

Probiotic formulation comprising *Lactiplanti bacillus plantarrum* and *Pedio coccus acidilactici* strain in studies of 293 COVID-19 outpatients with mild symptoms on being administered for 30 days showed reduction of nasopharyngeal and serum biomarkers [17]. Probiotic supplementation reduced viral load of nasopharyngeal and lung penetrating particles. The time span of digestive and non-digestive symptoms was compared to the inert medication and were found to be well-tolerated [40]. The utilization of nutrients from natural reserves can give long-established augmented immune effect. These natural foods include fruits and vegetables, cereals, millets, spices, herbs, nuts, oilseeds, and considerable superfoods like chlorella and spirulina.

Plant based formulation of Ayush Kwath powder is an immunity boosting health drink product that constitutes *Ocimum tenuiflorum* (Tulsi), *Cinnamomum verum* (Dalchini), *Zingiber officinale* (Sunthi), *Piper nigrum* (Krishna Marich). These plant-based products have been developed by the Ayush Ministry, Government of India (GoI) for immunomodulatory and prophylactic intervention to prevent COVID-19 [63].

In England, eight different plants were studied for preventive effects of COVID-19. These included (Schott) *Dryopteris filixmas*, *Epilobium hissutum*, *Lysimachia Nummularia*, *Epilobium angustifolium*, *Sedums purium* M. bieb, *Lysimachia vulagaris*, *Saxifraga ligulatewall*, *Myrica gale* L. Extracts of these plants can inhibit the action of Influenza [35].

6.7.2 PRACTICES BASED ON NUTRIENTS AND DIETS

It is explicit that the diet-practice influences the immune system responses. As a whole, where sensible, a coherent way to lessen the risk of viral infections is to regulate the activity of the inflammatory mediators through flexible risk factors such as diet, exercise, and healthful state of habits. Plant based diet studies demonstrates that individuals had very mild illness compared to the ones who didn't follow such diets [46]. Some assessed works have emphasized the potential application of fungal beta-glucans in nutrition and medicine by evaluating their formulation and efficacy as well as safety profile, and immunomodulating effect.

6.7.3 VITAMIN-BASED PRACTICES

It is evident that Vitamin A subjoining reduces fatality and death rate in infections like HIV, malaria, measles, measles-associated pneumonia, and diarrhea (Table 6.3). Adverse effects during viral afflictions have been associated with low vitamin A.

An adequate intake of vitamins B6, B12, C, and E along with zinc and iron are underlined for the maintenance of immunity [3]. Many dieticians and clinicians have recommended the food formulated products (Table 6.4).

TABLE 6.3 Vitamin-based Formulated Products

Product Name	Ingredients	Effects	Recommendations
Blue-bonnet soft gel	Vitamin A from fish, liver oil, bovine, purified water.	Formation of epithelium, front line-defense.	333%
Citrus Vitamin-C powder	Energy, protein, carbohydrates, fats, PUFA, zinc.	Stimulates proliferation and function of T lymphocytes.	Energy: 0.62% RDA Zinc: 100% RDA
Plant-based iron	Iron, Vitamin C, folic acid.	Better oxygen binding.	100% RDA
Vitamin D_3 soft gel capsules.	Vitamin D inhibition or immunomodulatory ways.	Direct viral	600–800 IU

6.7.4 MINERAL-BASED PRACTICES

The general wellbeing and strong immune system are maintained by food products consisting of sodium, zinc, potassium, chloride, calcium, and phosphorus (Table 6.4) thereby reducing the chances of infections [38]. Sodium plays crucial role in regulating electrolyte balance and the expression of ACE2 in SARS-CoV-2. Jan Alexander and his group found that Zn status in serum values less than 0.7 mg/L represents a risk factor for pneumonia. Inflammations and inflammatory biomarkers increase when long-term zinc deficiency occurs and most angle of the immune system are affected by zinc deficiency, especially the T-cell function. Deficiency of zinc drives a Th17 response, which is co-related with increased inflammation. In venerable subjects, reduced concentrations of circulating zinc correlated with rise in levels of the cytokines-Interleukin-6, Interleukin-8, and TNF-α (Tumor necrosis factor-α).

TABLE 6.4 Mineral-based Formulated Foods

Product Name	Ingredients	Function	Recommended
Ayush Kwath	Mulethi, Haridra, Munakka, Kantakari	Promotes wellness, prevents respiratory tract illness.	10–20 g
GNC	Magnesium supplement	Prevents lung disease, muscle cramps, energy booster.	300–420 mg for adults.
Himalayan organics plant-based zinc.	Zinc, Vitamin C	Impairs viral replication, enhances cytotoxicity.	300–400 mg
Natural calcium capsules/tablets.	Calcium from milk.	Exerts antiviral activity.	Mg: 33% RDA Phosphorus: 8.6% RDA

6.7.5 PHARMACOLOGICAL IMPACTS OF FOOD FORMULATION

Food mediates effects on bioavailability of numerous medications and has remarkable culmination in drug development and clinical setting [66]. Food exerts influence on drug absorption and the intensity of food-drug interactions may be influenced by the type of the food [34]. Foods can have strong impact on the medication taken during COVID infections. With respect to clinical significance, the degree of significant aspect of food effect is generally considered to be changing in the extent of bioavailability, and on the 'positive food effect' and 'negative food effect' which have been termed to set out either an increase or decrease in the furthest extent of bioavailability.

6.7.6 CHALLENGES FOR FOOD FORMULATION IN RESPONSE TO COVID-19

The COVID-19 pandemic has justified more of the well-functioning and balanced immune system requirements in recent times. Numerous elemental factors have not yet been uncovered as own, i.e., in the unpredictability on immune system responses to COVID-19. The human cell machinery is the circuit on which viruses rely to replicate themselves, hence antiviral food and drug designers face a challenge in stopping the virus without damaging the inner mechanism of healthy cells. Whilst scientists have found several ways out to the problem, the pharmacopeia antiviral products and antiviral food formulation still drop backs the abundance of antibiotics and foods to consider bacterial infections. Though food manufacturers are developing some foods and new drinks with immune boosting and anti-inflammatory properties, the pandemic itself has affected the involvement of hundreds of individuals, from R&D scientists to sensory specialist, procurement teams to marketing [30]. Therefore, the COVID-19 combating food formulation framework still lags, and challenges revolves in the backdrop of accurate disease treatment and enhancing wellness to the people.

6.8 DEVELOPMENT OF FOODS FOR INFANTS

Infancy (0 to 2 years) is a period of rapid growth with an imperative need to optimize nutrition to ensure proper organ development and appropriate growth. World Health Organization (WHO) has recommended exclusive breastfeeding (EBM) up to six months to prevent 13% of infant mortality rate globally while appropriate complementary nutrition and its foods can result in an additional 6% reduction in infant mortality. Also, malnutrition causes another major health problem in developing countries which contributes to poor physical and cognitive development of infants; therefore, WHO has targeted to reduce malnutrition/stunting by 40% by 2025 [16]. Thereby to achieve this goal, a nutritionally adequate and balanced complementary foods or food premixes or variety of nutritious weaning foods need to be accomplished that can suffice nutritional needs of the infants. Basically, Infant food means foods with special nutrition and highly-balanced composition which resembles breast milk [23].

Traditionally, infants are fed with semolina porridge, sweetened rice porridge, mashed fruits and vegetables. There are various approaches like supplementation and fortification of infant foods with milk and cereal products increasing gradually to combat MN deficiency like anemia (iron deficiency) in the developing countries [21]. Now-a-days instant infant foods promote the market due to its convenience and safe aspects. Also, mothers have become more responsive for readily available infant foods, which contain adequate nutrients required by infants and get influenced by various advertisement and trust gained through the market policy.

6.8.1 HISTORY OF INFANT FOODS

During the preindustrial period infants were fed cow's milk (CM), pre-chewed foods that were contaminated caused high mortality rates and ultimately child become malnourished. Infant's survival completely dependent on only breastfeeding as there was lack of successful substitute feeding. In this way, children's health had been compromised as they were fed with only animal's milk or formulae diluted with powdered milk. History of infant foods is summarized as follows:

- In the year 1838, Simon observed cow's milk contains more protein than breast milk.
- In the year 1856, Gail Borden: Condensed milk.
- In the year 1867, Leibig developed First commercial formula.
- In the year 1872, High energy and low-fat content condensed milk.
- In the year 1874, First complete artificial formula.
- In the year 1880, Chlorinated water.
- In the year 1883 (Meyenberg): Evaporated milk contains more fat and no sugar.
- In the year 1885, Meigs and Biedert revealed the composition of breast milk.
- In the year 1890, Milk pasteurization.
- In the year 1895 (Rotch): The first mathematical formulae developed to resemble the breastmilk (BM).

- In the year 1912, Funk: Macro- and micronutrient requirements to reduce deficiency diseases.
- In the year 1920, Supplement feeding incorporating with juice and cod liver oil.
- In the year 1929, Soy formula developed to reduce allergic to cow's milk.
- In the year 1940, Fortification with Vitamin D in pasteurized milk.
- In the year 1960, Initially baby food contains monosodium glutamate, sugar, and starch and later on changes in formulae with lactic acid, lactose, fat, minerals, and vitamins.
- In the year 1962, Regulation of formula ingredients.
- In the year 1970, For non-breast-fed child fortified cereal formula with iron or cow's milk.
- In the year 1990, Innocenti Declaration, Baby friendly Hospital Initiative.
- In the year 1993, WHO observe the difference in growth of breastfed child.
- In the year 1997–2003, Multicenter Growth Reference Study.
- In the year 2006: New Growth Charts WHO were constructed.
- 21st Century: it was recommended to use exclusively breastmilk up to 6 months, then breast milk along with complementary foods until two years.

Therefore, Infant nutrition whether natural or synthetic always accustomed by social value attributed to breastfeeding. Also, WHO recommended exclusive breastfeeding up to six months along with the complementary nutritious weaning foods until two years of age.

6.8.2 NUTRITIONAL NEEDS OF INFANTS

Infancy is the crucial stage for the proper growth in their later stage therefore; infant nutrition plays an important role in cognitive and physical development [29, 48]. Balance nutrition provides the foundation of health. Infant foods or weaning foods plays a significant role in the initial stage of life by providing the infant with energy, macro- and micro-nutrients, bioactive compounds and non-digestible carbohydrates that can stimulate the gut microbiota (Table 6.5). As WHO recommended exclusive breastfeeding up to six months with the introduction of semi-solid foods or milk-based foods from six months to infant diets known as complementary feeding, therefore this enables to meet the nutritional demands and thereby decrease the risk of malnutrition.

Complementary feeding practices help to expose for a new tastes and textures which ultimately leads to healthy eating habits. Nowadays, there are various commercially available readymade or instant infant foods (Table 6.6), are typically nutritious, safe, pre-prepared form or minimal cooking required before feeding [29].

TABLE 6.5 Recommended Dietary Allowance (RDA) for Infants

RDA	Infant Age Groups		
	0–6 Months	**6–12 Months**	**1–3 Years**
Body weight (kg)	5.4	8.6	12.2
Net energy (kcal/d)	108/kg	98/kg	1,240
Ascorbic acid (mg/d)	25	NA	40
Calcium (mg/d)	500	500	400
Free folic acid (μg/d)	25	NA	30
Iron (mg/d)	NA	NA	1,200
Nicotinic acid (mg/d)	710 μg/kg	550 μg/kg	0.8
Protein (g/d)	2.05/kg	1.05/kg	22
Pyridoxine (mg/d)	0.1	0.4	0.9
Riboflavin (mg/d)	65 μg/kg	60 μg/kg	0.6
Thiamine (mg/d)	55 μg/kg	50 μg/kg	1,600
Visible fat (g/d)	NA	NA	25
Vitamin (μg/kg): Retinol-ß	350	350	12
Vitamin (μg/kg): Carotene	1,200	1,200	400
Vitamin B12 (μg/d)	0.2	NA	0.2–1.0

Note: NA: Data not available.

TABLE 6.6 Major Nutrient Rich Foods Required by Infants

Nutrients	Food Constituents
Carbohydrate content	Lactose, starch, corn syrup, sucrose, and corn syrup solids.
Functional ingredients content	Soy lecithin, mono, and diglycerides, starch, carrageenan.
Major mineral content	Calcium carbonate, mono, di, and tribasic calcium phosphates, dibasic magnesium phosphates, potassium citrate, magnesium chloride.
Minor mineral content	Zinc sulfate, potassium iodide, cupric sulfate, ferrous sulfate.
Oil content	Soy, safflower, corn, sunflower, copra, palm, colza.
Protein content	Non-fat dry milk, whey, sodium, and calcium caseinates, soy protein isolate, condensed skim milk and soy milk.
Vitamin content	Niacin, folic acid, pantothenic acid, biotin, choline, inositol A, E, K, C, B1, B2, B6, B12.

6.8.3 CURRENT SCENARIO OF INFANT FOODS

In this modern era, consumption of infant formula-based foods is increasing rapidly and at the same time, its adverse health outcome is also considerable. Concerns for commercial infant foods are listed below:

- Adequate macro- and micro-nutrient;
- Amounts of fish and meat content;
- Energy content;
- Predominantly sweet and salty in taste;
- Safety of the ingredients used;
- Variety of food ingredients.

Therefore, to overcome the health related issues and for improvement of the food scientists and technologists have adopted some latest technology (Table 6.7) to add different components in the infant formula including human milk oligosaccharides (HMO), milk fat globule membranes (MFGM), hormones and high pressure processing techniques, resulting less allergy causing and disease free formula as compare to traditional formula milk [28, 60] and thereby became the most challenging task for the inclusion of human milk components in infant formula almost resembles BM in the near future.

TABLE 6.7 Methods and Benefits to Develop Risk-Free Infant Formulas

Methods	Benefits
Addition of docosahexaenoic acids (DHA).	For proper growth and development.
Addition of MFGM.	For proper neuro and intestinal development.
Addition of proteins.	Weight management and allergy free.
Enriched with prebiotic substrates.	Proper digestion and intestinal function.
Enrichment of polydextrose (PDX) and galacto-oligosaccharides (GOS).	For proper absorption of iron, magnesium, and calcium.
High pressure pasteurization.	Reduction in allergy prevalence.
Inclusion of corn sugar.	Reduces lactose intolerance.
Synthesis of HMOs.	Required for the prevention of toxins and pathogen.

6.8.4 TYPES OF NUTRITION-ENRICHED INSTANT INFANT FOODS

Globally infant food market has progressed due to the fact that nutritional qualities of infant foods have improved and are closer to that of breast milk.

As an infant requires an adequate amount of energy-dense and higher nutrients content food from the period of six months along with the breast milk and thereby complementary feeding is essential to fill the nutrition gap. If complementary foods are inappropriate or not introduced in the early period, may result in poor physical growth and leads to malnutrition. Therefore, to meet the nutritional requirement of infants, complementary foods need to be energy-dense, safe and enriched with various nutrients [67]. Keeping all these views in mind, nutritionists, scientists, and commercial manufacturers have attempted various studies and have developed value added nutritious infant foods or formulas (Table 6.8).

TABLE 6.8 List of Nutritionally Enriched Instant Infant-Foods

Instant Infant Foods	Ingredients	References
Complementary food formulation.	Maize, soyabean, peanut fortified with *Moringa olifera* leaf powder.	[59]
Complementary foods using fruits and vegetables.	Papaya, potato, carrot, apple, apricot, strawberry, banana, mango, and guava.	[5]
Fortified supplemented foods.	Sorghum-Ogi and cray fish.	[49]
Infant food premix.	Pearl millet flour, finger millet flour, wheat flour, pumpkin powder, carrot powder, roasted Bengal gram flour, Niger seeds, milk powder.	[58]
Porridge from germinated multigrain.	Germinated wheat, buckwheat, and green gram.	[14]
Prebiotics mix from fermented cereal legume blends.	Underutilized crops (finger millet, barnyard millet, black soyabean, amaranth grain) and *Lactobacillus plantarum.*	[57]
Weaning food formulation	Sorghum, finger millet, amaranth grain, moth bean, skim milk powder.	[6]
Weaning foods	Germinated wheat flour, mung bean flour, skim milk powder and sucrose.	[43]
Weaning foods	Pigeon pea and millet flour.	[50]
Weaning foods: Instant	Rice flakes, skim milk powder, butter, vitamin premix and sugar.	[60]
Weaning foods: Instant	Fermented maize, beans, fishmeal, pawpaw, and sugar.	[64]
Weaning foods: Low-cost nutritious	Whole wheat flour, linseed flour, lentil flour, potato flour, carrot flour.	[4]

6.9 FOOD FORMULATIONS FOR SPORTS PERSONS

Fortification is a process where essential ingredients like minerals, vitamins, omega-3 fatty acids, prebiotics, probiotics, and phytoconstituents are being incorporated to the dairy, plant based and various other products. Fortified foods are the foods with various nutritional constituents and increased in the reference to the usual constituents. Supplemented foods are not the whole foods constituting instead of concentrated nutrients sources. They have specific nutritional and physiological effects which are commercialized in dosed form such as capsules, tablets, pills, powder sachets, liquid ampoules, or powders. In recent years, the intake of fortified foods and dietary supplementation becomes a strategy to achieve health benefits.

With new trends, people are accepting a healthy lifestyle, which is rendering into rise of exercise, longevity, and persistence activities. Scientific evidence has confirmed that particular constituents enhance the performance of the athletes. People started recognizing the benefits of sports nutritive products and complementing the exercise sessions with the respective products [2, 41]. Sports health products were primarily used by athletes and bodybuilders but with days these has trending among the mainstream consumers. With the increase in awareness about health, consumers are interested on animal free, natural, minimally processed food, plant-based food, seasonal, and local foods are shaping sports nutrition products. Besides being a nutritional trend, it has become an important sector influencing nutritional trends, which being the main constituent for the protein surge. This chapter explores the fortification of food products for sports persons, its significance, functional ingredients, role of active compounds and challenges faced during their formulation.

6.9.1 SIGNIFICANCE OF FORTIFICATION OF SPORTS FOODS

Sports foods are the specialized foods which are designed for athletes and body builders to enhance their nutritional uptake, health, performance, wellness, and muscle growth. The main significance of fortification is to reduce the nutritional deficiency rate. Also, fortification helps to reduce related diseases in sports people. Dairy products like whole milk, yogurt drinks, buttermilk which are high source of proteins for sports persons favors the fortification ingredients and a good choice for the important nutrient fortification. The fortification process helps to fulfill the nutrients loss (too low iron) in dairy products. Several heat processes like homogenization, pasteurization, spray drying contribute to loss of some essential nutrients. Thus, these processes help to deliver the required MNs and other functional ingredients and serves as a tool to minimize the deficiency of nutrients.

Despite the food-based advantages, there are many other benefits of interventions which aims at controlling and preventing the food storages provided by the fortification of foods. It helps to fulfill the iron, folic acid, iodine, vitamin A and D deficiency to the staple foods and also targets the malnutrition diseases [23]. The process of implementing the food fortification schemes on large scale will help in decreasing the losses caused in food during the storage [26]. The regular intake of fortified foods like probiotic liquids with the essential MNs like vitamin A, D will be a promising and realistic approach to maintain body's reserve foods than its disintegration [27].

6.9.2 FUNCTIONAL INGREDIENTS FOR SPORTS PERSONS

Functional products provide several benefits, such as: better immune system, heart functioning, etc. Functional beverages for sports person provide available nutrients or other substances such as flavonoids, antioxidants, specific amino acids or phytonutrient into the ready to consume drinks. Principal drinks are yogurt drinks, functional drinks (calcium, vitamin fortified and omega-3) Principal functional beverage products are yogurt drinks, functional waters (vitamin and mineral fortified, sports drinks wellness and herbal drinks). The significance of functional ingredients is to recover various applications of health and immunity supporting drinks in the food industry such as organic functional and naturally healthy foods. Below are few examples of functional ingredients in different foods [18].

An example of functional ingredients in different nuts, seeds, and other food is Finger millet, which has functional foods as proteins including essential amino acids, leucine, methionine, and phenyl alanine, vitamins, riboflavin, niacin, and phosphorus. Flax seeds proteins have potent multi-functional ingredients for formulation of food. Flax seeds contain about 28% dietary fibers, 40% fat, 4% ash, 21% proteins and 6% carbohydrates, and lignins, phenolic acids, sugars, and hemicelluloses. Chia seeds have protein content of 15–24%. Jaggery is reported to purify blood, regulate liver function and keep the body healthy [56].

6.9.3 ROLE OF ACTIVE INGREDIENTS FOR SPORTS PERSONS

Active ingredients in functional food (Table 6.9) regulate the physiological activities of the body and to lead a healthy living. Maintenance of balanced function of each organ system is important for physical work and needs attention under extreme working conditions [42].

TABLE 6.9 List of Active Substance Content for Different Organ Systems and Its Functional Foods

Organ System	Active Substance Content	Functional Foods
Circulatory system.	Coenzyme Q10	Special enriched beverages with Q10.
Movement system: Muscle and skeleton system.	Magnesium, calcium, protein.	Milk and milk products, vegetable drinks and milk substitute fortified.
System of sensory organs: minerals.	Vitamin A and E, magnesium.	Vitamin-enriched functional foods, vegetables, and fruits.
Systems of regulatory bodies including nervous system.	Omega-3 fatty acid; colin.	Sea fish enriched with active ingredient.
Visceral organs: digestive and respiratory systems.	Conjugated linoleic acids, probiotics, prebiotics.	Dairy products, probiotic, and prebiotic enriched milk.

6.9.4 FOOD FORMULATIONS FOR SPORTS PERSONS

Due to their lack of time and the low supply of nutritive foods during practice, the sports persons are looking for such diets that can be consumed easily while at the same time supply the required nutrition. At the present days, demand of nutritionally balanced snacks and drinks are increasing among the athletes [56].

Energy bars are the cereal containing supplementary bars and other high energy foods which aims at individuals who requires quick energy but do not have time. The three main sources of energy in food comes from fats, proteins, and carbohydrates. Taking the advantages of nutritional values of ingredients, energy bars were developed by using pumpkin seeds, chia seeds, finger millets, flax seeds, honey, and jaggery in different proportions.

6.10 CHALLENGES OF FORTIFICATION OF FORMULATED FOOD

Iron formulations are in major trends these days. Addition of iron to the iodized salt (double fortified salt (DFS)) sometimes results in severe changes in DFS. These changes include black spots or losses in the iodine content and yellow discoloration. DFS needs salt with added iodine of highest quality iron formulation to minimize iodine losses and changes in sensory. Thus, proper packaging of iodized salt is important to prevent losses. Prevention of moisture in salt is very important to prevent the mentioned challenging issues. Encapsulated formulations decrease the sensory changes, but the integrity of encapsulation process is the most important and can be affected due to low quality assurance while production of iron formulation [15].

6.11 SUMMARY

Food fortification is an alternative strategy to reduce the MN deficiency diseases especially in low- and middle-income countries where common foods available cannot fulfill the nutritional requirements of the population. Fortification is considered as one of the effective strategies to combat MN deficiency diseases. In today's world, the consumption of fortified foods and dietary supplementation has become an alternative to attain a distinct health benefit and hence, fortification of food requires inclusion of various nutrients in commonly consumed foods to reduce malnutrition and mortality rate. In recent years, many food-based approaches, MN supplementation and MN powder are available. Thus, the highly nutritive instant weaning food for infants, nutritious foods for geriatric population, anti-diabetic foods and premixes, health drinks and energy bars for the sports persons, immune-regulatory foods to combat COVID patients as well as foods for vulnerable groups are developing day by day to meet the nutritional requirements, help to reduce malnutrition and provide extra nourishment. Therefore, Indian fortified food market may be highly urban centric and in the coming years market will progressed for fortified and supplemented food using functional ingredients. The currents trend of lifestyle and food habits has a significant contribution towards global health crisis including different section of population such as infants, diabetics, geriatric people, etc. This review primarily focuses on food habits, dietary need and approaches for different group of population as well as development of new functional food by means of fortification to meet dietary requirements, overcome health related issues and this may lead to good health as well as better nation.

ACKNOWLEDGMENTS

The authors express their gratitude to: (i) Dr. Kandarpa Kumar Saikia, Associate Professor, and Dr. Manab Deka, Professor, Department of Bioengineering and Technology, Gauhati University, Jalukbari, Guwahati-14, Assam for his immense support and kind assistance as well as constant motivation; (ii) Dr. Debabrat Baishya, Head of the Department, Department of Bioengineering and Technology, Gauhati University, Jalukbari, Guwahati-14, Assam for his valuable advice.

KEYWORDS

- **customized nutrition**
- **diabetic mellitus**
- **formulation**
- **fortification**
- **geriatric population**
- **micronutrient deficiency**
- **severe acute respiratory syndrome corona virus**

REFERENCES

1. Asif, M., (2014). The prevention and control type-2 diabetes by changing lifestyle and dietary pattern. *Journal of Education and Health Promotion, 3*(1), 1–8.
2. Rodriguez, N. R., Di Marko, N. M., & Langley, S., (2009). American college of sports medicine position stand: Nutrition and athletic performance. *Medicine & Science in Sports & Exercise, 41*(3), 709–731.
3. Aman, F., & Masood, S., (2020). How nutrition can help to fight against COVID-19 pandemic. *Pakistan Journal of Medical Science, 36*(4), 121–123.
4. Bala, N., Verma, A., & Singh, S., (2014). Development of low-cost malted cereal and legume-based nutritious weaning food to combat malnutrition in rural areas. *International Journal of Food and Nutritional Sciences, 3*(6), 209–212.
5. Bahlol, H. E. M., Sharoba, A. M., & El-Desouky, A. I., (2007). Production and evaluation of some food formulas as complementary food for babies using some fruits and vegetables. *Annals of Agricultural Science, 45*(1), 147–168.
6. Bhoge, N. Y., & Wankhade, P. R., (2020). Study and development of weaning food formulations prepared from germinated grains. *International Journal of Creative Research Thoughts, 8*(3), 21–33.

7. Colberg, S. R., Sigal, R. J., Yardley, J. E., Riddell, M. C., Dunstan, D. W., & Dempsey, P. C., (2016). Physical activity exercise and diabetes: A position statement of the American Diabetes Association. *Diabetes Care, 39*(11), 2065–2079.
8. Clarke, G., (2014). Mini review on gut microbiota: The neglected endocrine organ. *Molecular Endocrinology, 28*(8), 1221–1238.
9. Chow, N., Fleming-Dutra, K., Gierke, R., Hall, A., Hughes, M., & Pilishvili, T., (2020). Preliminary estimates of the prevalence of selected underlying health conditions among patients with COVID-19 in USA. *Morbidity and Mortality Weekly Report, 69*(13), 382–386.
10. Chee, Y. J., Ng, S. J. H., & Yeoh, E., (2020). Diabetic ketoacidosis precipitated by COVID-19 in a patient with newly diagnosed diabetes mellitus. *Diabetes Research and Clinical Practice, 164*, 1, 2.
11. Chaudharya, N., Kumar, V., Sangwanc, P., Pant, N. V., Saxenae, A., Joshi, S., & Yadavg, A. N., (2021). Personalized nutrition and omics. In: Cifuentes, A., (ed.), *Comprehensive Foodomics* (Vol. *3*, pp. 495–507). Amsterdam – Netherlands: Elsevier Inc.
12. Chantrill, B. H., Coulthard, C. E., Dickinson, L. O., Inkley, G. W., Morris, W., & Pyle, A. H., (1952). The action of plant extracts on a bacteriophage of *Pseudomonas pyocyanea* and on influenza-A virus. *Journal of General Microbiology, 6*(1, 2), 74–84.
13. Castilho, S. D., & Filho, A. A. B., (2010). The history of infant nutrition. *Journal de Pediatria, 86*(3), 179–188.
14. Dhital, A., (2021). *Preparation of Porridge from Germinated Multigrain and its Nutritional Evaluation* (p. 78). B.Sc. (Nutrition & Dietetics) project, Tribhuvan University, Kirtipur, Nepal.
15. Drewnowski, A., Garrett, G. S., Kansagra, R., Khan, N., Kupka, R., Kurpad, A. V., Mannar, V., et al., (2021). Key considerations for policymakers-iodized salt as a vehicle for iron fortification: Current evidence, challenges, and knowledge gaps. *The Journal of Nutrition, 151*(Suppl 1), 64S–73S.
16. Diana, A., Mallard, S. R., Haszard, J. J., Purnamasari, D. M., Nurulazmi, I., Herliani, P. D., Nugraha, G. I., et al., (2017). Consumption of fortified infant foods reduces dietary diversity but has a positive effect on subsequent growth in infant from Sumedang district – Indonesia. *PloS One, 12*(4), E-article ID e01759521.
17. Davis, J. L., Paris, H. L., Beals, J. W., Binns, S. E., Blair, E., & Bell, C., (2016). Liposomal-encapsulated ascorbic acid – influence on vitamin C: Bioavailability and capacity to protect against ischemia-reperfusion injury. *Nutrition & Metabolic Insights, 9*(9), 25–30.
18. Dimitriou, M. E., & Dedoussis, G. V. Z., (2012). Gene – diet interactions in cardiovascular disease. *Current Nutrition Report, 1*(3), 153–160.
19. Ejtahed, H., (2011). Effect of probiotic yogurt containing *Lactobacillus acidophilus* and *Bifidobacterium lactis* on lipid profile in individuals with type 2 diabetes mellitus. *Journal of Dairy Science, 94*(7), 3288–3294.
20. Evans, A., & Matthews, S., (2017). *The Customized Diet: Opportunities in Next-Generation Personalized* Nutrition (Vol. XIX, No. 22, pp. 1–4). L.E.K. Consulting / Executive Insights.
21. Eichler, K., Wieser, S., Ruthemann, I., & Brugger, U., (2012). Effects of micronutrient fortified milk and cereal food for infants and children: A systematic review. *BMC Public Health, 12*(1), E-article 506.

22. Flores, R., & Gillespie, S., (2001). *Health and Nutrition: Emerging and Reemerging Issues in Developing Countries* (p. 23). Washington, D.C.- USA: International Food Policy Research Institute (IFPRI).
23. Floris, R., Lambers, T., Alting, A., & Kiers, J., (2010). Trends in infant formulas: A dairy perspective. In: Griffiths, M. W., (ed.), *Improving the Safety and Quality of Milk: Improving Quality in Milk Product* (Vol. 2, pp. 454–474). Cambridge – UK: Woodhead Publishing Ltd.
24. Ferrocino, I., & Cocolin, L., (2017). Current perspectives in food-based studies exploiting multi-omics approaches. *Current Opinion in Food Science, 13*, 10–15.
25. *Federation IDF Diabetes Atlas*. International Diabetes Federation (IDF), (2020). https://www.diabetesatlas.org/en/resources/ (accessed on 2 November 2023).
26. Horton, S., (1999). Opportunities for investment in nutrition in low-income Asia. *Asian Development Review, 17*(1, 2), 246–273.
27. Hurrell, R. F., Reddy, M. B., Juillerat, M. A., & Cook, J. D., (2003). Degradation of phytic acid in cereal porridges improves iron absorption by human subjects. *American Journal of Clinical Nutrition, 77*(5), 1213–1219.
28. Hussain, M., Li, X., Wang, L., Qayum, A., Liu, L., Zhang, X., Hussain, A., et al., (2021). Recent approaches and methods for the formulation of a risk – free infant formula: Review. *Food Reviews International, 2021*, E-article ID 1901113; doi: 10.1080/87559129.2021.1901113.
29. Hajiabdolbaghi, M., (2020). Efficacy and safety of interferon β-1in the treatment of severe COVID-19: A randomized clinical agents and chemotherapy, antimicrobial agents. *Chemotherapy, 64*(9), 1061–1120.
30. Sharma, N., (2018). *RDA as per ICMR.* https://eduvark.com/rda-per-icmr-528090.html (accessed on 2 November 2023).
31. Michali, N., (2020). *New Product Development in a COVID-Restricted World.* https://insights.figlobal.com/new-product-development/new-product-development-covid-restricted-world (accessed on 2 November 2023).
32. Pray, L., Boon, C., Miller, E. A., & Pillsbury, L., (2010). *Providing Healthy and Safe Foods as We Age: Workshop Summary* (Vol. 4, p. 192). Washington (DC): US National Academies Press.
33. Imran, A., Imran, M., Khalid, M., & Khalid, N., (2021). Phenolic and flavonoid contents in *Malva Sylvestris, Journal of Saudi Chemical Society, 25*(8), 2–6.
34. Jshea, P. O., Holm, R., O'Driscoll, C. M., & Griffin, B. T., (2019). Food for thought: Formulating away the food effect – a PEARRL review. *Journal of Pharmacy and Pharmacology, 71*(4), 510–535.
35. King, H., Aubert, R., & Herman, E., (1998). WHO Global burden of diabetes, 1995–2025: Prevalence, numerical estimates, and projections. *Diabetes Care, 21*(9), 1414–1431.
36. Kazakou, P., Paschou, S. A., Psaltopoulou, T., Gavriatopoulou, M., Korompoki, E., & Stefanaki, K., (2021). Early and late endocrine complications of COVID-19. *Endocrinology Connect, 10*(9), 229–239.
37. Kaur, D., Rasane, P., Singh, J., Kaur, S., Kumar, V., Mahato, D. K., Dey, A., Dhawan, K., & Kumar, S., (2019). Nutritional interventions for elderly and considerations for the development of geriatric foods. *Current Aging Science, 12*(1), 15–27.
38. Kumar, P., & Kumar, M., (2021). Role of vitamins and minerals as immunity boosters in COVID-19. *Inflammo Pharmacology, 29*(4), 1001–1016.

39. Li, B., Yang, J., Zhao, F., Zhi, L., Wang, X., & Liu, L., (2020). Prevalence and impact of cardiovascular metabolic diseases on COVID-19 in China. *Clinical Research in Cardiology, 109*(5), 531–538.
40. Llie, P., Stefanescu, S., & Smith, L., (2020). The role of vitamin D in the prevention of coronavirus disease – 2019 infection and mortality. *Aging Clinical and Experimental Research, 32*(7), 1195–1198.
41. Maughan, R. J., & Shirreffs, S. M., (2012). Nutrition for sports performance: Issues and opportunities. *Proceedings of the Nutrition Society, 71*(1), 112–119.
42. Molnar, J., & Pal, M., (2021). Analysis of functional foods and active ingredients used in sports. *Journal of Nutrition and Food Processing, 4*(5), 1–3.
43. Maslin, K., & Venter, C., (2017). Nutritional aspects of commercially prepared infant foods in developed countries: A narrative review. *Nutrition Research Reviews, 30*(1), 138–148.
44. McCabe-Sellers, B. J., Chenard, C. A., Lovera, D., Champagne, C. M., Bogle, M. L., & Kaput, J., (2009). Readiness of food composition database and food component analysis systems for nutrigenomics. *Journal of Food Composition and Analysis, 22*(3), S57-S62.
45. Macready, A. L., Fallaize, R., Butler, L. T., Ellis, J. A., Kuznes, S., & Frewer, L. J., (2018). Application of behavior change techniques in a personalized nutrition electronic health intervention study: Protocol for the web-based food4me randomized controlled trial. *JMIR Research Protocol, 7*(4), 87–92.
46. Micha, R., Penalvo, J. L., Cudhea, F., Imamura, F., Rehm, C. D., & Mozaffarian, D., (2017). Association between dietary factors and mortality from heart disease, stroke, and type 2 diabetes in the United States. *Journal of American Medical Association, 317*(9), 912–924.
47. Maestre, M. M., Arias, A., Mata-Vazquez, E., Martın-Toledano, M., Ruiz-Chicote, A. M., (2021). Long-term outcomes of patients with coronavirus disease – 2019 at one year after hospital discharge. *Journal of Clinical Medicine, 10*(13), 29–45.
48. Nasirpour, A., Scher, J., & Desobry, S., (2006). Baby foods: Formulations and interactions (a review). *Critical Reviews in Food Science and Nutrition, 46*(8), 665–681.
49. Oluseyi, A. K., Oluwafunmilola, A., Oluwasegan, S. T., Abimbola, A. A., Oluwatoyin, A. C., & Olubusola, O., (2013). Dietary fortification of sorghum-ogi using crayfish (*Paranephropsplani frons*) as supplements in infancy. *Food Science and Quality Management, 15,* 1–13.
50. Onweluzo, J. C., & Nwabugwu, C. C., (2009). Development and evaluation of weaning foods from pigeon pea and millet. *Pakistan Journal of Nutrition, 8*(6), 725–730.
51. Ordovas, J. M., Ferguson, L. R., Tai, E. S., & Mathers, J. C., (2018). Personalized nutrition and health. *British Medical Journal, 361*, 1–7.
52. Page, G., Laight, D., & Cummings, M., (2011). Thiamine deficiency in diabetes mellitus and the impact of thiamine replacement on glucose metabolism and vascular disease. *International Journal of Clinical Practice, 65*(6), 684–690.
53. Puiggros, F., Sola, R., Blade, C., Salvado, M. J., & Arola, L., (2011). Nutritional biomarkers and foodomic methodologies for qualitative and quantitative analysis of bioactive ingredients in dietary intervention studies. *Journal of Chromatography A, 1218*(42), 7399–7414.
54. Pradeepa, V., (2018). Consumers' expectations in baby food products with reference to Indian market. *Paripex – Indian Journal of Research, 7*(1), 60, 61.

55. Ramachandran, A., Snehalatha, C., Shetty, A. S., & Nanditha, A., (2012). Trends in prevalence of diabetes in Asian countries. *World Journal of Diabetes, 3*(6), 110–117.
56. Sharma, D., Sood, S., Verma, R., & Thakur, A., (2021). Development and storage stability of multi seed bars for sports persons. *Himachal Pradesh Journal of Agricultural Research, 47*(1), 66–76.
57. Semwal, A., Singh, A., Khan, C., & Shahi, N. C., (2015). Quality assessment of prebiotics mix from fermented cereal- legume blends. *International Journal of Agriculture, Environment and Biotechnology, 8*(1), 49–59.
58. Singh, S., & Dunkwal, V., (2020). Development and nutritional evaluation of infant food premix to improve the nutritional status of the malnourished infants (06-24 months). *The Indian Journal of Nutrition and Dietetics, 57*(4), 439–449.
59. Shiriki, D., Igyor, M. A., & Gernah, D. I., (2015). Nutritional evaluation of complementary food formulations from maize, soybean and peanut fortified with *Moringa oleifera* leaf powder. *Food and Nutrition Sciences, 6*(5), 494–500.
60. Satter, M. A., Jabin, S. A., Abedin, N., Arzu, T., Mitra, K., Abdullah, A. M., & Paul, D. K., (2013). Development of nutritionally enriched instant weaning food and its safety aspects. *African Journal of Food Science, 7*(8), 238–245.
61. Smith, J., & Charter, E., (2010). *Functional Food Product Development* (1st edn., Vol. 1, pp. 178–355). Chichester – UK: Wiley-Blackwell Publishing Ltd.
62. Tiwari, S., Pratyush, D., Gupta, B., Dwivedi, A., & Chaudhary, S., (2013). Prevalence and severity of vitamin D deficiency in patients with diabetic foot infection. *British Journal of Nutrition, 109*(1), 99–102.
63. Takehiko, Y., & Hiromu, K., (2012). The effect of food on the oral bioavailability of drugs: A review of current developments and pharmaceutical technologies for pharmaco kinetic control. *Therapeutic Delivery, 3*(1), 81–90.
64. Tiencheu, B., Achidi, A. U., Fossi, B. T., Tenyang, N., Ngongang, E. F. T., & Womeni, H. M., (2016). Formulation and nutritional evaluation of instant weaning foods processed from maize (*Zea mays*), pawpaw (*Carica papaya*), red beans (*Phaseolus vulgaris*) and mackerel fish meal (*Scomber scombrus*). *American Journal of Food Science and Technology, 4*(5), 149–159.
65. United Nations (UN), (2019). *World Population Ageing 2019: Highlights* (pp. 1–5). https://www.un.org/en/development/desa/population/publications/pdf/ageing/WorldPopulationAgeing2019-Report.pdf (accessed on 13th May 2022).
66. Van, S. H. P., Bast, A., & DeBoer, A., (2021). Immuno-modulating effects of fungal beta-glucans: From traditional use to medicine. *Nutrients, 13*(4), E-article ID 1333.
67. World Health Organization (WHO)/ Food Agriculture Organization (FAO), (2005). *Vitamin and Mineral Requirements in Human Nutrition* (2nd edn., p. 341). Bangkok – Thailand: WHO/ FAO.
68. Webster, A., (2018). *Personalized Nutrition: Ready for Prime-time*? https://foodinsight.org/personalized-nutrition-ready-for-prime-time (accessed on 14 April 2022).
69. Yen, C., Rei, L. K., & Shin-Ru, S., (2020). COVID-19: The first documented coronavirus pandemic in history. *Biomedical Journal, 43*(4), 328–333.

PART III

ANTIDIABETIC ACTIVITIES OF PLANTS

CHAPTER 7

Potential Therapeutics of Cultivated and Wild Mushrooms Against Diabetes

Anup Kumar Sarkar and Mallika Mazumder

ABSTRACT

Diabetes is a potentially fatal multi-factorial metabolic condition characterized by hyperglycemia induced by insulin insufficiency and insulin malfunction. Presently most of the researchers have set their goal to formulate natural anti-diabetic drugs that are easy to use and without any side effects. Among the natural sources of anti-diabetic food edible mushrooms have gained much of attention by the researchers of several corners of the world. They have been acknowledged as a nutritious food since ancient times, and as civilization progressed, some of them have appreciated as therapeutic as well as pharmacological reinforcing agents. An array of physiologically active chemicals with anti-diabetic properties can be found in various edible and medicinal mushrooms. Clinical trials of many cultivated mushroom extracts have been shown to reduce the level of fasting plasma glucose level significantly. Among the huge active components of mushroom only a few have been identified and many of them are still to be characterized. Thus, there is a need of gradual clinical trial of different mushroom extractions. In this chapter, we have enlightened importance of mushrooms as anti-diabetic agents along with the future prospects, the obstacles and limits that can impact the development of mushroom based anti-diabetic medications.

7.1 INTRODUCTION

Globally diabetes is now considered as the most problematic health issues resulting from several unhealthy practices like intake of junk food, lack

The Functional Foods: Nutrient and Health Benefits.
Megh R. Goyal, Junaid Ahmad Malik, and Anu Kumari (Eds.)

of exercise along with some environmental and genetic factors. Diabetes develops as a result of autoimmune destruction of β cells of pancreas, subsequent insufficiency of insulin secretion and associated abnormalities in metabolic activities, resulting in aberrant glucose, lipid, and protein metabolism [63]. Postprandial rise in glucose is the predominant feature of a diabetic patient. Uncontrolled diabetes can cause both short-term as well as long lasting problems like neuropathy, retinopathy, amputation, malfunctioning of organs, and injury in vascular system, which increases the risk of cardiovascular disease (CVD) [35]. Diabetes along with high blood glucose levels are connected with a variety of cardiovascular problems, which together responsible for significant amount of morbidity and death among diabetic patients [13]. Comprehensive evidence indicating that dietary habits of people are important determinants of their health-related problems. Proper dietary ingredients can prevent a person against diabetes and similar disorders and the difficulties which might arise from the diseases. Diabetic patients are unable to make or proper utilization of insulin in the body, and as a result of this disparity in insulin level, the amount of glucose in their blood is too high. Diabetes can be controlled by consuming those foods that retain significant amounts of α-amylase and α-glycosidase inhibitors. Diabetes and obesity may be easier to treat if the enzymes α-amylase and α-glycosidase are stopped from working overtime [78]. In human system the α-amylase enzyme catalyzes the hydrolysis of α-1-4 glycosidic bonds from the non-reducing ends of polysaccharides (amylase, amylopectin, and glycogen) to produce maltose units [101]. α-amylase inhibitors and α-glycosidase inhibitors bind with the enzyme and form enzyme-inhibitor complex to inhibit α-amylase activity and α-glycosidase activity, respectively, which in turn resulted the slow digestibility of starch and reduces rate of glucose absorption. Many wild and cultivated mushrooms contain such inhibitors along with some other anti-diabetic substances. α-Glycosidase is situated in the brush border cells of intestine that converts oligosaccharides and disaccharides in to monosaccharides and it improves carbohydrate absorption as well as contributes to the increase in blood sugar content [9]. α-Glycosidase inhibitors can limit the breakdown of disaccharides and the absorption of glucose by delaying carbohydrate absorption through competitive inhibition [138]. The use of α-glycosidase inhibitors not only limit carbohydrate breakdown and but also results rise in blood glucose after meals. The use of α-glycosidase inhibitors from natural sources, such as mushrooms, has recently received a lot of attention.

In present era mushrooms are widely regarded as the treasure of natural resources. Mushrooms or macrofungi are the group of organisms from the division Ascomycota and Basidiomycota of the kingdom Fungi which, under certain circumstances, produce a fruiting body large enough to be seen with the naked eye or to be touched, that grow beneath the ground [121, 125]. Some of them are cultivable but some are still uncultivable but all of them have significant pharmaceutical values. In most cases mushroom dry matter is comprising roughly 10–40% protein, 2–8% fat, 3–28% carbohydrate, and 3–32% fiber [15]. Edible mushrooms have been appreciated as a valuable food for human due to their low-calorie content and immune-boosting properties [119]. In addition to dietary applications, mushroom derived products have specific functions in the ailments of several health issues including diabetes [67]. Since a long time, mushroom and their products have been recommended as anti-diabetic in several regions of the world [112, 120]. Several *in vivo* and *in vitro* studies have reported that many edible and wild mushrooms contain antioxidant mycochemicals [59]. Creditable research on this topic reveal that different extract and isolated components of various mushrooms can reduce diabetes and allied symptoms. The bulk of investigations on the anti-diabetic efficacy of mushrooms have utilized either the fruiting bodies or the mycelia, with a few focusing on components isolated from the culture medium during submerged fermentation. In a randomized placebo control study, it was discovered that therapy with mushroom derived natural compounds for three months decreased resistance of insulin in diabetic individuals of age group 20 to 75 [49].

7.2 DIABETES: A LIFE-THREATENING DISORDER

Diabetes represents a group of physiological illness distinguished by abnormally elevated sugar level in blood caused by insulin imbalances. Through a sequence of pathological alterations, the disease also promotes to a number of complications in circulatory system, excretory system and sense organs [133, 152]. According to epidemiological research, risk factors for the disease as well as its consequences include hypercholesterolemia and hyperglycemia, both of which are of which are primarily correlated with food habit [57, 135]. Delayed and disturbed wound healing in diabetes mellitus patients is a rising global issue linked with considerable morbidity and economic burden. This disease is usually categorized into two sub-categories: Type 1 diabetes Melitus (T1D) and Type 2 Diabetes Mellitus (T2D). Former one is characterized by insulin insufficiency, which occurs due to malfunctioning of the pancreas more specifically β-cells and patients suffering from this condition are completely reliant on exogenous insulin sources [105]. These groups were classified primarily according to the presence (in case of T1D) or lack (in case of T2D) of auto-antibodies against pancreatic islet-cell antigens [153]. T1D can be treated in a variety of medications including injection of insulin. In comparison, T2D is insulin-independent and may be managed with dietary modifications, exercise, and medication. In T2D, the insulin signaling pathways become dysregulated and insulin losing its ability to inhibit gluconeogenesis or stimulate glycogenesis. Insulin resistance is a major pathophysiological component in T2D, and such kind of resistance develops primarily in adipocytes, hepatocytes, and skeletal muscle over the course of T2D development [117]. With the advancement of science researchers have found a third category of diabetes that comprised of adult individuals with latent autoimmune diabetes [153]. In present condition five types of diabetes have been categorized namely – severe autoimmune diabetes, juvenile diabetes (formerly known as T1D), Severe Insulin-Deficient Diabetes (T2D), Severe Insulin Resistant Diabetes, Mild Obesity-related Diabetes and Mild age-related diabetes [2]. Hyperglycemia and alteration of lipoprotein metabolism are the basic features of all types of diabetes [37]. International Diabetes Federation clearly indicated the status of diabetes in the 9th edition of Global Diabetes Map and according to this, adults with diabetes (20–79 years) globally reached 415 million in 2017 and moreover estimates predict that the number will be reach 629 million within 2045 [153, 179]. Diabetes mellitus is presently treated with a variety of hypoglycemic medications that produce anti-diabetic effects via diverse mechanisms [39]. Diabetes has been intimately correlated with the activities of free radical, as these free radicals produce lipid per-oxidation, which is critical in atherosclerosis, aging, and late complications of diabetes mellitus [129]. In the diabetic patient endothelial cells and the cells that are significantly affected by hyperglycemia, are unable to decrease glucose absorption or alter the pace of glucose transport, exposing them to excessively high glucose circumstances and resulting in diabetic problems linked with vascular injury [37]. In the present scenario it is necessary to create innovative dietary adjuncts and anti-diabetic medications that restore a normal metabolic milieu, therefore decreasing the long-term consequences of the disease.

Undoubtedly diabetes is a serious concern about public health issue, and urgent solutions are needed to reduce or even reverse this trend by investing in controllable variables such as food, physical activity, and weight management [27]. On the basis of several epidemiological studies, World Health Organization (WHO) has confirmed that tobacco smoking is one of the preventable risk factors for T2D and recommends smoking cessation/avoidance as part of their lifestyle [100]. Some researchers have also enlightened the impact of pollution, especially of air pollution on impaired glucose metabolism along with insulin resistance [80, 90, 114]. Before the introduction of therapeutic use of insulin, diet was the principal way of treatment of the diabetes, and dietary treatment for the disease included the use of traditional remedies mainly derived from natural resources such as plants and mushrooms. Mushrooms play a significant role in decreasing stress created by internal and external factors and the stress relief process is a key determinant of long and healthy life [116]. Advancement of biochemistry and molecular biology of mushroom derived secondary metabolites can provide the information regarding underlying mechanisms for their anti-diabetic efficacy (Figure 7.1).

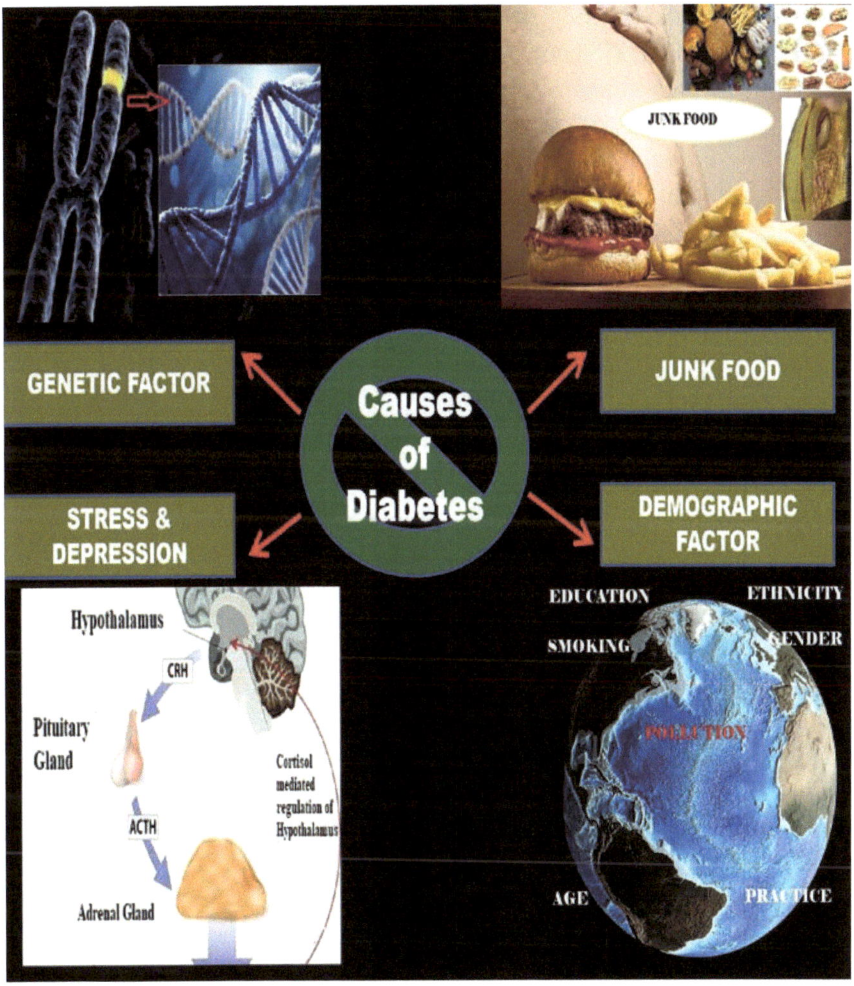

FIGURE 7.1 Several factors responsible for diabetes.

7.3 MYCOLOGICAL, NUTRITIONAL, AND COMMERCIAL ATTRIBUTES OF MUSHROOM

The term mushroom is popularly representing the fungal species that are taxonomically belongs to the orders Basidiomycetes and Ascomycetes under kingdom Fungi. Mushrooms have long been regarded as the most valuable natural resources of the globe due to their capacity to convert necessary input into nutrient-dense, high-protein diet. Technically these are fruiting bodies or spore producing structure of these fungal species which have the capability to degrade organic remains like wood, fallen leaves, dead bodies, compound organic debris and through the decomposition nutrients are recycled back in to the system [38]. The edible mushrooms are widely recommended as healthy food for humans of all age groups due to their nutritional and culinary values [125]. Mushroom diet is also beneficial for diabetic patients because of the low-fat content with higher amounts of unsaturated fatty acids and complete absence of cholesterol [137]. Clinical investigations also indicate that nutraceuticals derived from wild and cultivated mushrooms should be regarded as natural medications with low or no negative effects.

Although mushrooms have been used by humans for generations due to their low calorific value along with the presence of salt and fat content in appropriate amount, but large-scale commercialization of mushroom emerge out around the early 1900s [26, 123, 136]. Numerous mushrooms have been utilized by the Indian subcontinent and China for centuries, not only as food but also to cure diabetes, with the majority of these mushroom and their extracts containing compounds that inhibit Human Pancreatic Amylase [56]. Mushrooms have therapeutic qualities as they retain huge number of bioactive compounds having low molecular weight (LMW). Among the bioactive anti-diabetic compounds of mushrooms dietary polysaccharides, enzyme inhibitor proteins, phenolics, and some terpenoids are widely recognized. Dietary polyphenols derived from several mushrooms have been extensively studied as anti-diabetic drugs in cell, animal, human, and clinical trials. The molecular processes underpinning the relationship between mushroom consumption and a reduced incidence of diabetes might be attributed to the mushrooms' inherent antioxidant qualities, which are attributed to the mushroom components ergothioneine and glutathione [4]. Evidence shows that ergothioneine, vitamin D, α-glucan, and selenium have beneficial impacts on immunological system, intestinal function, and weight control.

Mushroom polysaccharides are a novel class of anti-diabetic medicines. Dietary polysaccharides derived from mushrooms offer a wide range of pharmacological effects and are almost free from toxicity [32]. Dietary polysaccharides include the naturally occurring polysaccharide molecules that are recommended for edible purposes in daily life, and they have favorable effects on glucose homeostasis. Remarkable number of *in vivo* as well as *in vitro* experiments confirmed that these molecules have lipid-lowering effect and anti-diabetic complications (Table 7.1). Additionally, dietary non-digestible polysaccharides derived from mushroom foods are well-known modulators of the gut microbiota, nourishing specific beneficial microorganisms in the human gut. *In vivo* studies on animal model indicated that these bioactive polysaccharides could help to prevent T2D through reducing oxidative stress. Some oligosaccharides present in mushrooms are also reported as effective anti-diabetic medication since they stimulate insulin synthesis in pancreatic islet cells in diabetic model animals [6]. Numerous investigations on the digested protein of the common mushroom demonstrated that it contained natural functional characteristics for reducing oxidative stress and indicated that it may be used in the food sector as an alternative natural antioxidant [76]. Presently it was well established that some proteins like AMP-activated protein kinase (AMPK) and cAMP-responsive element binding protein (CREB) critically regulate energy homeostasis through increasing cellular glucose uptake via an insulin-independent pathway and thus they are being investigated as a novel drug target for diabetes treatment. Many mushrooms have found to play significant role on AMPK and CREB-mediated insulin biosynthesis [102]. PTP1B, which plays a prominent role in the negative feedback of insulin signaling, is a recently identified molecular target for the treatment of diabetes, and certain mushrooms are discovered to dramatically inhibit PTP1B [8]. In diabetic rats, mushroom derived phenolic and terpenoid compounds were also found to exhibit anti-hyperglycemic and antioxidant activities [36] (Figure 7.2).

7.4 WILD AND CULTIVATED MUSHROOMS WITH POTENTIAL ANTIDIABETIC ACTIVITIES

Mushrooms have long been part of the human diet and several ancient texts published in oriental nations explain the curative qualities of fungus species. The therapy of diabetes relies heavily on methods for controlling and preventing hyperglycemia. Some of them have medicinal properties due to the presence of secondary metabolites such as polysaccharides, lectins, lactones, terpenoids, alkaloids, antibiotics, and metal-chelating agents. They have been found to contain many compounds that aid in the proper

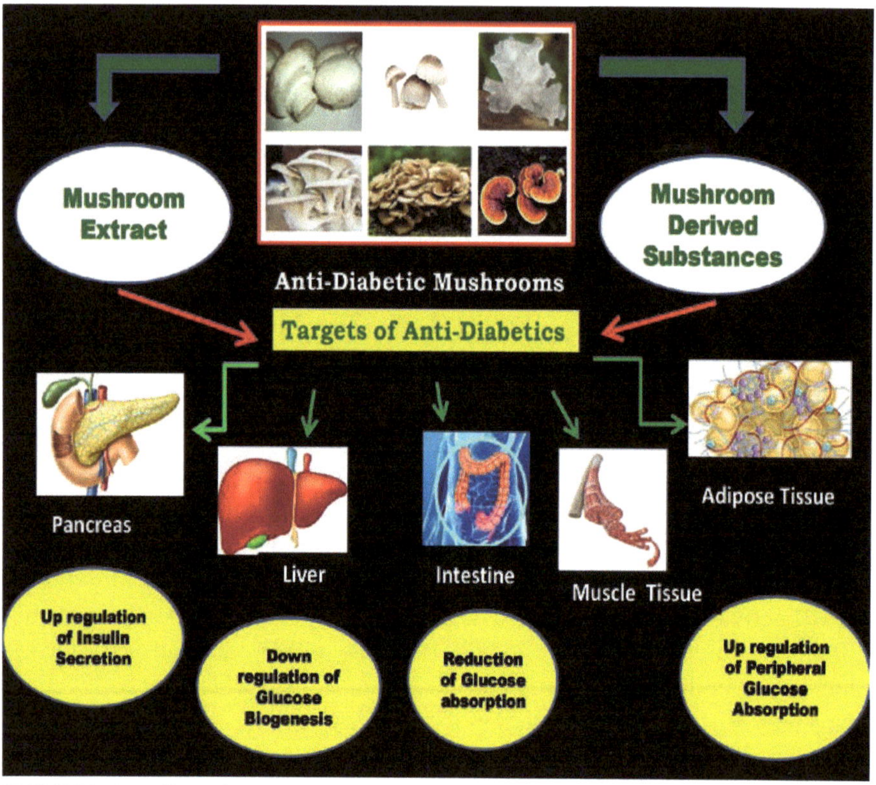

FIGURE 7.2 Effect of anti-diabetic mushrooms on glucose metabolism and diabetes.

functioning of the body parts like liver, pancreas, and other endocrine glands, stimulating the production of insulin and other hormones required for regular metabolic function [16, 149, 151, 173]. Mushroom polysaccharides appear to have a vital role in the treatment of diabetic problems, according to a large body of evidence [63]. The majority of carbohydrate is made up of dietary and digestible fibers and it contains no starch and only a trace of sugar [68]. Mushroom and their metabolite derived prospective anti-diabetic compounds represent a promising source of novel drugs with significant pharmacological and clinical potential. Glucans, dietary fibers, chitin, and their derivatives are primarily responsible for these functional qualities [125]. Consumption of mushrooms significantly reduces total cholesterol, total triglyceride, and low-density lipoproteins, while increasing high-density lipoproteins [84]. Remarkable numbers of experiments on model animals have confirmed that consuming nutraceuticals, particularly those with strong antioxidant capacity, has an inverse connection with CVD, different malignancies, and diabetes (Table 7.1). The importance of animal models in experimental research in answering numerous biological and biomedical problems and increasing human understanding cannot be overstated. Some animals like mice, rabbit, etc., evolutionary closeness to humans, and thus they are utilized in the area of biology. Mushrooms are also high in the antioxidants ergothioneine and glutathione, which contribute significantly to the prevention of chronic illnesses and death [22]. Mushroom derived glucans are the physiologically active homopolysaccharide that are remain unaffected by pancreatic enzymes [29]. According to majority of researchers, it is conceivable to determine the potential role of these polysaccharides on diabetes and some of these are – (i) up-regulation of insulin level in plasma and down regulation of pancreatic glucagon; (ii) acceleration of sensitivity of insulin receptors, which in turn decreases insulin resistance; (iii) enhancement of glycogen synthesis in liver; (iv) optimizing the utilization of glucose in peripheral tissues; (v) anti-inflammatory activity; and (vi) antioxidant and antiradical action [91, 139, 146]. Mushroom polysaccharides may reduce diabetes by reducing glucose absorption efficacy and postprandial glycemia via gastrointestinal (GI) viscosity mechanisms [66]. Mushroom derived polysaccharides, chiefly β-glucans can effectively restore the activity of several tissues of pancreas and stimulate β-cell to release insulin, thereby bringing down sugar level in blood as well as enhance the sensitivity of the peripheral tissue to circulate insulin [14]. Additional benefits include a reduced lipid profile devoid of cholesterol and a greater amount of polyunsaturated fatty acids [68]. Consumption of mushrooms significantly enhances high density lipoprotein and simultaneously reduces cholesterol, triglycerides, and low-density lipoprotein [62, 149]. Neyrinck et al. [106] found that fugal derived chitin-glucan complex regulate rebalancing of gut microbiota [106]. The complex also modulates glucose and lipid metabolism in obese mice fed a high fat diet. Jayasuriya et al. [61] did extensive research on the anti-diabetic activities of *Pleurotus ostreatus* and *Pleurotus cystidiosus* and identified a possible mechanism of action [61]. According to the findings, the hypoglycemic effect is linked to an increase in glucokinase activity and insulin secretion stimulation, which ultimately increase glucose consumption and glycogen synthesis by peripheral tissues. Certain compounds obtained from mushrooms found to lower postprandial blood glucose levels in db/db mice via increased insulin responsiveness, regardless of PPAR-γ [118].

Apart from that, they retain several important minerals in appropriate quantities and such minerals are utilized for the normal functioning of metabolic processes including diabetes related pathways when consume. Some mushrooms are rich in trace element and have been found to reduce sugar level and increase glycogen synthesis [43, 98]. Pronounced hypoglycemic effect is also noted for powder obtained from submerged culture medium of some mushrooms. Many studies found that these powders can regulate blood glucose levels and simultaneously maintain the level of glycosylated hemoglobin [139]. For example, *Coprinus comatus*, a vanadium-rich fermented fungus, was investigated for its hypoglycemic action in Alloxan- and adrenalin-induced hyperglycemic rats, where it was found to have a potent anti-hyperglycemic effect [11, 92]. Thus, at this time proper knowledge regarding mushroom and their proper use can solve many problems associated with diabetes. There are many mushroom derived remedies suggested for diabetes and diabetic complications. A list of edible and medicinal mushroom with anti-diabetic and related beneficial effects is given in Table 7.1. Proper knowledge regarding this natural treasure along with self-management education, monitoring of glycemic parameters, regular exercise and stress less life style can protect us from the risk of diabetes [21].

Apart from that, some of them include important minerals in appropriate amount and recommended as necessary food for the normal functioning of a variety of metabolic processes along with diabetes related pathways. Trace element-rich mushrooms have also been reported to reduce blood glucose and enhance glycogen levels [43, 98]. Powder obtained from submerged culture medium of some mushrooms also have a pronounced hypoglycemic effect manifesting in the normalization of blood glucose levels and content of glycosylated hemoglobin [139]. The hypoglycemic activity of fermented fungus, *Coprinus comatus*, was evaluated in Alloxan and adrenalin-induced hyperglycemic mice, and it was revealed that *C. comatus*, which is high in vanadium, had a strong anti-hyperglycemic impact [11, 92] (Figure 7.3).

7.5 ADVANTAGE MUSHROOM DERIVED ANTIDIABETIC DRUG OVER CONVENTIONAL DRUG

The paradigm of modern medicine is now shifting toward individualized care and the use of natural therapies such as biocompounds. As the nutritional value of mushroom exceeds the values present in most fruits and vegetables, they are effectively recommended for diabetes [116]. It is now seeking for natural remedies due to a variety of constraints associated with the usage of existing synthetic anti-diabetic medicines. Although chemical and biological hypoglycemic medications such as insulin, phenoformin, repaglinide, rosigitazone, tolbutamide, and troglitazone are recommended for treating

TABLE 7.1 Anti-Diabetic Potential of Mushroom and Mushroom Derived Substances

Mushroom	Active Compounds/Dose	Antidiabetic Effect	References
Agaricus bisporus	Dehydrated fruiting body.	Lowering of glucose and cholesterol in blood.	[62]
	Aqueous extracts.	Reduction of hyperglycemia; colonization of β-cells of pancreatic islets of Langerhans; enhances insulin level.	[159]
	Glycoprotein	Proliferation of pancreatic β-cells.	[148]
	β-glucan	Enhance the insulin secretion from pancreas and reduce the body weight of diabetic mice.	[162]
Agaricus brasiliensis	Powdered biomass during for 2 months.	Formation of mass of pancreatic β-cells in diabetic rats.	[107]
	β-glucans and their enzymatically hydrolyzed oligosaccharides.	Reduce the level of glucose cholesterol, triglyceride; anti-arterioscle-rotic activity.	[75]
Agaricus campestris	Filtrate suspension in dilute dose.	Down regulate hyperglycemia and enhance the transport of glucose.	[40]
Agaricus sylvaticus	Aquatic extract	Reduce cholesterol and triglyceride; antioxidant activity.	[28]
Agrocybe aegerita	Glucan	Modulate glucose in normal and diabetic mice.	[167]
Agrocybe chaxingu	Polysaccharides	Protect pancreatic β-cells.	[82]
Amillariella mella	Extract in NAOH solution.	Down regulate the fasting blood glucose, serum triglyceride and enhance glucose intolerance.	[165]
Antrodia cinnamomea	95% ethanolic extract.	Enhance glucose tolerance; lowered the total cholesterol and triglyceride; decrease lipid peroxidation; the expression of RAGE on plasma membrane and enhanced the superoxide dis-mutase activity.	[64, 170]
Antrodiella albocinnamomea	Antroalbol H	Up regulate GLUT4 translocation, albocinnamomea APMK activation by phosphorylation of LKB1 at Thr 189.	[143]

TABLE 7.1 *(Continued)*

Mushroom	Active Compounds/Dose	Antidiabetic Effect	References
Astraeus hygometricus	Ethanolic extract of fruiting bodies.	Reduce blood sugar level and enhance tolerance to glucose in alloxan induced diabetic mice.	[10]
Auricularia auricular-judae	Dried mycelial powder.	Reduce glucose, cholesterol, and triglyceride level in plasma.	[75]
	Polysaccharide	Modulation of antioxidative system as well as NF-kB-related signaling.	[50]
	Water soluble polysaccharide.	Down regulate plasma and urinary glucose and up regulate hepatic glycogen level. Reduction of overall body weight, urinary glucose, serum glucose, serum insulin and HbA.	[168]
Cerrena unicolor	Extracellular polysaccharides.	Down regulates glucose level in serum.	[156]
Calvatia gigantea	Dichloro-methane ethyl-acetate butanol aqueous.	Reduce 28% blood glucose; α-amylase inhibitory activity.	[109]
Catathelasma ventricosum	Selenium polysaccharide.	Normalize physiological damage in diabetic rats by protecting liver, kidney pancreas and also down regulates lipids.	[95]
Coprinus comatus	Methanolic extract.	Inhibition of α-amylase and α-glucosidase enzyme.	[134]
	Comatin	Reduces the level of triglycerides, cholesterol, fructosamine, etc., and maintains glucose level in blood.	[20]
Coriolus versicolor	Extract	Increases activity of antioxidant enzymes and reduces the amount of thiobarbituric acid reactive substances.	[169]
	Ternatin	Suppress biosynthesis of fatty acid in KK(Y) mice and 3T3L1 cell line by regulating PPAR-Y, SERBP-1C, FAS, ACC, and down regulate hyperglycemia.	[77]
Cordyceps militaris	Water and polysaccharidic extracts.	Down regulates diabetic symptoms in mice model.	[73, 93]
	Methanolic extract.	Inhibitory activity on α-glucosidase.	[134]

TABLE 7.1 *(Continued)*

Mushroom	Active Compounds/Dose	Antidiabetic Effect	References
	Acidic-extractable polysaccharides.	Antioxidant and hypoglycemic effects.	[178]
	Lipopolysaccharide and cordycepin.	Reduces diabetic symptom in LPS induced NFKB p^{65} in RAW 264.7 cells.	[130]
Cordyceps sinensis	Polysaccharide	Hypoglycemic activity in mice having inheritable diabetes.	[41, 45]
	CSP-1 fraction of polysaccharide.	Induce pancreatic cell to induce insulin and enhance insulin level in blood.	[86]
Cordyceps taii	Polysaccharides	Reduce lipid and glucose in case of streptozotocin induced T1D in mice model.	[93]
Ganoderma lucidum	Ganoderic acid C2, Ganoderenic acid A, Ganoderic acid Df.	Down regulates the enzyme aldose reductase.	[24, 99]
	Ganoderol B	Down regulates α-glucosidase.	[11]
	Ethanolic extract of powder obtained from fruiting body.	Reduce glucose level in serum of white mice model.	[122]
	Aquatic extract and extract in 80% ethanolic solvent.	Reduce the concentration of serum glucose in streptozotocin-induced diabetic mice.	[5]
Ganoderma lingzhi	Ethanolic extract	Down regulates α-glucosidase.	[128]
Ganoderma leucocontexum	Ganoderone A, Ganoleuconin E, M, N, P.	Down regulates α-glucosidase.	[146]
	Ganoleuconin A, C, F, J, K, L, M, N, Ganoderiol	Down regulates HMG-CoA reductase.	[146]
	Ganoleucin D, spiroapplanatumine L, (±)-Spirolingzhine A, Spirolingzhine	Down regulates aldose reductase.	[174]

TABLE 7.1 *(Continued)*

Mushroom	Active Compounds/Dose	Antidiabetic Effect	References
	Ganoleucin D, Ganomycin I	HMG-CoA reductase.	[145]
Grifola frondosa	Ethanolic extract	Regulate PPARS target genes in skeletal muscle, glucose intolerance, decrease the blood cholesterol level. Enhance glucose uptake under sodium palmitite induced insulin resistant cell line.	[3]
	Polysaccharide	Influence insulin synthesis; reduce the concentration of glucose and similar compounds in serum of diabetic rats; normalize inflammation regulating factors in diabetic rats.	[79]
	Polysaccharide	Reduce blood sugar level by suppression of glucose absorption from intestinal cell by creating abundant fiber volume.	[97]
	Glucose, mannose, galactose, xylose, arabinose, ribose	Control fasting serum insulin and fasting serum glucose level. Target PI3-AKT pathway.	[155]
	MT-α glucan	Increase the amount of low affinity insulin receptors on the liver membrane; decrease serum lipid level in mice. Improve peripheral insulin resistance and sensitivity to insulin.	[75]
Grifola gargal	Hot aqueous extraction.	Decrease blood glucose and body fat of diabetic mice.	[44]
Hericium erinaceus	Methanolic extract.	Reduces elevation rates of blood glucose levels.	[144]
Hypsizygus ulmarius	Aqueous extract.	Down regulate gluconeogenic enzymes in STZ treated diabetic mice model.	[103]
Inonotus obliquus	Water extracts and polysaccharidic fractions.	Inhibitory potential on α-amylase and α-glucosidase.	[134, 141]
	Hot aqueous extraction.	Enhance GLUT 4 expression in 3T3-L1 pre adipocyte cell.	[65]
	Extracts in hexane, chloroform, ethyl acetate, methanol.	Anti-diabetic activity.	[34]

TABLE 7.1 *(Continued)*

Mushroom	Active Compounds/Dose	Antidiabetic Effect	References
Lactarius deterrimus	Ethanolic extract	Improve hyperglycemic condition in STZ induced rats via protecting pancreatic cells and beta cell mass by activation of CXCL12/Akt pathway.	[54, 104]
Laetiporus sulphureus	Extracellular polysaccharides.	Increased the insulin antigenesity via proliferation or regeneration of diabetic islet β-cells.	[53]
Lentinula edodes	Mycellial exopolymer	Reduce the level of glucose, triglycerides, and cholesterol in blood.	[161, 162]
	Exopolymer	Improve plasma insulin level, repair damage of pancreatic β cells, up regulation of insulin synthesis, lowers plasma glucose, plasma cholesterol and triglycerides.	[60]
Lentinus strigosus	Crude exopolysaccharides from a submerged culture.	Dose-dependent reduction in serum glucose levels and regeneration in the pancreatic islets of Langerhans.	[158]
Lentinus tigrinus	Lyophilized hot water extraction.	Lowers blood glucose level in mice.	[23]
Morchella conica	Methanolic extract.	Inhibitory potential on α-amylase and α-glucosidase.	[8, 134]
Ophiocordyceps sinensis	Polysaccharide	Reduce 60–70% of blood glucose in STZ induced diabetic mice.	[172]
Paecilomyces japonica	Exo-biopolymer from submerged mycelial cultures.	Hypoglycemic effects.	[163]
Phellinus baumii	Oral administration of extracellular polysaccharide for two weeks.	Down-regulates AST, ALT, serum tri-glyceride in STZ induced diabetic rat. Decrease blood glucose by 32% in diabetic mice model.	[54, 166]
	Mushroom extract in polar solvent.	Protect the β-cell from damage.	[164]
Phellinus ribis	Polychlorinated compounds.	Stimulate PPAR-γ agonistic activity.	[83]

TABLE 7.1 *(Continued)*

Mushroom	Active Compounds/Dose	Antidiabetic Effect	References
Phellinus linteus	Polysaccharide	Hypoglycemic properties in alloxan-induced diabetic mice.	[134, 177]
		Inhibits autoimmuno diabetes through regulating cytokine expression.	[73]
	Hot aqueous extraction.	Down regulate serum glucose and up regulate number of Langerhans islets.	[158]
	α linked glycosyl residues.	Reduces blood sugar level in diabetic mice model.	[134]
	Polysaccharide	Significantly reduce sugar concentration in blood.	[96]
	Hispidin	Improve the condition of β-cell, enhance insulin secretion via cytoprotection against oxidative damage in H_2O_2 treated RINm5F cell line.	[60]
	Derivatives of hispidin.	Act as natural aldose reductase inhibitors.	[83, 85]
Pleurotus abalonus	$[\rightarrow 6)$-α-D-Glcp-$(1\rightarrow]_n$	Lowers blood glucose level in diabetic mice.	[142]
Pleurotus albidus	70% ethanolic extract.	Modulate complex 1 activity by regulating SIRT3 and decrease the ROS production, oxidative damage of lipid and protein.	[31]
Pleurotus citrinopileatus	Methanolic extract.	Regulate blood glucose level, insulin, catalase, stimulate insulin, production from pancreas in STZ induced rat.	[126]
	Water soluble polysaccharide.	Reduce fasting blood sugar level in diabetic mice.	[50]
Pleurotus cystidiosus	Freeze-dried powders.	Enhance insulin level T2D patient and down regulate glucose in serum.	[61]
Pleurotus djamor	Zinc polysaccharide.	Increase anti-diabetic activity in STZ induced diabetic mice.	[175]
Pleurotus eryngii	Aqueous extract.	Significantly decrease blood glucose and HbAlc in diabetic mice model.	[87]

TABLE 7.1 *(Continued)*

Mushroom	Active Compounds/Dose	Antidiabetic Effect	References
	Mycelial extract	Enhance insulin secretion and reduce blood sugar level; up regulate glycogen synthesis; restoration of pancreatic β-cells in alloxan treated mice model.	[12]
	Extracellular polysaccharide.	Inhibition of α-amylase and α glucosidase; decrease of blood sugar level.	[171]
Pleurotus florida	Polysaccharide	Down regulate glucose, cholesterol, and triglyceride in serum; restoration of SOD and catalase.	[33, 124]
Pleurotus giganteus	Methanolic extract.	Increase adipogenic differentiation and glucose uptake by GLUT1 and GLUT4 expression in 3T3 L1 cells.	[111]
Pleurotus ostreatus	Mycelial extract	Reduce the glucose level in blood of streptozotocin- induced mice model.	[127]
	Freeze-dried powders.	Increase insulin secretion and reduce glucose level in T2D patient.	[61]
	Ethanolic extract	Decrease glucose, triglyceride, low density lipoprotein (LDL), very low-density lipoprotein, high density lipoprotein, cholesterol, creatinine, SGPT, and SGOT in serum of diabetic mice.	[23]
	Polysaccharide	Reduces lipid and glucose level in blood. Up regulate the expression of GSK3 phosphorylation and GLUT 4 protein expression.	[176]
	Mushroom extract	Down regulate blood glucose, total cholesterol, Triglycerides, and low-density lipoproteins No impaired function was observed for liver, kidney, and hemopoietic tissue.	[1]
Pleurotus pulmonarius	Profilin-like protein, glyceraldehyde-3-phosphate dehydrogenase-like protein, trehalose phosphorylase-like (TP-like) protein, and catalase-like protein.	Down regulate glucose concentration in blood.	[7, 132, 140]

TABLE 7.1 *(Continued)*

Mushroom	Active Compounds/Dose	Antidiabetic Effect	References
Pleurotus sajor-caju	Aqueous extract.	Hypoglycemic activity in normal as well as diabetic mice.	[108]
	Glucan rich polysaccharide.	Up-regulate GLUT4, adiponectin, and down regulate the expression of IL-6, TNF-α, SAA2, CRP, and MCP1 via reduction of transcription factor NF-KB.	[69]
Pleurotus tuberregium	Mycellial extract.	Enhance insulin secretion; down regulate glucose in blood; up regulate antioxidant activity.	[54]
	Aqueous extract.	Reduce triglyceride and glucose in blood in diabetic rabbit.	[110]
	Extracellular polysaccharide.	Up-regulate PPAR-α; hyperglycemia and hyperlipidemia.	[54]
Pleurotus sajor-caju	Aqueous extraction.	Decrease blood glucose level and regulate body weight of diabetic mice.	[108]
Sparassis crispa	β-glucan	Promoter of wound healing in diabetic patients.	[81]
	Freeze dried fruiting body samples.	Decrease blood sugar level, serum triglyceride and total cholesterol.	[160]
Tremella aurantia	Acidic polysaccharide.	Down regulate glucose in serum and decrease cholesterol, triglyceride, and lipoperoxide in plasma.	[70]
Tremella fuciformis	Glucuronoxylomannan	Hypoglycemic activity in both normal and diabetic mice.	[71]
	Exopolysaccharide of submerged mycelial culture.	Enhances insulin sensitivity and hypoglycemic impact via PPAR-gamma-mediated lipid metabolism.	[17, 154]
Tremella mesenterica	Acidic polysaccharide glucuronoxylomannan.	Glucuronoxylomannan minimize high blood glucose in only streptozocin induced and nicotinamide induced diabetic rats.	[48]
Trametes pubescens	Hot water extract and methanolic extract of fruit body.	Inhibit the activity of α-amylase and α glucosidase.	[58]
Volvariella volvacea	Lyophilized extract.	Enhance insulin secretion and reduce blood glucose level in diabetic rats; down regulate LDH, AST, ALT, urea, and creatinine.	[42]

TABLE 7.1 *(Continued)*

Mushroom	Active Compounds/Dose	Antidiabetic Effect	References
Wolfiporia cocos	Dehydrotumulosic acid, polyporenic acid C, pachymic acid, dehydrotrametenolic acid, and dehydroeburicoic acid.	Anti-hyperglycemia; insulin sensitivity.	[30]
Wolfiporia extensa	Chloroform fractions.	Decrease postprandial blood glucose levels in db/db mice through enhanced insulin sensitivity irrespective of PPAR-γ.	[88]

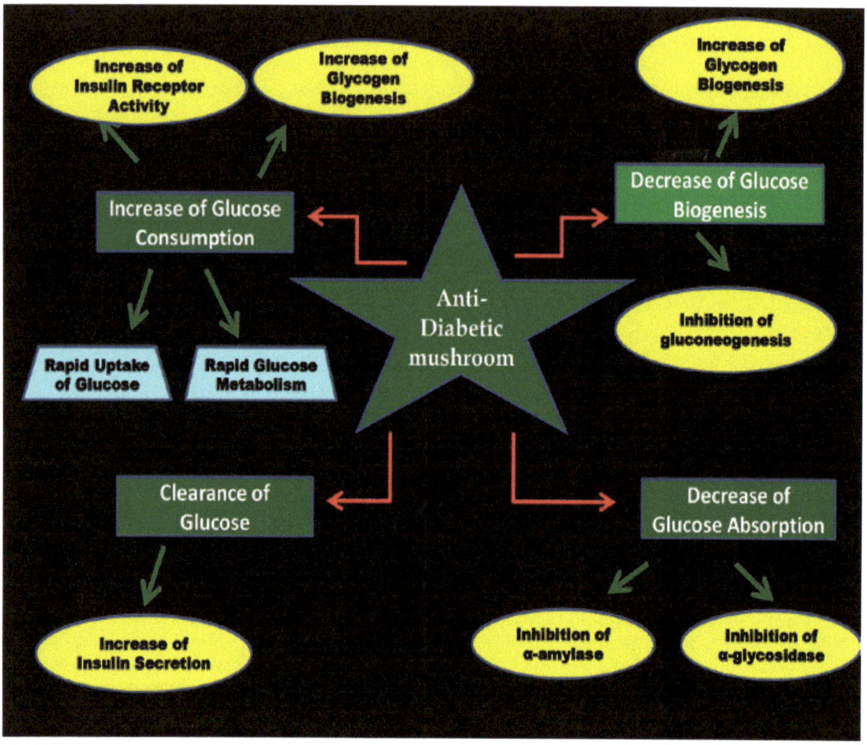

FIGURE 7.3 Mechanism of action of anti-diabetic mushroom.

hyperglycemia, however they impart unpleasant side effects and also, they do not appreciably control the diabetic complications [86]. Despite multiple prevention methods and arsenals of medicine, diabetes treatment remains woefully inadequate. Thus, it is critical to develop innovative nutraceuticals or medications for the treatment or prevention of diabetes that have fewer adverse effects. Numerous mycologists advocated for more study on functional mushrooms as preventative and therapeutic strategies for diabetes and its consequences. Mushroom based meals have been considered as healthy food for all age groups including children and aged people as all the nutrients are present in mushroom in appropriate proportion [68]. Some of them have delicate taste and that is why the consumption of such food supplements or therapeutic preparations is not so difficult. The bulk of mushroom-derived formulations and chemicals are used as dietary supplements or nutraceuticals and are either a processed or semi processed extract/dried biomass obtained from the fruiting body or from the cultured medium [131]. Nutraceuticals derived from mushrooms are currently eaten either as capsules or as tablets.

They are also shown to boost the immunological response of the human body, enhancing resistance to a variety of diseases, including diabetes, and in some instances, reversing the disease state [151]. Till date numerous mycochemicals have been shown to impart anti-diabetic effects by regulating multiple pathophysiological processes associated with diabetes without any disturbance in normal physiological process [20, 51]. The advantages of mushroom based anti-diabetics over conventional drug as mushroom is very affordable, since they may be cultivated on a variety of forest products and low-cost agricultural waste, such as straw of cereals, maize cobs, saw dust, etc. Mushrooms are also useful in treatment of some disease interlinked with diabetes. Some of them are utilized for both the management of blood glucose and the alteration of the course of diabetic problems without side-effects [92]. Many mushrooms have been proven in research to have the capacity to reduce raised blood sugar levels. However, with the exception of some mushrooms, the explanation for this effect is restricted. In order to obtain highly active mycelial extracts, the extraction protocols have to be optimized, specifically regarding extraction solvents, temperature, and duration of extraction. Current scientific study can now understand the rationale as well as extensive mechanisms underlying many biological functions of these bioactive substances. Various investigations found that some indigestible components in mushroom fruit bodies may act as prebiotics, modulating the formation and development of gut microbiota [19, 47, 113]. On the contrary at present moment, there is no pharmaceutical available that is capable of curing the ailment without side effects in all persons.

7.6 FUTURE PROSPECTS OF MUSHROOM DERIVED ANTIDIABETIC DRUG RESEARCH

The new discovery of bioactive components in mushrooms that may be used to improve human health will offer an extra boost to mushroom intake. Mushrooms are edible fungi that have been utilized as antioxidants since ancient times to combat metabolic diseases such as diabetes. Mushrooms, which already have traditionally been used in the diabetic remedies, are the appealing target for the development of innovative modern therapeutics to manage diabetes and its complications. As functional food, mushroom may have a greater potential influence on diabetes prevention as well as therapeutic capabilities for the illness than other species [115]. Comprehensive research showed that mild doses (500–1,000 mg-kg^{-1}) of all mushrooms had more or less a similar impact just like 150 mg-kg^{-1} dose of metformin both in diabetic and nondiabetic mice [108]. Long-term safety concerns about mushrooms, as well as interactions with other medicines, need to be addressed. Future research into these concerns is required to justify the use of mushrooms and their compounds as possible medicines or nutraceutical in diabetic therapy. In addition to these mushrooms are frequently employed in the manufacturing and consumption of food items due to their amazing nutritional content and anti-diabetic efficiency. There is a rising trend in protein content in both breads and muffins supplemented with 10% mushroom-powder and due to these breads and muffins lack the gluten network and viscosity-enhancing properties [46].

Finding more efficient and cost-effective methods for preparing and modifying the active chemicals from mushrooms can enhance the prospective use of mushroom derived anti-diabetic drug. Every year many known as well as novel mycochemicals have been reported from fruiting bodies or cultured mycelium of several wild as well as cultivated mushrooms, and some of them are found to have anti-hyperglycemic properties in clinical tests [66]. Because of the rising interest in mycotherapy, the scientific community must make a strong commitment to increase clinical studies and provide supplements of safe origin and genetic purity. However, as technology advances, researchers get a better grasp of the pathophysiology of diabetes.

7.7 DIFFICULTIES AND CONSTRAINTS OF MUSHROOM DERIVED ANTIDIABETIC DRUG

Considering the above discussion, it is mentionable that mushroom is the nutrient dense, versatile antidiabetic food supplement. Albeit many advances have been made in the research of mushroom based formulation with anti-diabetic effects, there are still significant gaps in their clinical application. Consuming high amounts of mushrooms to improve blood glucose levels may not be useful for diabetics and pre-diabetics. Emerging information from several mushrooms and their compounds implies the need of more randomized, controlled research with large sample sizes, particularly for long-term use [18]. The connection between the structural and functional aspect of certain polysaccharides as well as triterpenoids present in mushroom is still not precisely understood. In contradiction to the enormous number of experimental results, there are currently just a few clinical trials of different mushroom-based formulations that were recommended for the treatment of diabetes in human [66]. In addition, few anti-diabetic mushrooms are epiphytic and thus there is a need of suitable method for large scale production through culture medium. The discovery of physiologically active chemicals in mushrooms and there subsequent study could transform the trajectory of treatment system about diabetes and also explore many directions of medical research in the near future.

7.8 SUMMARY

Diabetes is a potentially fatal disease that has recently reached epidemic proportions around the world. In modern medical system, controlling diabetes is still hard to control without having side effects. Due to the negative effects and ineffectiveness of commercial hypoglycemic drugs, diabetes complications pose a threat to the health of a majority of diabetic people worldwide. On the other side, due to relative safety and affordability, a variety of edible and medicinal mushrooms have gained wide acceptance as a cure for diabetic conditions since last few years. Mushroom derivatives especially polysaccharides have demonstrated promising outcomes and are worthy of further investigation. However, scientific or clinical evidence of hypoglycemic effects of majority of wild mushroom is insufficient to support their usage as official anti-diabetic medications or nutraceuticals. Because of this, present communication recommend that more attention be given to cultivated and wild mushrooms, as both of which have potential for preventing and treating diabetes as well as its associated complications. In addition to this a lot of the mushrooms that are known to help people with diabetes have not yet been grown in the laboratory, more research should be done to find the best conditions and methods for growing these mushrooms.

KEYWORDS

- **anti-dyslipidemic**
- **anti-hyperglycemic activities**
- **blood glucose level**
- **diabetes**
- **macrofungi**
- **mushroom**

REFERENCES

1. Abu Sayeed, M., Banu, A., Khatun, K., Khanam, P. A., Begum, T., Mahtab, H., & Ashraful, H. J., (2014). Effect of edible mushroom (*Pleurotus ostreatus*) on type-2 diabetics. *Ibrahim Medical College Journal, 8*(1), 6–11.
2. Ahlqvist, E., Storm, P., Karajamaki, A., Martinell, M., Dorkhan, M., Carlsson, A., Vikman, P., Prasad, R. B., Aly, D. M., Almgren, P., et al., (2018). Novel subgroups of adult-onset diabetes and their association with outcomes: A data-driven cluster analysis of six variables. *The Lancet Diabetes and Endocrinology, 6*, 361–369.
3. Aoki, H., Hanayama, M., Mori, K., & Sato, R., (2018). *Grifola frondosa* (Maitake) extract activates PPARδ and improves glucose intolerance in high-fat-diet-induced obese mice. *Bioscience, Biotechnology, and Biochemistry, 82*(9), 1550–1559.
4. Ba, D. M., Gao, X., Muscat, J., Al-Shaar, L., Chinchilli, V., et al., (2021). Association of mushroom consumption with all-cause and cause-specific mortality among American adults: Prospective cohort study findings from NHANES III. *Nutrition Journal, 20*(1), 38. https://doi.org/10.1186/s12937-021-00691-8.
5. Bach, E., Hi, E., Martins, A., Nascimento, P., & Wadt, N., (2018). Hypoglicemic and hypolipedimic effects of *Ganoderma lucidum* in streptozotocin-induced diabetic rats. *Medicines, 5*(3), 78. https://doi.org/10.3390/medicines5030078.
6. Bahrami, G., Miraghaee, S. S., Mohammadi, B., Bahrami, M. T., Taheripak, G., Keshavarzi, S., Babaei, A., et al., (2020). Molecular mechanism of the anti-diabetic activity of an identified oligosaccharide from *Rosa canina*. *Research in Pharmaceutical Sciences, 15*(1), 36–47. https://doi.org/10.4103/1735-5362.278713.
7. Balaji, P., Madhanraj, R., Rameshkumar, K., Veeramanikandan, V., Eyini, M., Arun, A., Thulasinathan, B., et al., (2020). Evaluation of antidiabetic activity of *Pleurotus pulmonarius* against streptozotocin-nicotinamide induced diabetic Wistar albino rats. *Saudi Journal of Biological Sciences, 27*(3), 913–924. https://doi.org/10.1016/j.sjbs.2020.01.027
8. Begum, N., Nasir, A., Parveen, Z., Muhammad, T., Ahmed, A., Farman, S., et al., (2021). Evaluation of the hypoglycemic activity of *Morchella conica* by targeting protein tyrosine phosphatase 1B. *Frontiers in Pharmacology, 12*, 661803. https://doi.org/10.3389/fphar.2021.661803.
9. Bischoff, H., (1995). Pharmacology of α-glucosidase inhibitors. In: Vasselli, J., Maggio, C., Scriabine, A., (eds.), *Drugs in Development: α-Glycosidase Inhibition: Potential Use in Diabetes* (Vol. 1, pp. 1–13). Branford, CT: NEVA Press.
10. Biswas, G., & Acharya, K., (2013). Hypoglycemic activity of ethanolic extract of *Astraeus hygometricus* (Pers.) Morg. in alloxan induced diabetic mice. *International Journal of Pharmacy and Pharmaceutical Sciences, 5*(1), 391–394.
11. Cao, C., Yang, S., & Zhou, Z., (2020). The potential application of *Cordyceps* in metabolic-related disorders. *Phytotherapy Research, 34*, 295–305.
12. Cao, Y., Zou, S., Xu, H., Li, M., Tong, Z., Xu, M., & Xu, X., (2016). Hypoglycemic activity of the Baker's yeast β-glucan in obese/type 2 diabetic mice and the underlying mechanism. *Molecular Nutrition & Food Research, 60*(12), 2678–2690. https://doi.org/10.1002/mnfr.201600032.
13. Ceriello, A., Davidson, J., Hanefeld, M., Leiter, L., Monnier, L., Owens, D., Tajima, N., & Tuomilehto, J., (2006). Postprandial hyperglycaemia and cardiovascular complications of diabetes: An update. *Nutrition, Metabolism, and Cardiovascular Diseases, 16*, 453–456.

14. Cerletti, C., Esposito, S., & Iacoviello, L., (2021). Edible mushrooms and beta-glucans: Impact on human health. *Nutrients, 13*(7), 2195. https://doi.org/10.3390/nu13072195.
15. Chakraborty, N., Banerjee, A., Sarkar, A., Ghosh, S., & Acharya, K., (2021). Mushroom polysaccharides: A potent immune-modulator. *Bio Interface Research in Applied Chemistry, 11*(2), 8915–8930. https://doi.org/10.33263/BRIAC112.89158930.
16. Chen, J., Mao, D., Yong, Y., Li, J., Wei, H., & Lu, L., (2012). Hepatoprotective and hypolipidemic effects of water-soluble polysaccharidic extract of *Pleurotus eryngii. Food Chemistry, 130*(3), 687–694.
17. Cho, E. J., Hwang, H. J., Kim, S. W., Oh, J. Y., Baek, Y. M., Choi, J. W., et al., (2007). Hypoglycemic effects of exopolysaccharides produced by mycelial cultures of two different mushrooms *Tremella fuciformis* and *Phellinus baumii* in ob/ob mice. *Applied Microbiology and Biotechnology, 75*, 1257–1265.
18. De Silva, D. D., Rapior, S., Hyde, K. D., & Bahkali, A. H., (2012). Medicinal mushrooms in prevention and control of diabetes mellitus. *Fungal Diversity, 56*, 1–29. https://doi.org/ 10.1007/s13225-012-0187-4.
19. Delzenne, N. M., & Bindels, L. B., (2015). *Ganoderma lucidum*, a new prebiotic agent to treat obesity? *Nature Reviews Gastroenterology & Hepatology, 12*, 553. https://doi.org/10.1038/nrgastro.2015.137
20. Ding, Z. Y., Lu, Y. J., Lu, Z. X., lv, F. X., Wang, Y. H., Bei, X. M., Wang, F., & Zhang, K. C., (2010). Hypoglycaemic effect of comatin, an antidiabetic substance separated from *Coprinus comatus* broth, on alloxan induced-diabetic rats. *Food Chemistry, 121*, 39–43.
21. Dixit, S., & Alahmari, A. F., (2014). Pharmacological and nonpharmacological therapies in the management of diabetic peripheral neuropathy in type 2 diabetes: A comprehensive review. *Journal of Cardiovascular Disease Researches, 5*, 37e48.
22. Dubost, N., Ou, B., & Beelman, R., (2007). Quantification of polyphenols and ergothioneine in cultivated mushrooms and correlation to total antioxidant capacity. *Food Chemistry, 105*(2), 727–735.
23. Dulay, R. M. R., Arenas, M. C., Kalaw, S. P., Reyes, R. G., & Cabrera, E. C., (2014). Proximate composition and functionality of the culinary-medicinal tiger sawgill mushroom, *Lentinus tigrinus* (higher basidiomycetes), from the Philippines. *International Journal of Medicinal Mushrooms, 16*(1), 85–94.
24. Fatmawati, S., Shimizu, K., & Kondo, R., (2010). Ganoderic acid Df, a new triterpenoid with aldose reductase inhibitory activity from the fruiting body of *Ganoderma lucidum. Fitoterapia, 81*, 1033–1036.
25. Fatmawati, S., Shimizua, K., Kondoa, R., & Ganoderol, B., (2011). A potent α-glucosidase inhibitor isolated from the fruiting body of *Ganoderma lucidum. Phytomedicine, 18*, 1053–1055.
26. Feeney, M. J., Dwyer, J., Hasler-Lewis, C. M., Milner, J. A., Noakes, M., Rowe, S., et al., (2014). Mushrooms and health summit proceedings. *The Journal of Nutrition, 144*(7), 1128S–1136S. https://doi.org/10.3945/jn.114.190728.
27. Forouhi, N. G., Misra, A., Mohan, V., Taylor, R., Yancy, W., (2018). Dietary and nutritional approaches for prevention and management of type 2 diabetes. *BMJ, 361*, k2234.
28. Fortes, R. C., & Carvalho, G. N. M. R., (2011). The effects of *Agaricus sylvaticus* fungi dietary supplementation on the metabolism and blood pressure of patients with colorectal cancer during post-surgical phase. *Nutrición Hospitalaria, 26*, 176–186.
29. Friedman, M., (2016). Mushroom polysaccharides: Chemistry and antiobesity, antidiabetes, anticancer, and antibiotic properties in cells, rodents, and humans. *Foods, 5*(4), 80. https://doi.org/10.3390/foods5040080

30. Fu, M., Wang, L., Wang, X., Deng, B., Hu, X., & Zou, J., (2018). Determination of the five main terpenoids in different tissues of *Wolfiporia cocos*. *Molecules, 23*(8), 1839. https://doi.org/10.3390/molecules23081839.
31. Gambato, G., Pavao, E. M., Chilanti, G., Fontana, R. C., Salvador, M., & Camassola, M., (2018). *Pleurotus albidus* modulates mitochondrial metabolism disrupted by hyperglycaemia in EA.hy926 endothelial cells. *BioMed Research International, 2018*, 2859787. https://doi.org/10.1155/2018/2859787.
32. Ganesan, K., & Xu, B., (2019). Anti-diabetic effects and mechanisms of dietary polysaccharides. *Molecules, 24*(14), 1–28. https://doi.org/10.3390/molecules24142556.
33. Ganeshpurkar, A., Kohli, S., & Rai, G., (2014). Antidiabetic potential of polysaccharides from the white oyster culinary- medicinal mushroom *Pleurotus florida* (higher basidiomycetes). *International Journal of Medicinal Mushrooms, 16*(3), 207–217.
34. Geng, Y., Lu, Z. M., Huang, W., Xu, H. Y., Shi, J. S., & Xu, Z. H., (2013). Bioassay-guided isolation of DPP-4 inhibitory fractions from extracts of submerged cultured of *Inonotus obliquus*. *Molecules, 18*(1), 1150–1161.
35. Genuth, S., Alberti, K. G., Bennett, P., et al., (2003). Follow-up report on the diagnosis of diabetes mellitus. *Diabetes Care, 11*, 3160–3167.
36. Germoush, M. O., Elgebaly, H. A., Hassan, S., Kamel, E. M., Bin-Jumah, M., & Mahmoud, A. M., (2020). Consumption of terpenoids-rich *Padina pavonia* extract attenuates hyperglycemia, insulin resistance and oxidative stress, and upregulates PPARγ in a rat model of type 2 diabetes. *Antioxidants, 9*, 1–24.
37. Giacco, F., & Brownlee, M., (2010). Oxidative stress and diabetic complications. *Circulation Research, 107*(9), 1058–1070. https://doi.org/10.1161/CIRCRESAHA.110.223545.
38. Girma, W., & Tasisa, T., (2018). Application of mushroom as food and medicine. *Advances in Biotechnology and Microbiology, 11*(4), 555817. https://doi.org/10.19080/0099AIBM.2018.11.555817.
39. Gloyn, A. L., & Drucker, D. J., (2018). Precision medicine in the management of type 2 diabetes. *The Lancet Diabetes and Endocrinology, 6*(11), 891–900.
40. Gray, A. M., & Flatt, P. R., (1998). Insulin-releasing and insulin-like activity of *Agaricus campestris* (mushroom). *The Journal of Endocrinology, 157*(2), 259–266. https://doi.org/10.1677/joe.0.1570259.
41. Guo, W. L., Shi, F. F., Li, L., Xu, J. X., Chen, M., et al., (2019). Preparation of a novel *Grifola frondosa* polysaccharide-chromium (III) complex and its hypoglycemic and hypolipidemic activities in high fat diet and streptozotocin-induced diabetic mice. *International Journal of Biological Macromolecules, 131*, 81–88.
42. Gupta, P. P., Chandrakar, S., Roy, A., Verma, S., Gupta, N., & Sahu, R. K., (2021). Assessment of antidiabetic activity of hydroalcoholic extract of *Volvariella volvacea*. *European Journal of Molecular & Clinical Medicine, 8*(3), 1499–1509.
43. Han, C., & Liu, T., (2009). A comparison of hypoglycemic activity of three species of basidiomycetes rich in vanadium. *Biological Trace Element Research, 127*, 177–182.
44. Harada, E., Morizono, T., & Saito, M., (2017). Blood glucose-reducing and fat- reducing effects of a novel medicinal mushroom, *Grifola gargal* (agaricomycetes). *International Journal of Medicinal Mushrooms, 19*(2), 1071–1081.
45. He, Y., Zhang, L., & Wang, H., (2019). The biological activities of the antitumor drug *Grifola frondosa* polysaccharide. *Progress in Molecular Biology and Translational Science, 163*, 221–261.

46. Ho, L. H., Zulkifli, N. A., & Tan, T. C., (2020). Edible mushroom: Nutritional properties, potential nutraceutical values, and its utilization in food product development. In Passari, A. K., & Sanchez, S., (eds.), *An Introduction to Mushroom*. Intech Open, UK. http://dx.doi.org/10.5772/intechopen.91827.
47. Holmes, D., (2015). Medicinal mushroom reduces obesity by modulating microbiota. *Nature Reviews Endocrinology, 11*, 504. https://doi.org/10.1038/nrendo.2015.114.
48. Hong, L., Xun, M., & Wutong, W., (2007). Anti-diabetic effect of an alpha-glucan from fruit body of maitake (*Grifola frondosa*) on KK-Ay mice. *Journal of Pharmacy and Pharmacology, 59*(4), 575–582.
49. Hsu, C. H., Liao, Y. L., & Lin, S. C., (2007). The mushroom *Agaricus Blazei* Murill in combination with metformin and gliclazide improves insulin resistance in type 2 diabetes: A randomized, double-blinded, and placebo-controlled clinical trial. *The Journal of Alternative and Complementary Medicine, 13*(1), 97–102. https://doi.org/10.1089/acm.2006.6054.
50. Hu, S. H., Wang, J. C., Lien, J. L., Liaw, E. T., & Lee, M. Y., (2006). Anti-hyperglycemic effect of polysaccharide from the fermented broth of *Pleurotus citrinopileatus*. *Applied Microbiology and Biotechnology, 70*(1), 107–113.
51. Huang, G. J., Hsieh, W. T., Chang, H. Y., Huang, S. S., Lin, Y. C., & Kuo, Y. H., (2011). α-Glucosidase and aldose reductase inhibitory activities from the fruiting body of *Phellinus merrillii*. *Journal of Agricultural and Food Chemistry, 59*, 5702–5706.
52. Hwang, H. J., Kim, S. W., Lim, J. M., Joo, J. H., Kim, H. O., Kim, H. M., & Yun, J. W., (2005). Hypoglycemic effect of crude exopolysaccharides produced by a medicinal mushroom *Phellinus baumii* in streptozotocin induced diabetic rats. *Life Sciences, 76*(26), 3069–3080.
53. Hwang, H. S., Yun, J. W., (2010). Hypoglycemic effect of polysaccharides produced by submerged mycelial culture of *Laetiporus sulphureus* on streptozotocin-induced diabetic rats. *Biotechnology and Bioprocess Engineering, 15*, 173–181.
54. Huang, H. Y., Korivi, M., Chaing, Y. Y., et al., (2012). *Pleurotus tuber-regium* polysaccharides attenuate hyperglycemia and oxidative stress in experimental diabetic rats. *Evidence-Based Complementary and Alternative Medicine, 2012*, 856381. https://doi.org/10.1155/2012/856381.
55. Huang, H. Y., Korivi, M., Yang, H. T., Huang, C. C., Chaing, Y. Y., & Tsai, Y. C., (2014). Effect of *Pleurotus tuber-regium* polysaccharides supplementation on the progression of diabetes complications in obese-diabetic rats. *Chinese Journal of Physiology, 57*(4), 198–208.
56. Iddrisu, I., & Chikwere, P., (2020). Plants and mushrooms in the management of diabetes. *World Nutrition, 11*(1), 190–209.
57. Igel, L. I., Powell, A. G., Apovian, C. M., & Aronne, L. J., (2012). Advances in medical therapy for weight loss and the weight-centric management of type 2 diabetes mellitus. *Current Atherosclerosis Reports, 14*, 60–69.
58. Im, K. H., Nguyen, T. K., Choi, J., Lee, T. S., (2016). *In vitro* antioxidant, anti-diabetes, anti- dementia and inflammation inhibitory effect of *Trametes pubescens* fruiting body extracts. *Molecules, 21*(5), 639.
59. Islam, T., Ganesan, K., & Xu, B., (2019). New insight into mycochemical profiles and antioxidant potential of edible and medicinal mushrooms: A review. *International Journal of Medicinal Mushrooms, 21*(3), 237–251. https://doi.org/10.1615/IntJMedMushrooms.2019030079.

60. Jang, J. S., Lee, J. S., Lee, J. H., Kwon, D. S., Lee, K. E., Lee, S. Y., Hong, E. K., (2010). Hispidin produced from *Phellinus linteus* protects pancreatic beta-cells from damage by hydrogen peroxide. *Archives of Pharmacal Research, 33*(6), 853–861.
61. Jayasuriya, W. B. N., Wanigatunge, C. A., Fernando, G. H., Abeytunga, D. T. U., & Suresh, T. S., (2015). Hypoglycaemic activity of culinary *Pleurotus ostreatus* and P. *cystidiosus* mushrooms in healthy volunteers and type 2 diabetic patients on diet control and the possible mechanisms of action. *Phytotherapy Research, 29*(2), 303–309. https://doi.org/10.1002/ptr.5255.
62. Jeong, S. C., Jeong, Y. T., Yang, B. K., Islam, R., Koyyalamudi, S. R., Pang, G., Cho, K. Y., & Song, C. H., (2010). White button mushroom (*Agaricus bisporus*) lowers blood glucose and cholesterol levels in diabetic and hypercholesterolemic rats. *Nutrition Research, 30*, 49–56.
63. Jiang, X., Meng, W., Li, L., Meng, Z., & Wang, D., (2020). Adjuvant therapy with mushroom polysaccharides for diabetic complications. *Frontiers in Pharmacology, 11*, 168. doi: 10.3389/fphar.2020.00168.
64. Johnson, A., Cheng, S. C., Tsou, D., & Kong, Z. L., (2019). Attenuation of reproductive dysfunction in diabetic male rats with timber cultured *Antrodia cinnamomea* ethanol extract. *Biomedicine and Pharmacotherapy, 112*, 1–13.
65. Joo, J. I., Kim, D. H., & Yun, J. W., (2010). Extract of chaga mushroom (*Inonotus obliquus*) stimulates 3T3-L1 adipocyte differentiation. *Phytotherapy Research, 24*(11), 1592–1599.
66. Jovanovic, J. A., Mihailovic, M., Uskokovic, A., Grdovic, N., Dinic, S., & Vidakovic, M., (2021). The effects of major mushroom bioactive compounds on mechanisms that control blood glucose level. *Journal of Fungi, 7*(1), 58. https://doi.org/10.3390/jof7010058.
67. Jovanovic, J. A., Mihailovic, M., Uskokovic, A. S., Grdovic, N., Dinic, S., Poznanovic, G., Mujic, I., & Vidakovic, M., (2017). Evaluation of the antioxidant and antiglycation effects of *Lactarius deterrimus* and *Castanea sativa* extracts on hepatorenal injury in streptozotocin-induced diabetic rats. *Frontiers in Pharmacology, 8*, 793.
68. Kakraliya, S. S., (2020). Economic importance of mushroom and their uses. *Just Agriculture, 3*(1), 025.
69. Kanagsabapathy, G., Kuppusamy, U. R., Malek, S. N. A., Abdulla, M. A., Chua, K. H., & Sabaratnam, V., (2012). Glucan-rich polysaccharides from *Pleurotus sajor-caju* (Fr.) Singer prevents glucose intolerance, insulin resistance and inflammation in C57BL/6J mice fed a high fat diet. *BMC Complementary Medicine and Therapies, 12*, 261.
70. Kiho, T., Kochi, M., Usui, S., Hirano, K., Aizawa, K., & Inakuma, T., (2001). Antidiabetic effect of an acidic polysaccharide (TAP) from *Tremella aurantia* and its degradation product (TAP-H). *Biological & Pharmaceutical Bulletin, 24*(12), 1400–1403. https://doi.org/10.1248/bpb.24.1400.
71. Kiho, T., Tsujimura, Y., Sakushima, M., Usui, S., & Ukai, S., (1994). Polysaccharides in fungi. XXXIII. Hypoglycemic activity of an acidic polysaccharide (AC) from *Tremella fuciformis*. *Yakugaku Zasshi, 114*, 308–315.
72. Kim, D. J., Kang, Y. H., Kim, K. K., Kim, T. W., Park, J. B., & Choe, M., (2017). Increased glucose metabolism and alpha-glucosidase inhibition in *Cordyceps militaris* water extract-treated HepG2 cells. *Nutrition Research and Practice, 11*, 180–189.

73. Kim, H. M., Kang, J. S., Kim, J. Y., Park, S. K., Kim, H. S., Lee, Y. J., Yun, J., et al., (2010). Evaluation of antidiabetic activity of polysaccharide isolated from *Phellinus linteus* in non-obese diabetic mouse. *International Immunopharmacology*, *10*, 72–78.
74. Kim, S. K., Hong, U. P., Kim, J. S., Kim, C. H., Lee, K. W., Choi, S. E., Park, K. H., & Lee, M. W., (2007). Antidiabetic effect of *Auricularia auricula* mycelia in streptozotocin-induced diabetic rats. *Natural Product Sciences, 13*, 390–393.
75. Kim, Y. W., Kim, K. H., Choi, H. J., Lee, D. S., (2005). Anti-diabetic activity of beta-glucans and their enzymatically hydrolyzed oligosaccharides from *Agaricus blazei*. *Biotechnology Letters*, *27*(7), 483–487.
76. Kimatu, B. M., Zhao, L., Biao, Y., Ma, G., Yang, W., Pei, F., & Hu, Q., (2017). Antioxidant potential of edible mushroom (*Agaricus bisporus*) protein hydrolysates and their ultrafiltration fractions. *Food Chemistry*, *230*, 58–67.
77. Kobayashi, M., Kawashima, H., Takemori, K., Ito, H., Murai, A., Masuda, S., Yamada, K., et al., (2012). Ternatin, a cyclic peptide isolated mushroom and it's derivative suppress hyperglycemia and hepatic fatty acid synthesis in spontaneously diabetic KK-A(y) mice. *Biochemical and Biophysical Research Communications*, *427*(2), 299–304.
78. Koike, D., (2005). Effect of a wheat amylase inhibitor on canine carbohydrate digestion, gastrointestinal function and pancreatic growth. *Gastroenterology, 10*, 1221–1229.
79. Kou, L., Du, M., Liu, P., Zhang, Y., Yang, P., Shang, M., & Wang, X., (2019). Anti-diabetic and anti-nephritic activities of *Grifola frondosa* mycelium polysaccharides in diet-streptozotocin- induced diabetic rats via modulation on oxidative stress. *Applied Biochemistry and Biotechnology*, *187*(1), 310–322.
80. Kramer, U., Herder, C., Sugiri, D., et al., (2010). Traffic related air pollution and incident type 2 diabetes: Results from the SALIA cohort study. *Environ Health Perspect., 118*, 1273–1279.
81. Kwon, A. H., Qiu, Z., Hashimoto, M., Yamamoto, K., & Kimura, T., (2008). Effects of medicinal mushroom (*Sparassis crispa*) on wound healing in streptozotocin-induced diabetic rats. *The American Journal of Surgery*, *197*(4), 503–509.
82. Lee, B. R., Lee, Y. P., Kim, D. W., Song, H. Y., Yoo, K. Y., et al., (2010). Amelioration of streptozotocin-induced diabetes by *Agrocybe chaxingu* polysaccharide. *Molecules and Cells 29*(4), 349–354. https://doi.org/10.1007/s10059-010-0044-9.
83. Lee, I. K., Lee, J. H., & Yun, B. S., (2008a). Polychlorinated compounds with PPAR-gamma agonistic effect from the medicinal fungus *Phellinus ribis*. *Bioorganic and Medicinal Chemistry Letters, 18*(16), 4566–4568.
84. Lee, K. H., Morris-Natschke, S. L., Yang, X., Huang, R., Zhou, T., Wu, S. F., et al., (2012). Recent progress of research on medicinal mushrooms, foods, and other herbal products used in traditional Chinese medicine. *Journal of Traditional and Complementary Medicine*, *2*(2), 84–95.
85. Lee, Y. S., Kang, Y. H., Jung, J. Y., Lee, S., Ohuchi, K., Shin, K. H., Kang, I. J., et al., (2008b) Protein glycation inhibitors from the fruiting body of *Phellinus linteus*. *Biological and Pharmaceutical Bulletin, 31*, 1968–1972.
86. Li, S. P., Zhang, G. H., Zeng, Q., Huang, Z. G., Wang, Y. T., Dong, T. T., & Tsim, K. W., (2006). Hypoglycemic activity of polysaccharide, with antioxidation, isolated from cultured *Cordyceps* mycelia. *Phytomedicine, 13*(6), 428–433.
87. Li, J. P., Lei, Y. L., & Zhan, H., (2014). The effects of the king oyster mushroom *Pleurotus eryngii* (higher basidiomycetes) on glycemic control in alloxan-induced diabetic mice. *International Journal of Medicinal Mushrooms, 16*(3), 219–225.

88. Li, T. H., Hou, C. C., Chang, C. L. T., & Yang, W. C., (2011). Anti-hyperglycemic properties of crude extract and triterpenes from *Poria cocos*. *Evidence-Based Complementary and Alternative Medicine: ECAM*, 1–8.
89. Li, W. L., Zheng, H. C., Bukuru, J., & Kimpe, N., (2004). Natural medicines used in the traditional Chinese medical system for therapy of diabetes mellitus. *Journal of Ethnopharmacology*, *92*, 1–21.
90. Li, Y., Xu, L., Shan, Z., Teng, W., & Han, C., (2019). Association between air pollution and type 2 diabetes: An updated review of the literature. *Therapeutic Advances in Endocrinology and Metabolism, 10*, 2042018819897046. https://doi.org/10.1177/2042018819897046.
91. Lima, A. T., Santos, M. N., De Souza, L. A., Pinheiro, T. S., et al., (2016). Chemical characteristics of a heteropolysaccharide from *Tylopilus ballouii* mushroom and its antioxidant and anti-inflammatory activities. *Carbohydrate Polymers, 144*, 400–409 https://doi.org/10.1016/j.carbpol.2016.02.050.
92. Lindequist, U., & Haertel, B., (2020). Medicinal mushrooms for treatment of type 2 diabetes: An update on clinical trials. *International Journal of Medicinal Mushrooms, 22*, 845–854.
93. Liu, C., Song, J., Teng, M., Zheng, X., Li, X., Tian, Y., Pan, M., et al., (2016). Antidiabetic and antinephritic activities of aqueous extract of *Cordyceps militaris* fruit body in diet-streptozotocin-induced diabetic Sprague Dawley rats. *Oxidative Medicine and Cellular Longevity, 2016*, 9685257.
94. Liu, R. M., Dai, R., Luo, Y., & Xiao, J. H., (2019). Glucose-lowering and hypolipidemic activities of polysaccharides from *Cordyceps taii* in streptozotocin-induced diabetic mice. *BMC Complementary and Alternative Medicine, 19*(1), 230. https://doi.org/10.1186/s12906-019-2646-x.
95. Liu, Y., Sun, J., Rao, S., Su, Y., & Yang, Y., (2013). Antihyperglycemic, antihyperlipidemic and antioxidant activities of polysaccharides from *Catathelasma ventricosum* in streptozotocin-induced diabetic mice. *Food and Chemical Toxicology, 57*, 39–45. https://doi.org/10.1016/j.fct.2013.03.001.
96. Liu, Y., Wang, C., Li, J., Mei, Y., & Liang, Y., (2019). Hypoglycemic and hypolipidemic effects of *Phellinus Linteus* mycelial extract from solid-state culture in a rat model of type 2 diabetes. *Nutrients, 11*(2), 296. https://doi.org/10.3390/nu11020296.
97. Lo, H. C., Hsu, T. H., & Chen, C. Y., (2008). Submerged culture mycelium and broth of *Grifola frondosa* improve glycemic responses in diabetic rats. *The American Journal of Chinese Medicine, 36*(2), 265–285.
98. Lv, Y., Han, L., Chao, Y., & Guo, J., (2009). Comparison of hypoglycemic activity of trace elements absorbed in fermented mushroom of *Coprinus comatus*. *Biological Trace Element Research, 10*, 8352–8357.
99. Ma, H. T., Hsieh, J. F., & Chen, S. T., (2015). Anti-diabetic effects of *Ganoderma lucidum*. *Phytochemistry, 114*, 109–113.
100. Maddatu, J., Anderson-Baucum, E., & Evans-Molina, C., (2017). Smoking and the risk of type 2 diabetes. *Translational Research: The Journal of Laboratory and Clinical Medicine, 184*, 101–107. https://doi.org/10.1016/j.trsl.2017.02.004.
101. Maktoof, S. A., Alrubayae, I. M. N., & Almansorii, N. A., (2019). Evaluation of *in vitro* antidiabetic effect and phytochemical screening of some wild mushroom extracts isolated in Basrah, Iraq. *Drug Invention Today, 11*(11), 2797–2803.

102. Mandal, M., Rakibuzzaman, Rokeya, B., Ali, L., Hassan, Z., Faruque, M. O., (2018). Anti-diabetic effect of oyster mushroom mediates through increased AMP-activated protein kinase (AMPK) and cyclic AMP-response element binding (CREB) protein in type 2 diabetic model rats. *Bangladesh Journal of Medical Science, 17*(4), 661–668.
103. Meera, K. S., Sudha, G., Rajathi, K., & Manjusha, G. V., (2011). Antidiabetic effect of aqueous extract of *Hypsizygus ulmarius* on streptozotocin-nicotiinamide induced diabetic rats. *Asian Journal of Pharmaceutical and Biological Research, 1*, 151–157.
104. Mihailovic, M., Jovanovic, J. A., Uskokovic, A., Grdovic, N., Dinic, S., Vidovic, S., Poznanovic, G., et al., (2015). Protective effects of the mushroom *Lactarius deterrimus* extract on systemic oxidative stress and pancreatic islets in streptozotocin-induced diabetic rats. *Journal of Diabetes Research, 2015*, 576726, https://doi.org/10.1155/2015/576726.
105. Modak, M., Dixit, P., Londhe, J., Ghaskadbi, S., & Devasagayam, T. P. A., (2006). Indian herbs and herbal drugs used for the treatment of diabetes. *Journal of Clinical Biochemistry and Nutrition, 40*, 163–173.
106. Neyrinck, A. M., Possemiers, S., Verstraete, W., De Backer, F., Cani, P. D., Delzenne, N. M., (2012). Dietary modulation of clostridial cluster XIVa gut bacteria (*Roseburia* spp.) by chitin-glucan fiber improves host metabolic alterations induced by high-fat diet in mice. *The Journal of Nutritional Biochemistry, 23*(1), 51–59. https://doi.org/10.1016/j.jnutbio.2010.10.008.
107. Niwa, A., Tajiri, T., & Higashino, H., (2011). *Ipomoea batatas* and *Agarics blazei* ameliorate diabetic disorders with therapeutic antioxidant potential in streptozotocin-induced diabetic rats. *Journal of Clinical Biochemistry and Nutrition, 48*(3), 194–202. https://doi.org/10.3164/jcbn.10-78.
108. Ng, S. H., Mohd, Z. M. S., Zakaria, F., et al., (2015). Hypoglycemic and antidiabetic effect of *Pleurotus sajor-caju* aqueous extract in normal and streptozotocin-induced diabetic rats. *BioMed Research International, 2015*, 214918. https://doi.org/10.1155/2015/214918.
109. Ogbole, O. O., Nkumah, A. O., Linus, A. U., Falade, M. O., (2019). Molecular identification, *in vivo* and *in vitro* activities of *Calvatia gigantea* (macrofungus) as an antidiabetic agent. *Mycology, 10*(3), 166–173.
110. Onvekwuzu, I. M., Catherine, C. I., & Jude, C. I., (2019). Anti-diabetic effect of a flavonoid and sitosterol-rich aqueous extract of *Pleurotus tuberregium* sclerotia in alloxan-induced diabetic rabbits. *Endocrine, Metabolic & Immune Disorders Drug Targets, 19*(8), 1148–1156.
111. Paravansivam, P., Heng, C. K., Malek, S. N., Sabaratnam, V., M, R. R., & Kuppusamy, U. R., (2016). Giant oyster mushroom *Pleurotus giganteus* (*Agaricomycetes*) enhances adipocyte differentiation and glucose uptake via activation of PPARy and glucose transporters 1 and 4 in 3T3-L1 cells. *International Journal of Medicinal Mushrooms, 18*(9), 821–831.
112. Park, Y. J., Park, S. Y., Arasu, M. V., Al-Dhabi, N. A., Ahn, H. G., Kim, J. K., Park, S. U., (2017). Accumulation of carotenoids and metabolic profiling in different cultivars of tagetes flowers. *Molecules, 22*, 313.
113. Parnell, J. A., & Reimer, R. A., (2012). Prebiotic fiber modulation of the gut microbiota improves risk factors for obesity and the metabolic syndrome. *Gutmicrobes, 3*(1), 29–34. https://doi.org/10.4161/gmic.19246

114. Pearson, J. F., Bachireddy, C., Shyamprasad, S., et al., (2010). Association between fine particulate matter and diabetes prevalence in the U.S. *Diabetes Care, 33*, 2196–2201.
115. Perera, P. K., & Li, Y., (2011). Mushrooms as a functional food mediator in preventing and ameliorating diabetes. *Functional Foods in Health and Diseases, 1*(4), 161–171. https://doi.org/10.31989/ffhd.v1i4.133
116. Qadir, S., (2021). Ethnomycology of mushroom and their antioxidant properties. In: Prakash, V., (ed.), *Significance of Indian Medicinal Plants and Mushrooms* (pp. 434–458). Krishna Publication House, Bhavnagar Gujrat, India.
117. Qian, S., Zhang, M., He, Y., Wang, W., & Liu, S., (2016). Recent advances in the development of protein tyrosine phosphatase 1B inhibitors for type 2 diabetes. *Future Medicinal Chemistry, 8*(11), 1239–1258. https://doi.org/10.4155/fmc-2016-0064.
118. Rahman, M. A., Abdullah, N., Aminudin, N., & Aminudin, N., (2017). *Ganoderma lucidum* ameliorates spatial memory and memory-related protein markers in hypercholesterolemic and Alzheimer's disease model rats. *Archives of Neurology & Neurological Disorders, 3*(2), 117.
119. Ramos, M., Burgos, N., Barnard, A., Evans, G., Preece, J., Graz, M., Ruthes, A. C., et al., (2019). *Agaricus bisporus* and its by-products as a source of valuable extracts and bioactive compounds. *Food Chemistry, 292*, 176–187. https://doi.org/10.1016/j.foodchem.2019.04.035
120. Rajkumari, J., Magdalane, C. M., Siddhardha, B., Madhavan, J., Ramalingam, G., Al-Dhabi, N. A., Arasu, M. V., et al., (2019). Synthesis of titanium oxide nanoparticles using *Aloe barbadensis mill* and evaluation of its antibiofilm potential against *Pseudomonas aeruginosa* PAO1. *Journal of Photochemistry and Photobiology, B. Biology, 201*, 111667. https://doi.org/10.1016/j.jphotobiol.2019.111667.
121. Rathore, H., Prasad, S., Kapri, M., Tiwari, A., & Sharma, S., (2019). Medicinal importance of mushroom mycelium: Mechanisms and applications. *Journal of Functional Foods, 56*, 182–193.
122. Ratnaningtyas, N. I., Hernayanti, H., Andarwanti, S., Ekowati, N., Purwanti, E. S., & Sukmawati, D., (2018). Effects of *Ganoderma lucidum* extract on diabetic rats. *Biosaintifika Journal of Biology & Biology Education, 10*, 642–647.
123. Roncero-Ramos, I., & Delgado-Andrade, C., (2017). The beneficial role of edible mushrooms in human health. *Current Opinion in Food Science, 14*, 122–128. https://doi.org/10.1016/j.cofs.2017.04.002.
124. Roy, S., (2020). Evaluation of antioxidant and antidiabetic activity of white oyster mushroom (*Pleurotus florida*) cultivated in North Bengal. In Sarkar, A. K., (ed.), *Organisms and Environment* (pp. 136–142). Educreation Publishing, New Delhi, India.
125. Roy, S., & Sarkar, A. K., (2021). *Macrofungal Diversity in Terai-Dooars Region of West Bengal, India* (pp. 1–123). Akinik Publication, New Delhi, India.
126. Rushita, S., Vijayakumar, M., Noorlidah, A., Abdulla, M. A., & Vikineswary, S., (2013). Effect of *Pleurotus citrinopileatus* on blood glucose-insulin and catalase of streptozotocin-induced type 2 diabetes mellitus rats. *The Journal of Animal and Plant Sciences, 23*(6), 1566–1571.
127. Sangi, A. S. M., Bawadekji, A., & Al-Ali, M., (2018). Comparative effects of metformin, *Pleurotus ostreatus, Nigella sativa* and *Zingiber officinale* on the streptozotocin-induced diabetes mellitus in rats. *Pharmacognosy Magazine, 14*, S268–S273.

128. Satria, D., Amena, Y., Niwa, Y., Ashoura, A., Allama, A. E., & Shimizua, K., (2019). Lucidumol D, a new lanostane-type triterpene from fruiting bodies of Reishi (*Ganoderma lingzhi*). *Natural Product Research*, *33*, 189–195.
129. Sharma, A., Kaur, R., Kaur, J., Garg, S., Bhatti, R., & Kaur, A., (2021). An endophytic *Schizophyllum commune* Fr. exhibits *in-vitro* and *in-vivo* antidiabetic activity in streptozotocin induced diabetic rats. *AMB Express, 11*(1), 58. https://doi.org/10.1186/s13568-021-01219-3.
130. Shin, S., Lee, S., Kwan, J., Moon, S., Lee, S., Lee, C. K., Cho, K., et al., (2009). Cordycepin suppresses expression of diabetes regulating genes by inhibition of lipopolysaccharide-induced inflammation in macrophages. *Immune Network*, *9*(3), 98–105.
131. Singh, R., (2017). A review on different benefits of mushroom. *IOSR Journal of Pharmacy and Biological Sciences*, *12*(1), 107–111. https://doi.org/10.9790/3008-120102107111
132. Smiderle, F. R., Olsen, L. M., Ruthes, A. C., Czelusniak, P. A., Santana-Filho, A. P., Sassaki, G. L., Gorin, P. A. J., & Iacomini, M., (2012). Exopolysaccharides, proteins, and lipids in *Pleurotus pulmonarius* submerged culture using different carbon sources. *Carbohydrate Polymers*, *87*, 368–376.
133. Sobngwi, E., Ndour-Mbaye, M., Boateng, K. A., Ramaiya, K. L., Njenga, E. W., Diop, S. N., Mbanya, J. C., & Ohwovoriole, A. E., (2012). Type 2 diabetes control and complications in specialized diabetes care centers of six sub-Saharan African countries: The Diabcare Africa study. *Diabetes Research and Clinical Practice*, *95*, 30–36.
134. Stojkovic, D., Smiljkovic, M., Ciric, A., Glamoclija, J., Van, G. L., Ferreira, I. C. F. R., & Sokovic, M., (2019). An insight into antidiabetic properties of six medicinal and edible mushrooms: Inhibition of α-amylase and α-glucosidase linked to type- 2 diabetes. *South African Journal of Botany, 120*, 100–103.
135. Tourlouki, E., Matalas, A. L., & Panagiotakos, D. B., (2009). Dietary habits and cardiovascular disease risk in middle-aged and elderly populations: A review of evidence. *Clinical Interventions in Aging, 4*, 319–330.
136. Valverde, M. E., Hernandez-Perez, T., & Paredes-Lopez, O., (2015). Edible mushrooms: Improving human health and promoting quality life. *International Journal of Microbiology*, *2015*, 376387.
137. Veter, J., (2019). Biological values of cultivated mushrooms-a review. *Acta Alimentaria, 48*(2), 229–240. https://doi.org/10.1556/066.2019.48.2.11.
138. Vichayanrat, A., Ploybutr, S., Tunlakit, M., & Watanakejorn, P., (2002). Efficacy and safety of volibose in comparison with acarbose in type 2 diabetic patients. *Diabetes Research and Clinical Practice*, *55*, 99–103.
139. Vitak, T., Yurkiv, B., Wasser, S., Nevo, E., & Sybirna, N., (2017). Effect of medicinal mushrooms on blood cells under conditions of diabetes mellitus. *World Journal of Diabetes, 8*(5), 187–201. http://dx.doi.org/10.4239/wjd.v8.i5.187
140. Wahab, N. A., Abdullah, N., & Aminudin, N., (2014). Characterization of potential antidiabetic-related proteins from *Pleurotus pulmonarius* (Fr.) Quel. (grey oyster mushroom) by MALDI-TOF/TOF mass spectrometry. *BioMed Research International*, *2014*, 131607. https://doi.org/10.1155/2014/131607.
141. Wang, C., Gao, X., Santhanam, R. K., Chen, Z., Chen, Y., Xu, L., Wang, C., Ferri, N., & Chen, H., (2018). Effects of polysaccharides from inonotus obliquus and its chromium (III) complex on advanced glycation end-products formation, α-amylase, α-glucosidase

activity and H_2O_2-induced oxidative damage in hepatic L02 cells. *Food and Chemical Toxicology, 116*(Pt B), 335–345. https://doi.org/10.1016/j.fct.2018.04.047.
142. Wang, C. R., Ng, T. B., Li, L., Fang, J. C., Jiang, Y., Wen, T. Y., Qiao, W. T., Li, N., & Liu, F., (2011). Isolation of a polysaccharide with antiproliferative, hypoglycemic, antioxidant and HIV-1 reverse transcriptase inhibitory activities from the fruiting bodies of the abalone mushroom *Pleurotus abalones*. *JPP, 63*, 825–832.
143. Wang, F., Yang, X., Lu, Y., Li, Z., Xu, Y., Hu, J., Liu, J., & Xiong, W., (2019). The natural product antroalbol H promotes phosphorylation of liver kinase B1 (LKB1) at threonine 189 and thereby enhances cellular glucose uptake. *The Journal of Biological Chemistry, 294*(27), 10415–10427. https://doi.org/10.1074/jbc.RA118.007231.
144. Wang, J. C. Y., Hu, S. H., Wang, J. T., Chen, K. S., & Chia, Y. C., (2005). Hypoglycemic effect of extract of *Hericium erinaceus*. *Journal of the Science of Food and Agriculture, 85*, 641–646.
145. Wang, K., Bao, L., Ma, K., Zhang, J., Chen, B., Han, J., Ren, J., Luo, H., & Liu, H., (2017). A novel class of α-glucosidase and HMG-CoA reductase inhibitors from *Ganoderma leucocontextum* and the anti-diabetic properties of ganomycin I in KK-Ay mice. *European Journal of Medicinal Chemistry, 127*, 1035–1046. https://doi.org/10.3389/fphar.2017.00793.
146. Wang, K., Bao, L., Xiong, W., Ma, K., Han, J., Wang, W., Yin, W., & Liu, H., (2015). Lanostane triterpenes from the Tibetan medicinal mushroom *Ganoderma leucocontextum* and their inhibitory effects on HMG-CoA reductase and α-glucosidase. *Journal of Natural Products, 78*, 1977–1989.
147. Wang, P. C., Zhao, S., Yang, B. Y., Wang, Q. H., & Kuang, H. X., (2016). Anti-diabetic polysaccharides from natural sources: A review. *Carbohydrate Polymers, 148*, 86–97. https://doi.org/10.1016/j.carbpol.2016.02.060.
148. Wang, Y., Liu, Y., Wang, H., Li, C., Qi, P., & Bao, J., (2012). *Agaricus bisporus* lectins mediates islet β-cell proliferation through regulation of cell cycle proteins. *Experimental Biology and Medicine, 237*(3), 287–296. https://doi.org/10.1258/ebm.2011.011251.
149. Wani, B. A., Bodha, R. H., & Wani, A. H., (2010). Nutritional and medicinal importance of mushrooms. *Journal of Medicinal Plants Research, 4*, 2598–2604.
150. Wasser, S. P., Sokolov, D., Reshetnikov, S. V., & Timor-Tismenetsky, M., (2000). Dietary supplements from medicinal mushrooms: Diversity of types and variety of regulations. *International Journal of Medicinal Mushrooms, 2*(1), 1–19. https://doi.org/10.1615/IntJMedMushr.v2.i1.10.
151. Wasser, S. P., & Weis, A. L., (1999). Medicinal properties of substances occurring in higher basidiomycetes mushrooms: Current perspectives (review). *International Journal of Medicinal Mushrooms, 1*, 31–62.
152. Winkley, K., Sallis, H., Kariyawasam, D., Leelarathna, L. H., Chalder, T., Edmonds, M. E., Stahl, D., & Ismail, K., (2012). Five-year follow-up of a cohort of people with their first diabetic foot ulcer: The persistent effect of depression on mortality. *Diabetologia, 55*, 303–310.
153. Winska, K., Mączka, W., Gabryelska, K., & Grabarczyk, M., (2019). Mushrooms of the genus *Ganoderma* used to treat diabetes and insulin resistance. *Molecules (Basel, Switzerland), 24*(22), 4075. https://doi.org/10.3390/molecules24224075.
154. Wu, T., & Xu, B., (2015). Antidiabetic and antioxidant activities of eight medicinal mushroom species from China. *International Journal of Medicinal Mushrooms, 17*(2), 129–140.

155. Xiao, C., Wu, Q., Xie, Y., Zhang, J., & Tan, J., (2015). Hypoglycemic effects of *Grifola frondosa* (Maitake) polysaccharides F2 and F3 through the improvement of insulin resistance in diabetic rats. *Food and Function Journal, 6*(11), 3567–3575.
156. Yamac, M., Zeytinoglu, M., Kanbak, G., Bayramoglu, G., Senturk, H., (2009). Hypoglycemic effect of crude exopolysaccharides produced by *Cerrena unicolor, Coprinus comatus*, and *Lenzites betulina* isolates in streptozotocin-induced diabetic rats. *Pharmaceutical Biology, 47*, 168–174.
157. Yamac, M., Zeytinoglu, M., Senturk, H., Kartkaya, K., Kaubak, G., Bayramoglu, G., Oglakci, A., & Griensuen, L. J. L. D. V., (2016). Effects of black hoof medicinal mushroom, *Phellinus linteus* (*Agaricomycetes*), polysaccharide extract in streptozotocin-induced diabetic rats. *International Journal of Medicinal Mushrooms, 18*(4), 301–311.
158. Yamac, M., Kanbak, G., Zeytinoglu, M., Bayramoglu, G., Senturk, H., Uyanoglu, M., (2008). Hypoglycemic effect of *Lentinus strigosus* (Schwein.) Fr. crude exopolysaccharide in streptozotocin-induced diabetic rats. *Journal of Medicinal Food, 11*(3), 513–517. 10.1089/jmf.2007.0551.
159. Yamac, M., Kanbak, G., Zeytinoglu, M., Senturk, H., Bayramoglu, G., Dokumacioglu, A., & Van, G. L. J. L. D., (2010). Pancreas protective effect of button mushroom *Agaricus bisporus* (JE Lange) imbach (*Agaricomycetidae*) extract on rats with streptozotocin-induced diabetes. *International Journal of Medicinal Mushrooms, 12*(4), 379–389. https://doi.org/10.1615/IntJMedMushr.v12.i4.50.
160. Yamamoto, K., & Kimura, T., (2010). Dietary *Sparassis crispa* (Hanabiratake) ameliorates plasma levels of adiponectin and glucose in type 2 diabetic mice. *Journal of Health Sciences, 56*, 541–546.
161. Yang, B. K., Kim, D. H., Jeong, S. C., Das, S., Choi, Y. S., Shin, J. S., Lee, S. C., & Song, C. H., (2002). Hypoglycemic effect of a *Lentinus edodes* exopolymer produced from a submerged mycelial culture. *Bioscience, Biotechnology, and Biochemistry, 66*, 937–942.
162. Yang, B. K., Kim, D. H., Jeong, S. C., Das, S., Choi, Y. S., Shin, J. S., Lee, S. C., & Song, C. H., (2014). Hypoglycemic effect of a *Lentinus edodes* exo-polymer produced from a submerged mycelial culture. *Bioscience, Biotechnology and Biochemistry, 66*(5), 937–942.
163. Yang, B. K., Kim, G. N., Jeong, Y. T., Jeong, H., Mehta, P., & Song, C. H., (2008). Hypoglycemic effects of exo-biopolymers produced by five different medicinal mushrooms in STZ-induced diabetic rats. *Mycobiology, 36*(1), 45–49. https://doi.org/10.4489/MYCO.2008.36.1.045
164. Yang, K., Zhang, S., Ying, Y., Li, Y., Cai, M., Guan, R., Hu, J., & Sun, P., (2020). Cultivated fruit body of *Phellinus baumii*: A potentially sustainable antidiabetic resource. *ACS Omega, 5*(15), 8596–8604.
165. Yang, S., Meng, Y., Yan, J., Wang, N., Xue, Z., Zhang, H., & Fan, Y., (2019). Polysaccharide- enriched fraction from *Amillariella mellea* fruiting body improves insulin resistance. *Molecules, 24*(1), 46. https://doi.org/10.3390/molecules24010046.
166. Yayeh, T., Oh, W. J., Park, S. C., et al., (2012). *Phellinus baumii* ethyl acetate extract inhibits lipopolysaccharide-induced iNOS, COX-2, and proinflammatory cytokine expression in RAW264.7 cells. *Journal of Natural Medicines, 66*, 49–54. https://doi.org/10.1007/s11418-011-0552-8.
167. Yong, T., Chen, S., Xie, Y., Shuai, O., Li, X., Chen, D., Su, J., et al., (2018). Hypouricemic effects of extracts from *Agrocybe aegerita* on hyperuricemia mice and virtual prediction

of bioactives by molecular docking. *Frontiers in Pharmacology, 9*, 498. https://doi.org/10.3389/fphar.2018.00498.
168. Yuan, Z., He, P., & Takeuchi, H., (1998). Ameliorating effects of water-soluble polysaccharides from the woody ear (*Auricularia auricular-judae* Quel.) in genetically diabetic KK-Ay mice. *Journal of Nutritional Science and Vitaminology, 44*(6), 829–840.
169. Yurkiv, B., Wasser, S. P., Nevo, E., & Sybirna, N. O., (2015). Antioxidant effects of medicinal mushrooms *Agaricus brasiliensis* and *Ganoderma lucidum* (higher basidiomycetes): Evidence from animal studies. *International Journal of Medicinal Mushrooms, 17*(10), 943–955. https://doi.org/10.1615/intjmedmushrooms.v17.i10.40.
170. Zhang, B. B., Guan, Y. Y., Hu, P. F., Chen, L., Xu, G. R., Liu, L., & Cheung, P., (2019). Production of bioactive metabolites by submerged fermentation of the medicinal mushroom *Antrodia cinnamomea*: Recent advances and future development. *Critical Reviews in Biotechnology, 39*(4), 541–554. https://doi.org/10.1080/07388551.2019.1577798.
171. Zhang, C., Zhang, L., Liu, H., Zhang, J., Hu, C., & Jia, L., (2018). Antioxidation, anti-hyperglycemia and renoprotective effects of extracellular polysaccharides from *Pleurotus eryngii* SI-04. *International Journal of Biological Macromolecules, 111*, 219–228.
172. Zhang, G., Huang, Y., Bain, Y., Wang, J. H., & Ng, T. B., (2006). Hypoglycemic activity of the fungi *Cordyceps militaris, Cordyceps sinensis, Tricholoma mongolicum* and *Omphalia lapidescens* in streptozotocin-induced diabetic rats. *Applied Microbiology and Biotechnology, 72*(6), 1152–1156.
173. Zhang, H. N., & Lin, Z. B., (2004). Hypoglycemic effect of *Ganoderma lucidum* polysaccharides. *Acta Pharmacologica Sinica, 25*, 191–195.
174. Zhang, J., Ma, K., Chen, H., Wang, K., Xiong, W., Bao, L., & Liu, H., (2017). A novel polycyclic meroterpenoid with aldose reductase inhibitory activity from medicinal mushroom *Ganoderma leucocontextum*. *The Journal of Antibiotics, 70*, 915–917.
175. Zhang, J., Meng, G., Zhang, C., Liu, L., Xu, N., Liu, M., Cui, M., et al., (2015). The antioxidative effects of acidic-, alkalic- and enzymatic-extractable mycelium zinc polysaccharides by *Pleurotus djamor* on liver and kidney of streptozocin-induced diabetic mice. BMC Complement. *BMC Complementary and Alternative Medicine, 15*, 440.
176. Zhang, Y., Hu, T., Zhou, H., Zhang, Y., Jin, G., & Yang, Y., (2016). Antidiabetic effect of poly saccharides from *Pleurotus ostreatus* in streptozotocin-induced diabetic rats. *International Journal of Biological Macromolecules, 83*, 126–132.
177. Zhao, C., Liao, Z., Wu, X., Liu, Y., Liu, X., Lin, Z., Huang, Y., & Liu, B., (2014). Isolation, purification, and structural features of a polysaccharide from *Phellinus linteus* and its hypoglycemic effect in alloxan-induced diabetic mice. *Journal of Food Science, 2014*, H1002–H1010.
178. Zhao, H., Lai, Q., Zhang, J., Huang, C., & Jia, L., (2018). Antioxidant and hypoglycemic effects of acidic-extractable polysaccharides from *Cordyceps militaris* on type 2 diabetes mice. *Oxidative Medicine and Cellular Longevity*, 1–15.
179. Zheng, Y., Bai, L., Zhou, Y., Tong, R., Zeng, M., Li, X., & Shi, J., (2019). Polysaccharides from Chinese herbal medicine for anti-diabetes recent advances. *International Journal of Biological Macromolecules, 121*, 1240–1253.

CHAPTER 8

Antidiabetic and Nutritional Properties of Guava (*Psidium guavaja* L.) Pulp (Fermented with Lactococcus lactis)

Moushumi Ghosh and Harshpinder Kaur

ABSTRACT

This study investigated the functional and nutritional properties of guava (Psidium guavaja L., Punjab – pink) pulp fermented with a probiotic strain of Lactococcus lactis especially with reference to diabetes. Deseeded guava pulp was fermented at ambient temperature for 8 h with overnight grown culture of L. lactis that exhibited positive prebiotic effects, production of organic acids in the guava pulp. The fermented extracts of guava pulp possessed α-glucosidase (IC50 values of 3.40 ± 0.08) and high antioxidant activity (81%), vitamin C (289 mg/g), and total polyphenolics (712 mg/GAE/g). There was significant increase in the antioxidant activity, polyphenolic content, inhibitory activities of α-glucosidase after 6 h. Proximate analysis indicated the decrease in carbohydrates, lipids, and moisture contents; and increase in protein content. Sensorial analysis (9-point hedonic scale) indicated good acceptability of the fermented pulp in the 6-h fermented product. Sensory properties and bioactive content in the fermented product were retained on storage. The results of this study indicated a sustainable alternative for utilizing guava pulp for nutritional and antidiabetic benefits.

8.1 INTRODUCTION

The importance of functional foods has assumed importance especially in the post pandemic scenario. As a consequence, the demand for nutritionally

The Functional Foods: Nutrient and Health Benefits.
Megh R. Goyal, Junaid Ahmad Malik, and Anu Kumari (Eds.)

enriched horticultural produces and their products have accelerated the efforts of food industries to develop innovative products. Guava (Psidium guajava L.) is an affordable choice because of its nutritional qualities and ease of availability [17]. Also, beneficial properties of guava fruit (such as: antioxidants, high content of vitamin-C and polyphenols) have been reported [13]. The pulp of guava fruits can be utilized as a viable source of nutrients; since after processing, it contains an appreciable quantity of bioactives (high dietary fibers, polyphenols, and antioxidants) [32]. Significantly enhanced dietary fibers and other beneficial properties in fermented guava pulp have been reported [27]. Fermentation is the most desirable process for modifying the primary food matrix [33]. During fermentation, metabolic processes of microorganisms can enhance both bioaccessibility and bioavailability of nutrients from agro-produces; and can render acceptable sensory properties [18].

The guava pulp is an attractive food matrix and fermentation may render enhanced bioactive compounds, in addition to improving palatability. Considerable quantity of guava is wasted following the harvest due to the lack of adequate technologies for preservation; and sustainable utilization of guava fruits is desired.

The aim of this study was to develop a functional formulation comprising of guava pulp and to investigate the inhibitory properties of α-glucosidase of the formulated fermented product. Amongst non-communicable diseases (NCDs), diabetes has assumed alarming proportions and dietary interventions assume significance. Besides, the proximate composition and other important nutritional parameters of the fermented formulation were also evaluated.

8.2 MATERIALS AND METHODS

8.2.1 MICROORGANISM AND FERMENTATION

The Lactococcus lactis subsp. lactis strain (Accession no: JN618456) isolated from nondairy sources for this study has been reported for its probiotic and technological properties [4]. For maintaining cultures in MRS (deMann Rogosa Sharpes medium), agar (Himedia, India) (4–6°C) was used.

- Influence of Prebiotic Components on Growth of L. lactis: Prior to fermentation experiments, overnight grown L. lactis culture was inoculated in MRS alone (control) and MRS supplemented with fructo-oligosaccharides (FOS), inulin (Sigma, USA) and guava extracts; and was incubated for 48 h at 37°C with shaking. Cell counts were determined by plating appropriately with diluted aliquots on MRS plates, which were incubated at 37°C for 24 h. The fermentation matrix was comprised of ripe and over-ripe guava (Psidium guavaja L.) of Punjab pink variety, which was first pulped and deseeded. Guava pulps were then inoculated in a ratio of 1.5: 1 (based on previous optimization studies) to form a batter along with overnight grown cultures of Lactococcus lactis subsp. lactis (MRS grown cultures were centrifuged, pellets were washed with sterile PBS (phosphate buffered saline)). The pulp was incubated at ambient temperature (35°C) for 8 h. Drop in the pH value confirmed the fermentation. The fermented pulp was steamed for 20–30 minutes. Formulated food samples were analyzed for sensory parameters (color, smell, consistency, texture, overall acceptability) using a 9-point hedonic scale [21]. All samples were kept in airtight containers at 4°C.

8.2.2 α-GLUCOSIDASE INHIBITORY ACTIVITY

Acarbose stocks (1 mg/ml from Sigma, MO–USA) were prepared and were used within 3 h. Briefly, α-glucosidase enzyme from Saccharomyces cerevisiae (EC3.2.1.20, Sigma-Aldrich, St Louis, MO–USA) was dissolved in 100 mM of PBS (pH 7.0) containing 2 g/liter BSA and 0.2 g/liter and NaN3; 5 mM PNPG (para nitrophenyl alpha d glucopyranoside)) (Sigma, MO, USA) was dissolved in the same buffer (pH 7.0) and was used as a substrate solution; 500 microliters (μl) of the enzyme solution and 100 microliters of the fermented extract was mixed, and the mixture was incubated for 5 minutes in 2 ml Eppendorf tubes at room temperature, then 500 μl of substrate PNPG solution was added to the reaction mixture and it was further incubated for another 5 minutes at room temperature. For positive control, 100 μl of the fermented extract was replaced with same volume of Acarbose stock (1 mg/ml); and for negative control, PBS replaced fermented extract. Absorbance (405 nm) was measured spectrophotometrically (Shimadzu). Inhibitory activity was expressed as follows:

$$\text{Inhibitory activity} = \{100 - [(\text{Relative absorbance (\%) of test compounds}) - (\text{Absorbance of the control (C)})\} \quad (1)$$

where; test solution was comprised of PBS [11].

8.2.3 DPPH RADICAL SCAVENGING ACTIVITY

In the modified method [20], 2 ml of 0.15 mM DPPH was mixed with 1 ml extract in various dilutions. Control sample consisted of 2 ml of DPPH added to 1 ml of methanol. After mixing, the tubes were kept in the standing position for 30 min; and absorbance was measured spectrophotometrically at 517 nm (Shimadzu, Japan). Samples were triplicated. Scavenging ability (%) was calculated as follows:

$$\text{Superoxide radical scavenging activities (\%)}$$
$$= 100 \times \{1 - [(\text{absorbance of sample})/(\text{absorbance of control})]\} \quad (2)$$

8.2.4 TOTAL PHENOLIC CONTENT (TPC) OF FERMENTED GUAVA PULP

The 4 mL aliquots of methanol extract were added into 0.25 mL of Folin-Ciocalteau reagent after dilution with water, followed by addition of 0.75 mL sodium carbonate (20%). The magnetically stirred solution was placed in the dark for 2 h; and absorbance was recorded spectrophotometrically at 765 nm (Shimadzu, Japan). Controls were comprised of water and reagent. TPC was expressed as (GAE)/g [29].

8.2.5 ESTIMATION OF ASCORBIC ACID

Ascorbic acid was determined by 2,6-dichlorophenol indophenol (DCPIPH) dye method with metaphosphoric acid – acetic acid reagent. Titration of 5 mL extract was performed against dye solution according to the method described by AOAC [3]. Changes in color to pink as a result of DCPIPH reduction was indicative of the end point. The ascorbic acid content was calculated in mg/100 g.

8.2.6 PROXIMATE ANALYSIS

The sample was dried to a constant weight at 105°C for measuring the moisture level. Carbohydrate content (Anthrone method) was analyzed by standard method [31]. The standard Kjeldahl method was used to determine crude protein content; and results were converted using a nitrogen factor of 6.25. The Soxhlet extraction method was used to determine total lipid content [3].

8.2.7 STATISTICAL ANALYSIS

Experiments were performed in triplicates. Data are represented as "mean values ± standard deviation (S. D.)." ANOVA was used to estimate the level of significance ($p<0.05$).

8.3 RESULTS

In MRS media supplemented with FOS, Inulin, and Guava extracts, excellent growth was observed in comparison with MRS alone. The cell counts of the probiotic L. lactis are presented in Table 8.1(a) and these indicated that growth was enhanced in prebiotic supplemented media. To determine whether L. lactis utilized fructooligosaccharides as sole source of carbon [6], medium was supplemented with 1% inulin, FOS, and guava pulp. Glucose was used as a control. A strain of Lactobacillus acidophilus was evaluated in parallel for comparison. Table 8.1(a) summarizes the prebiotic activity scores and acidification parameters.

TABLE 8.1(A) Metabolic Response of L. lactis in MRS Supplemented with FOS, Inulin, and Guava Extracts After 24 h

Media Used for *L. lactis* Growth	Cell Counts (CFU/g)
Fructooligiosaccharides (FOS) + MRS	6.5×10^9 ($\pm 5.97 \times 10^9$)
Inulin + MRS	4.3×10^8 ($\pm 3.7 \times 10^8$)
Guava extract + MRS	5.7×10^9 ($\pm 5.0 \times 10^9$)
MRS	3.1×10^6 ($\pm 2.3 \times 10^6$)
FOS as a sole carbon source	0.38
Inulin	0.49
Guava extract	0.54

The assimilation of fermentable carbohydrates and acidification was notable for L. lactis (higher prebiotic score) than for L. acidophilus. The formulations comprising of guava pulp in the ratio of 1:5 for fermentation time of 4 h with 2% inoculum was judged as the optimum for guava pulp product (results not shown). Increase in sensory attributes and α-glucosidase inhibitory activities after this period was not significant ($P>0.05$). Lactic acid was the principal organic acid produced upon fermentation. In the fermented sample, pH of 4.2 was noted after 6 h, while the pH was 3.5 after 8 h.

Growth was recorded as CFU/g in "average ±SEM" of three determinations (Table 8.1(a)). E. coli was used as the reference enteric strain. Prebiotic score was calculated considering the growth of L. lactis during fermentation [14] as follows:

$$N: Nt = 10\ h - N0 = 0\ h \tag{3}$$

8.3.1 THE INHIBITORY ACTIVITIES OF α-GLUCOSIDASE

The IC50 for α-glucosidase was 3.40 mg/mL. Commercial α-glucosidase inhibitor acarbose had an IC50 value of 8.19 (Table 8.1(b)). Furthermore, enzyme inhibitory properties showed a marked increase at 6 h of fermentation against α-glucosidase, the IC50 values being much lower than that observed after 2 h of fermentation. The inhibitory profile for both enzymes during the course of fermentation is depicted in Figure 8.1. Pancreatic lipase inhibition (Red) is also shown. Highest inhibition was observed at 6 hours; inhibition remained constant after this period. None of the inhibitory activities were increased notably after 6 h. Interestingly, concomitant increase in pancreatic lipase inhibitory activity was also observed.

TABLE 8.1(B) The IC50 of α-Glucosidase Inhibition (μg/mL) for Fermented Guava Pulp

Sample	Concentration	The IC50 of α-Glucosidase Inhibition		
		4 h	6 h	8 h
Fermented guava pulp	200 μg/mL	7.23 ± 1.23	3.40 ± 0.08	3.34 ± 0.07
Guava pulp	200 μg/mL	9.21 ± 2.14	8.33 ± 2.10	8.29 ± 1.11
Acarbose	200 μg/mL	8.19 ± 1.75	NA	NA

Note: Values in parenthesis indicate "mean ± (S.D.);" ND: Not detected; and NA: Not applicable.

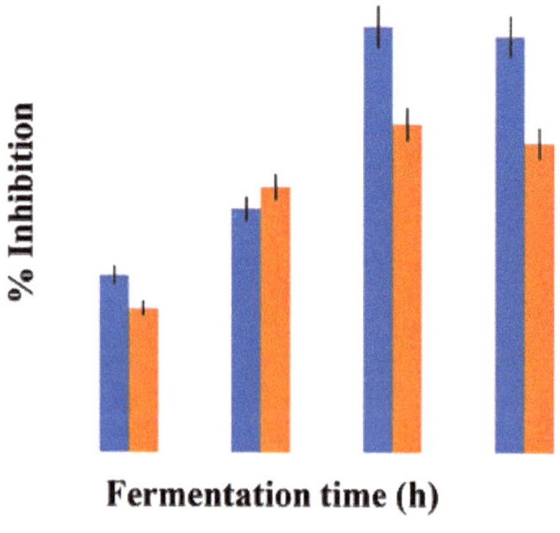

FIGURE 8.1 Evolution of αglucosidase inhibitory activities (blue) in guava pulp during the course of L. lactis fermentation (30°C).

8.3.2 ANTIOXIDANT ACTIVITY AND TOTAL POLYPHENOLS IN FERMENTED GUAVA PULP

Enhanced levels of polyphenols and antioxidant activities were noted in the fermented guava in comparison to the unfermented samples (Table 8.2). Increase in antioxidant and polyphenolic content did not differ significantly ($p>0.05$) after the fermentation period of 6 h.

TABLE 8.2 Total Phenolic Content, Antioxidant Activity, and Vitamin C Content of Fermented Guava Pulp

Fermentation Time (h)	Total Phenolic Content (mg/GAE/g)	Antioxidant Activity (%)	Vitamin C (mg/100 g)
0	369 ± 11.32	35 ± 7.06	119.17 ± 12.20
2	658 ± 10.21	61.2 ± 11.20	258.4 ± 10.09
4	712 ± 9.77	81.0 ± 10.33	289.3 ± 9.08
6	703 ± 8.78	80.4 ± 11.09	298.7 ± 10.02

Note: Values in parenthesis are "mean ± (S.D.)."

8.3.3 ASCORBIC ACID CONTENT

The fermented guava pulp revealed enhanced level of Ascorbic acid in comparison to the unfermented product (Table 8.2). Storage of samples at 4°C for 48 h did not change the amount of Vitamin-C significantly ($p>0.05$).

8.3.4 PROXIMATE COMPOSITION

The carbohydrate, fat, and moisture levels in guava fermented samples declined with the progress of fermentation. An increase in protein content was observed (Table 8.3).

8.3.5 SENSORY PROFILE

Scores for all attributes gradually decreased with higher fermentation time (more than 6 h) as indicated in Table 8.4. During fermentation, LAB population levels increased, accompanied by changes in titratable acidity, pH, and VFAs (volatile fatty acids) responsible for influencing the sensory parameters. Sensory attributes of 6 h-fermented samples after storage of 48 h under refrigerated conditions possessed overall acceptability (Table 8.5).

TABLE 8.3 Proximate Composition of Fermented Guava Pulp

Parameter	Proximate Composition (%)			
	0 h	2 h	4 h	6 h
Carbohydrate	68.3	51.1	40.03	31
Fat	10.41	7.4	5.8	6.0
Protein	10.64	21.4	22.0	23.6
Moisture	53.04	44.7	37.6	33.6

Note: All values are in "mean ± (S.D.)."

TABLE 8.4 Sensory Evaluation of Fermented Guava Pulp at Different Fermentation Time

Sensory Attributes	Fermentation Time (Hours)			
	0*	4	6	10
Aroma (9)	4.6 ± 1.02	7.8 ± 0.07	7.6 ± 0.03	5.8 ± 0.02
Consistency (9)	5.2 ± 0.05	7.5 ± 0.02	7.4 ± 0.05	6.5 ± 0.04
Texture (9)	5.0 ± 0.09	7.7 ± 0.03	7.3 ± 0.05	6.6 ± 0.03
Color and appearance (9)	4.8 ± 0.04	7.3 ± 0.02	7.3 ± 0.03	6.4 ± 0.02
Overall acceptability (9)	4.8 ± 0.87	7.6 ± 0.02	7.4 ± 0.07	6.6 ± 0.05

Note: *: Unfermented sample; values are means ± (S.D.). A prior consent of panelists was obtained for the study.

TABLE 8.5 Functional and Sensory Characteristics of Fermented Guava Pulp After 48 h of Storage at 4°C

Attributes	Refrigerated Sample (After 48 h of Storage)	Freshly Fermented
α-glucosidase inhibitory activity (%)	27.6 ±1.76	29.7 ± 2.11
Vitamin C	297.6 ± 11.03	298.7 ± 9.33
Antioxidant activity	800.3 ± 9.51	804 ± 8.65
Overall acceptability	7.2 ± 0.87	7.4 ± 0.07

Note: Values are means ± (S.D.). Prior consent of panelists was obtained for the sensory study.

8.4 DISCUSSION

An important trend of consumers is the demand for functional foods; in this category, such effective foods for diabetes have been intensely sought. The antidiabetic potential of guava has been established. Fermentation of guava pulps offered an innovative approach in utilizing the fruits intrinsic properties, which can be stored. The rapid senescence in guava leads to spoilage and loss of alpha-glucosidase functions.

High nutritional value, pleasant palatability, and promising healthy properties [1, 8] are hallmarks of fermentation; therefore, the latter is an emerging source for developing innovative functional food products. Considering the nutritional benefits vis-a-vis, the need for innovations could promote utilization, due to novel alternative of guava as a combination for palatability and functional benefits.

The Punjab pink variety of guava was selected because of its high nutritional content (carotenoids: 5.72–6.84 mg/100 g) apart from crude protein, fiber, carbohydrates, Vitamin C, potassium, calcium, and phosphorous. It is envisaged that a simple fermentation process using guava can be subsequently scaled up and has commercial feasibility, besides the approach can balance the nutrition, resource utilization, and sustainable food processing.

It was desirable to evaluate the effects of prebiotics on L. lactis. Both, inulin and fructooligosaccharides are well established prebiotics and are widely used for investigating prebiotic effects of starter lactic acid bacteria (LAB) [9] that had significant positive effects on L. lactis. The L. lactis strain survived well in MRS media supplemented with FOS, inulin and guava extracts indicating the positive influence of prebiotic components; and these results are in agreement with other investigators [15, 24]. It has been suggested that ability to utilize prebiotic is dependent on strains- and is related to its depolymerizing functions [35].

In the study by authors of this chapter, maximum specific growth rates were noted with guava pulp (di- and oligo-saccharides) in line with other reports [26]. The elaboration and capability to co-metabolize indigestible di- and oligo-saccharides simultaneously present a selective advantage of candidate strains for the utilization of the intestinal fermentable carbon sources [2].

The 6-h fermentation period was chosen as optimal based on the sensory score. Sensory analysis is an indispensable tool for judging the consumer acceptability of any food product. Food products must possess the basic attributes of pleasure and taste, which have been described as the most meaningful indicators to be successful among consumers [5, 36]. The sensory score obtained for the product was balanced and acceptable to panelists; however due to absence of similar product(s), a more critical evaluation was not possible. Although similar products have not been investigated, yet pH at the end of fermentation affects inhibitory activity of milk (fermented), due to hydrolysis of the inhibitory peptides [22]. During fermentation, structural changes render insoluble proteins more accessible to the attack of pepsin;

such changes have a marked effect on protein digestibility besides generating short chain peptides.

Alpha glucosidase inhibitors reduce postprandial absorption of glucose (and other mono-saccharides) by reducing polysaccharide hydrolysis in the gut. In view of limited choice and safety of synthetic α-glucosidase inhibitors (e.g., acarbose, etc.), potential molecules from natural origin are being extensively explored and their subsequent activities in both dairy and non-dairy food products have been reported [34]. For instance, supplementation of fermented camel milk with Allium sativum has been shown to be effective in inhibiting α-glucosidase in cow and camel milk yogurt.

Storage under refrigerated condition (4°C) for 48 did not lead to significant alteration of the inhibitory activities. Extracts from plant foods have been suggested as good sources of anti-diabetic compounds. The ethanol extracts of popular Korean fruits and vegetables was evaluated for α-glucosidase inhibitory activity. Variable activity (5% to 30%) was observed for various vegetables. Highest inhibition was noted for ethanol extracts [25]. Both polyphenols and flavonoids are abundant in guava and can potentially act as inhibitor for both the enzymes especially α-glucosidase.

Hydrolytic reactions and biochemical changes brought about by micro-organisms play dominant role during fermentation leading to an increase of total phenolic compounds. During such processes, bioactivity [16, 30] of pre-existing functional components are elevated, and digestibility is promoted while new compounds are also synthesized [12]. It has been suggested that degradation of phenolic acid esters and tannins by LAB esterase can increase the antioxidant activity [12, 30]. The high levels of total phenolics along with antioxidant activities observed in fermented guava pulp may be explained on this basis. The antioxidant activities can also be accounted by the increased vitamin-C levels in the fermented samples. Both Lactococcus lactis and Streptococcus thermophilus can synthesize folate (vitamin B11) during fermentation thus enhancing the products' "natural" folate levels [19]. High production of Vitamin B12 in fermented guava pulp was also reported recently [27]. The L. lactis was specifically chosen in this chapter, due to its ability to grow and ferment non-dairy matrices and its capability to produce bioactives from these fermented products [4].

Enhanced protein content was observed in the fermented guava. This may be attributed to the utilization of carbohydrates and due to the hydrolysis of protein by extra cellular enzymes of L. lactis, such as: proteases and also growth of the microorganism. The decrease of con-commutant in carbohydrate indicated its utilization. Both lipid content and moisture content were decreased in the course of fermentation. However, no further decline in moisture content occurred beyond that observed at 6 h. Similar observations were reported for fermented soymilk [23].

Overall, the guava pulp fermented product possessed α-glucosidase inhibitory activity in addition to high levels of polyphenolics, Vitamin C and antioxidants. Although not a part of this study, inhibitory activities were also noted for the pancreatic lipase. Both these enzymes play an important role in diabetes. These properties render possibilities of dietary interventions for both diabetes and CVD by beneficially utilizing the guava pulp. Further studies should shed light on other functional attributes in this product.

8.5 SUMMARY

The potential functional benefits of guava pulp fermented with L. lactis in inhibiting α-glucosidase activity were demonstrated in this study. The important benefits of the product lie in α-glucosidase inhibitory properties, high antioxidants, increased levels of vitamin C. The enhanced level of polyphenolics after fermentation of guava pulp possibly has a potential effect in inhibition of α-glucosidase. Functional properties were retained after storage under refrigerated conditions. The approach can be useful in minimizing wastage of guava fruits in a way of developing ready-to-eat high value product with antidiabetic properties acceptable by all ages.

Credit Author Statement

Harshpinder Kaur performed the experiments and analyzed the results. Moushumi Ghosh wrote the manuscript and provided necessary funds. The authors declare that they have no conflicts of interest. Neither human subjects nor animal experiments were part of this study.

ACKNOWLEDGMENTS

The authors are thankful to Food Security Center of Excellence, TIET for the financial assistance.

KEYWORDS

- **anti-diabetic compounds**
- **guava pulp**
- **Lactococcus lactis**
- **probiotic strain**
- **ready-to-eat**
- **α-glucosidase inhibitory activity**

REFERENCES

1. Adebo, O. A., Njobeh, P. B., Adebiyi, J. A., Gbashi, S., Phoku, J. Z., & Kayitesi, E., (2017). Fermented Pulse-Based Food Products in Developing Nations as Functional Foods and Ingredients in Functional Food – Improved Health Through Adequate Food. Intech Open; online; DOI: 10.5772/intechopen.69170. Available from: https://www.intechopen.com/chapters/55507 (accessed on 2 November 2023).
2. Amaretti, A., Tamburini, E., Bernardi, T., Pompei, A., Zanoni, S., Vaccari, G., Matteuzzi, D., & Rossi, M., (2006). Substrate preference of Bifidobacterium adolescentis MB 239: Compared growth on single and mixed carbohydrates. Applied Microbiology & Biotechnology, 73. 654–662.
3. Association of Official analytical chemists (AOAC), (2000). Official Methods of Analysis. (17th edn.). Association of Official Analytical Chemists, Washington, DC -USA.
4. Bhanwar, S., Singh, A., & Ganguli, A., (2013). Probiotic characterization of hydrolases producing Lactococcus lactis Subsp lactis isolated from pickled yam. International Journal of Food Science &. Nutrition, 65(1), 53–61.
5. Brunso, K., Grunert, K. G., & Fjord, T. A., (2002). Consumers Food Choice and Qualitative Perception (Vol. 77). MAPP, Centre for markedsovervagning – vurdering ogbearbejdning til- Aarhus School of Business; MAPP – Centre for Research on Customer relations in Food Sector.
6. Bustamante, P., Mayorga, L., Ramírez, H., Martínez, P., Barranco, E., & Azaola, A., (2006). Evaluación microbiológica de compuestos conactividad prebiótica (Evalaution of probiotic activity of microbiological compounds). Revista Mex. de Ciencias Farma., 37, 5–10.
7. Capurso, A., Crepaldi, G., & Capurso, C., (2018). Legumes and Pulses: Benefits of Mediterranean Diet in the Elderly Patient (pp. 285–325). Springer International Publishing AG.
8. Cardinali, F., Osimani, A., Milanović, V., Garofalo, C., & Aquilanti, L., (2021). Innovative fermented beverages made with red rice, barley, and buckwheat. Foods, 10(3). Online; DOI: 10.3390/foods10030613.

9. Choudhary, S., Singh, M., Sharma, D., Attri, S., Sharma, K., & Goel, G., (2019). Principal component analysis of stimulatory effect of synbiotic combination of indigenous probiotic and inulin on antioxidant activity of soymilk. Probiotics & Antimicrobial Proteins, 11, 813–819.
10. Church, F. C., Swaisgood, H. E., Porter, D. H., & Catignani, G. L., (1983). Spectrophotometric assay using ophthaldialdehyde for determination of proteolysis in milk and isolated milk proteins. Journal of Dairy Science, 66, 1219–1227.
11. Dong, Q., Hu, N., Yue, H., & Wang, H., (2021). Inhibitory activity and mechanism investigation of hypericin as novel α-glucosidase inhibitor. Molecules, 26, E-article ID: 4566; https://doi.org/10.3390/molecules26154566.
12. Filanino, P., Cagno, R. D., Tlais, A. Z. A., Cantatore, V., & Gobetti, M., (2019). Fructose-rich niches traced the evolution of lactic acid bacteria toward fructophilic species. Critical Reviews in Microbiology, 45(1), 65–81.
13. Gutiérrez, R. M., Mitchell, S., & Solis, R. V., (2008). Psidium guavaja: A review of its traditional uses, phytochemistry and pharmacology. Journal of Ethnopharmacology, 117(1), 1–27.
14. Huebner, J., Wehling, R. I., & Hutkins, R. W., (2007). Functional activity of commercial prebiotics. International Dairy Journal, 17, 770–775.
15. Hussein, H. S., Awad, S., El-Sayed, I., & Ibrahim, A., (2020). Impact of chickpea as prebiotic, antioxidant and thickener agent of stirred bio-yoghurt. Annals of Agricultural Science, 65, 49–58.
16. Hur, S. J., Lee, S. Y., Kim, Y. C., Choi, I., & Kim, G. B., (2014). Effect of fermentation on the antioxidant activity in plant-based foods. Food Chemistry, 160, 346–356.
17. Kamath, J. V., Rahul, N., Kumar, A. C. K., & Lakshmi, S. M., (2008). Psidium guajava L.: A Review. International Journal of Green Pharmacology, 2, 9–12.
18. Kohajdova, Z., & Karovicova, J., (2007). Fermentation of cereals for specific purposes. Journal of Food and Nutrition Research, 46(2), 51–57.
19. LeBlanc, J. G., Laiño, J. E., Del Valle, M. J., Vannini, V. D., Van, S. M. P., De Valdez, G. F., De Giori, G. S., & Sesma, F., (2011). B-group vitamin production by lactic acid bacteria: Current knowledge and potential applications. Journal of Applied Microbiology, 111(6), 1297–1309.
20. Lim, Y. Y., & Quah, E. P. L., (2007). Antioxidant properties of different cultivars of Portula caoleracea. Food Chemistry, 103, 734–740.
21. Moretti, V. M., Madonia, G., Diaferia, C., Paleari, M. A., Pirone, G., & Gandini, G., (2004). Chemical and micro biological parameters and sensory attributes of a typical Sicilian salami ripened in different conditions. Meat Science, 66, 845–854.
22. Nielson, M. S., Martinussen, T., Flambard, B., Sorensen, K. I., & Otte, J. A. H., (2009). Peptide profiles and angiotensin-I-converting enzyme inhibitory activity of fermented milk products: Effect of bacterial strain, fermentation pH, and storage time. International Dairy Journal, 19(3), 155–165.
23. Obadina, A. O., Akinola, O. J., Shittu, T. A., & Bakare, H. A., (2013). Effect of natural fermentation on the chemical and nutritional composition of fermented soymilk. Nigerian Food Journal, 31, 91–97.
24. Olievera, R. P. S., Perego, P., Olievera, M. N., & Converti, A., (2012). Growth, organic acid profile and sugar metabolism of Bifidobacterium lactis in co-culture with Streptococcus thermophilus: The inulin effect. Food Research International, 48, 21–27.

25. Park, K. Y., Jeong, J. K., Lee, Y. E., & Daily, J. W., (2014). Health benefits of kimchi (Korean fermented vegetables) as a probiotic food. Journal of Medicinal. Foods, 17, 6–20.
26. Parra-Matadamas, A., Mayorga-Reyes, L., & Pérez-Chabela, M. L., (2015). In vitro fermentation of agro-industrial by-products: Grapefruit albedo and peel, cactus pear peel and pineapple peel by lactic acid bacteria. International Food Research Journal, 22(2), 859–865.
27. Palachum, W., Choorit, W., Manurakchinakorn, S., & Chisti, Y., (2020). Guava pulp fermentation and processing to a vitamin B12-enriched product. Journal of Food Processing and Preservation, 44(8), 36–41.
28. Qiao, Q., Chen, L., Li, X., Lu, X., & Xu, Q., (2021). Molecular mechanisms of dietary bioactive compounds in redox balance and metabolic disorders. Oxidative Medicine and Cellular Longevity, 2021, 1–27.
29. Qureshi, M. N., Stecher, G., & Bonn, G. K., (2014). Determination of total polyphenolic compounds and flavonoids in Juglans regia leaves. Pakistan Journal of Pharmaceutical Science, 27(4), 865–869.
30. Rodrìguez, H., Curiel, J. A., Landete, J. M., Las Rivas, B., De Felipe, F. L., & Gòmez-Cordovés, C. G., (2009). Food phenolics and lactic acid bacteria. International Journal of Food Microbiology, 132, 79–90.
31. Sadasivam, S., & Manickam, A., (2008). Biochemical Methods (pp. 9–11). New Age International (P) Ltd.; third edition.
32. Shabbir, H., Kausar, T., Noreen, S., Rehman, H. U., Hussain, A., Hunag, Q., Gani, A., et al., (2020). In vivo screening and antidiabetic potential of polyphenol extracts from guava pulp, seeds, and leaves. Animals, 201, 1–14.
33. Serfaini, M., & Peluso, I., (2016). Functional foods for health: The interrelated antioxidant and anti-inflammatory role of fruits, vegetables, herbs, spices, and cocoa in humans. Current Pharmaceutical Design, 22(44), 6701–6715.
34. Watanabe, J., Kawabata, J., Kurihara, H., & Niki, R., (1997). Isolation and identification of alpha-glucosidase inhibitors from Tochu-cha (Eucommia ulmoides). Bioscience, Biotechnology and Biochemistry, 61(1), 177–178.
35. Yao, X., Mauldin, R., & Byers, L., (2003). Multiple sugar binding sites in α-glucosidase. Biochim et Biophy. Acta (BBA) – Prot. and Proteomics, 1645(1), 22–29.
36. Zieliński, H., Szawara-Nowak, D., Bączekand, N., & Wronkowska, M., (2019). Effect of liquid-state fermentation on the antioxidant and functional properties of raw and roasted buckwheat flours. Food Chemistry, 271, 291–297.

PART IV

ROLE OF PROBIOTICS AS FUNCTIONAL FOODS ON HUMAN HEALTH

CHAPTER 9

Role of Probiotics on Gut Microflora

Sangeeta Shukla and Anita Raisagar

ABSTRACT

Probiotics possess health benefits to the host when consumed in suitable amounts. This chapter focuses on history and concepts of probiotics, and therapeutic values of probiotics. There are different mechanisms by which probiotics positively benefit the health of the host by adhesion to the intestinal mucosa, enchaining epithelial barriers, modeling immunity, competitive exclusion of pathogens and producing antimicrobials like bacteriocin, organic acids, etc. The beneficial effects of probiotics include treatment of intestinal disorders, infectious diarrhea and *Helicobacter pylori* infection, prevention of osteoporosis, hypercholesteremia, hepatic encephalopathy, IBD, IBS, UTI, managing lactose intolerance and anticarcinogenic activity. There are no side effects noted by the use of probiotics *in vivo* study, thus it may be used as an alternative to antibiotics to enhance gut microflora.

9.1 INTRODUCTION

The term probiotic originates from the *probios* that means "for life." In 1953, this term was firstly introduced by the W. Kollath to represent all food complexes either organic or inorganic as probiotics. The purpose of such practice was to elevate these food complexes as supplements. In 1965, Lilly and Stillwell used this term in a completely different context. Lilly and Stillwell define probiotics as substances that is secreted by an organism and enhance the growth of another organism [54].

According to McFarland [61], "*probiotics are fed supplemented with live microbes that positively affect the consumer by improving gut microflora*

The Functional Foods: Nutrient and Health Benefits.
Megh R. Goyal, Junaid Ahmad Malik, and Anu Kumari (Eds.)

balance" [61]. Technically the term probiotics is defined by WHO and FAO, as "*live microorganisms which upon assimilation in a defined numbers provides health profits elsewhere inherent general nourishment*" [25, 115]. In other words, probiotics are defined as live microorganisms that positively affect the host's health when administered in appropriate amounts [10, 41].

The human gut contains approximately 10^{14} bacterial cells of more than 400 different bacterial species. Some of these bacteria are beneficial while others might be pathogenic for humans. The factors that affect the composition of the gut microflora are the age, immunological status, nutritional requirements, and alcohol consumption habits of the host. The other factors which determine the gut microflora composition are pH, transit time, and the presence of material in the gut. The amount and type of fermentable material or growth substances is also important to determine the bacterial species present in the gut.

The microbial colonization of the gut possesses health implications for humans. The gut microflora influences human health by maturing the gut, providing nutrition to the host, giving resistance to pathogens and maintaining host health [93].

This chapter highlights and discusses in detail about history and concept of probiotics, different mechanisms of probiotics, and the importance of probiotics for influencing gut microflora as a therapeutic agent.

9.2 HISTORY OF PROBIOTICS

The history of probiotics is closely linked with the use of fermented food; therefore, it is as old as human history [73]. It is theorized that around 10,000 years ago, man started to produce fermented food products. The domestication of animals occurred during neolithic period of the stone age. Sumerians were first people, who developed animal farming. Nevertheless, the first symbolic indication of the practice of milking appeared in the ancient center of Mesopotamia before 3100 BC. In that pictorial evidence, one man sited on a stool is squeezing a cow's nipples and collecting the milk in a bucket. A wall painting dated back to 2500 B.C. also showed the milking and fermentation habit of Sumerians [48]. Also, in ancient Indian Ayurvedic texts, it is noted that ingesting of milk or milk products is directly linked with the healthy and long life of consumers.

Milk fermentation originated in the ancient Egyptians, Phoenicians, and Eastern cultures. In ancient times, collected milk was kept in bottles. The bottle was made up of from stomach or skin of the animals from which the milk was collected. In that bottle, milk directly interacted with the bacterial strains which might be the dynasties of the bulgaricus or acidophilus. It is hypothesized that in the desert of Turk, one of the shepherds forgot a skin bag filled with goat milk, which converted into a creamy, thick, and tasty custard. That is now known as "yogurt." It is believed by Turks that yogurt gives physical and inner well-being and long life.

Prophet Muhammad gave kefir grains to the dynasties of the climbers of the Caucasus. This is the first recorded evidence of the symbiotic association of probiotic bacteria and yeast. In the book Naturalis Historia, Pliny, the Elder suggested the use of fermented milk for the treatment of intestinal infections. Hippocrates IV believed Olympic athletes got nutrition by consumption of cheese. In the 16th century, treatment of severe diarrhea by using yogurt was recorded. This treatment was prescribed by a physician in a Turkish court on the order of Suleiman the Magnificent.

Hippocrates and Galeno recognized high nutritional value of functional food along with its negative effects and advised the use of functional food only for medicinal use. These beliefs were the base of environment and climate contamination. In Italy, women prepare a drink made with milk, lemon, and water. This drink helps people to struggle with dust and heat and also avoid colonic infections. A probiotic therapy, called fecal bacteriotherapy, also known as stool or fecal microbiota transplantation, was first reported in the late 1950s when a team of surgeons effectively treated *Pseudomembranous enterocolitis* infection in four patients [21]. In this therapy, washed suspended feces from a healthy donor was permeated into a patient's colon via endoscopy.

Physician Ge Hong also used this therapy for the treatment of diarrhea. At the same time, the idea of transferring colonic samples in veterinary practice was in use in Europe also. F. Aquapendente, an Italian anatomist, designated the practice of "transfaunation." In this practice, the cow who lost the capacity to ruminate was inoculated with rumen fluid. By this inoculation, the gastrointestinal tract (GIT) of the animal was colonized with bacteria. Later, transplantation of fecal was also used in avian species for protection of chicken from *Salmonella* infections [38].

In Africa, German soldiers used fresh feces of camel for treatment of diarrhea. This therapy was successful only when the used fecal or dung was fresh in condition. The active ingredients of fecal or dung which are responsible for the treatment of infections were later identified as *Bacillus subtilis.* Kelly [44] also reported the use of this therapy for prevention of infection caused by *Clostridium difficile*. The popularity of this therapy is its effectiveness and ease of use.

Here historical viewpoints offer a very expressive background to the recent state of the research about gut microflora and its role in promoting human health. Nowadays, the mutual interaction of gut microflora with the host becomes an important research topic in field of biomedical and clinical research. However, its significance and precise role need more research and studies.

9.3 CONCEPT OF PROBIOTICS

The probiotic's modern history is started at the beginning of the 1900s with the revolutionary studies of Elie Metchnikoff. The microbes responsible for fermentation was identified by Louis Pasteur. Later, Metchnikoff, who worked at Pasteur institute, tried to find the influence of identified microorganisms on human health. In 1908, Russian scientist Ellie Metchnikoff studied the lifestyle of Bulgarians as they are healthy and long-lived people. He observed that in Bulgaria, consumption of fermented milk products, i.e., Bulgarian milk is a part of daily routine. For the production of Bulgarian milk rod-shaped *Lactobacillus* bacteria was used. Metchnikoff hypothesized that these bacteria positively affect the gut microflora as well as decrease the toxic activity of the microbes. Later he was awarded the Nobel prize for proposing the beneficial effect of probiotics on human health [15].

Bulgarian physician Stamen Grigorov discovered *Bulgarian bacillus*, which is the *Lactobacillus* bacteria used for the production of Bulgarian milk. According to Stamen Grigorov, *Bulgarian bacillus* prevents the putrefactive effects of gastrointestinal (GI) metabolism that may result in sickness and aging. Metchnikoff also supported these assumptions and described that due to bacterial putrefaction, toxins released in the large intestine cause aging. This putrefying bacterium is nowadays known as *Clostridia*. Metchnikoff specified that "*requirement of colonic microbes on food makes it possible to adopt actions to alter the flora in host bodies and to substitute the harmful microorganisms with beneficial microorganisms*." This statement clearly defines the "probiotic concept." Metchnikoff considered the *Lactobacilli* as probiotics [34].

9.4 MECHANISM OF IMPACT OF PROBIOTICS ON THE GUT MICROFLORA

9.4.1 INCREASED ADHESION TO THE INTESTINAL MUCOSA

Adhesion to the intestinal mucosa is a significant criterion for the selection of a strain as probiotics because it is directly associated with the beneficial effects of probiotics [36, 77]. Adhesion to intestinal mucosa is essential for colonization and interface between probiotic strains and the host. It is also vital for modulating immune system [77] and antagonistic activity against pathogens [36]. Interaction of probiotics with colonic epithelial cells and mucus is possible because of the presence of surface determinants in probiotics. Mucin, secreted by intestinal epithelial cells is a complex mixture of glycoprotein and the main constituent of mucous. Mucin prevents the bonding of bacterial pathogens [16, 30]. In mucous gel, immunoglobulins, free proteins, salts, and Lipids are present. This shows a conceivable association between competitive exclusion of pathogens from the mucus and probiotic bacterial surface proteins [31, 72, 106].

Probiotics also cause qualitative modifications in colonic mucins that avert pathogen adhesion [47]. Probiotics might also encourage release of defensins from epithelial cells which stabilize the function of gut barrier [26]. There are several mechanisms explained in the previous studies, still, there is a need to understand the factors that limit the mucosal adhesion of probiotics.

9.4.2 PRODUCTION OF INHIBITORY SUBSTANCES AND ANTIMICROBIALS

Probiotics produce inhibitory substances like organic acids and bacteriocins. Production of these antimicrobials also plays a significant role in influencing gut microflora. Organic acids are the key antimicrobic compounds that are responsible for the repressive or inhibitory action of probiotics against gut pathogens, as it shows a strong repressive action against Gram-negative bacteria [3, 19, 58].

Organic acid enters in a pathogenic cell in an undissociated form and inside the cell's cytoplasm, it is dissociated. This lowers the intracellular pH and leads to the pathogen's death by intracellular accumulation of organic acid in an ionized form. Probiotics also produce peptides having antimicrobial potentials, such as bacteriocins and small antimicrobial peptides.

The common mechanisms of antimicrobial activity of bacteriocin include inhibition of cell wall synthesis and destruction of target cells by pore formation [33].

9.4.3 COMPETITIVE EXCLUSION OF PATHOGENIC MICROORGANISMS

Greenberg in 1969 used the term, Competitive exclusion,' for the situation in which one intestinal bacterial species strongly competes with other bacteria for receptor sites. Several studies documented competitive exclusion of pathogens namely *Salmonella, Listeria monocytogenes, Escherichia coli, Helicobacter pylori,* Rotavirus, etc., by probiotic bacteria *Lactobacilli* and *Bifidobacteria* [12, 14, 66, 67, 87, 99].

The mechanisms of exclusion of one species of bacteria by another species include blocking of adhesion sites of pathogens, production of antimicrobial substances, secretion of selective metabolites, exclusion of available bacterial receptor sites, and competitive diminution of essential nutrients. In competitive exclusion, probiotics modify the gut microecology and make it less suitable for pathogens. An example of such modification is the production of organic acid as reported by Schiffrin & Blum [86].

9.4.4 STIMULATION AND MODULATION OF IMMUNITY

Probiotics can interact with epithelial cells, lymphocytes, dendritic cells, monocytes, and macrophages. Dendritic cells, also known as antigen-presenting cells, are significant in bacterial acknowledgment and determining the subsequent T-cell responses. These cells induce oral tolerance, generate regulatory T cells, and produce cytokines in the intestine [2, 11]. The effects of diverse probiotic bacteria on Dendritic cells have been documented in several experimental systems and different species (human and mouse).

Ulisse et al. [102] studied the treatment of pouchitis *in vitro* with probiotic combination by enhancement of IL-10 level and decrement of TNF-α, IL-1, matrix metalloproteinase and inducible nitric oxide synthase in patients but in *in-vivo* study cellular source of the IL-10 was not known. Pathmakanthan et al. [75] also reported the synthesis of IL-10 and secretion of macrophages by probiotics *Lactobacillus plantarum* derived from the inflamed colon. Probiotics also modulate immunity by the production of antibodies and B lymphocytes.

Martin et al. [60] also reported elevation of natural killing cells activity in the spleen by using probiotic *Lactobacillus casei* in combination with dextran in an *in vivo* study and suggested oral administration of this symbiotic combination for natural killing cells induction and IL-12 production in human peripheral blood.

9.4.5 AUGMENTATION OF THE EPITHELIAL BARRIER

In maintaining epithelial integrity and protection of organisms from the environment, the epithelial barrier is a major defense mechanism. It comprises epithelial junction adhesion complex, mucous layer, secretory IgA, and antimicrobial peptides [71]. Disruption or disturbance of this epithelial barrier function is reported in the condition of enteric infections, coeliac disease (CD), inflammatory bowel disease (IBD), and some autoimmune diseases like Type 1 diabetes [84, 111]. If the function of the epithelial barrier is disturbed, food and bacterial antigens may influence the submucosa which may cause intestinal disorders [59, 104, 117].

Several studies have reported the involvement of probiotics in the maintenance of epithelial barrier, but the exact mechanism is not known. In both *in vivo* and *in vitro* studies enhancement of barrier function by probiotic strain is documented. The possible mechanisms might be enhancement of gene expressions involved in tight junction signaling, promotion of mucous secretion, and prevention of epithelial damage induced by cytokine [69].

9.5 THERAPEUTIC VALUES OF PROBIOTICS

9.5.1 TREATMENT OF INTESTINAL DISORDERS

Several studies reported a decrement in occurrence, duration, and brutality of gastric and intestinal illness by using probiotics at a level of 10^9–10^{11} per day. In this section, treatment of intestinal disorders (i.e., antibiotic-associated Diarrhea and intestinal disease associated with *Clostridium difficile*) with use of probiotics are discussed. Antibiotic-associated diarrhea is a disorder characterized by imbalances in natural gut microflora because of the side effect of antibiotic therapy. Treatments/use of antibiotics may cause insignificant occurrences of diarrhea by defeating the normal gut microflora and stimulating the growth of pathogens.

Antibiotic-associated diarrhea associated with the overgrowth of *Clostridium difficile* is termed *Clostridium difficile*-associated diarrhea [85]. Many studies demonstrated the use of probiotics in antibiotic-associated diarrhea, but this practice is not found practical from the cost-benefit viewpoint. For treatment of antibiotic-associated diarrhea, *Enterococcus faecium, L. acidophilus, L. delbrueckii, L. rhamnosus* GG, *L. fermentum* and *Saccharomyces boulardii* showed positive effects. In the reduction of the onset of diarrhea probiotics [62], probiotic yeast *Saccharomyces boulardii* also showed a beneficial effect on the prevention of intestinal disease associated with *Clostridium difficile.*

Alberda et al. [4] suggested the use of *Lactobacillus casei* drink for the treatment of antibiotic-associated diarrhea and *Clostridium difficile*-associated diarrhea. Syngai et al. [96] recommended the use of probiotics *Leuconostoc cremoris* individually or in combination with *Saccharomyces, Bacillus, Lactobacillus, Lactococcus, Bifidobacterium,* or *Streptococcus* spp. for effective treatment of antibiotic-associated diarrhea.

9.5.2 TREATMENT OF INFECTIOUS DIARRHEA

WHO defines diarrhea as 3–4 watery stools in 24 h. It is very difficult to investigate the treatment of infectious diarrhea by using probiotics because there are several etiologic agents that cause diarrhea (e.g., rotavirus, *Shigella, Clostridium difficile, Escherichia coli, Salmonella*), and the bulk of the investigations do not try to framework the cause of diarrhea.

According to epidemiological data, approximately 20–60% of the people, traveling around the world are infected with traveler's diarrhea. It particularly impacts people who travel from industrial areas to remote areas or developed areas to backward areas. This type of diarrhea is reliant on the origin, destination, and mode of travel. In 60–85% of cases of traveler's diarrhea, Bacterial pathogens are responsible. The major cause of this diarrhea is *Campylobacter jejuni, Salmonella* spp., *Escherichia coli,* and *Shigella* [27, 35]. There are several studies performed for the treatment of traveler's diarrhea by use of probiotics. Bacterial Strains that are supposed to have a defensive consequence on the traveler's diarrhea includes *Lactobacilli, Enterococci, Bifidobacteria,* and *Streptococci.*

Bae [7] concluded the statistically significant effect of probiotics in the inhibition of traveler's diarrhea. The role of probiotics against bacterial and viral diarrhea was also studied by Sniffen et al. [91]. The study concluded the beneficial role of probiotics *Saccharomyces boulardii* and *Lactobacillus rhamnosus* GG against bacterial diarrhea and viral diarrhea, respectively.

9.5.3 REDUCTION OF HELICOBACTER PYLORI INFECTION

Helicobacter pylori infection is the most common and worst bacterial infection found in humans because it causes several diseases namely, peptic chronic gastritis, ulcers, lymphoma, gastric malignancies, and gastric adenocarcinoma [74]. Several investigations reported that probiotics prevent *Helicobacter pylori* infection. The main mechanisms of using probiotics include immunoregulation function enhancement that works as antagonism for infection caused by *Helicobacter pylori* [11, 24, 32]. Homan & Orel [37] found a reduction in *Helicobacter pylori* loads by the use of probiotic *Saccharomyces boulardii* and *Lactobacillus johnsonii* La1 but complete eradication of pathogen was not observed.

9.5.4 INFLAMMATORY BOWEL DISEASE (IBD)

This disease is instigated by GIT inflammation that might persuade severe watery and bloody diarrhea followed by intestinal discomfort. IBD includes ulcerative colitis, pouchitis, and Crohn's disease which affects both the colon and small intestine [20]. Ulcerative colitis fundamentally affects the rectum's internal layer and an inner layer of large intestines and can lead to colon cancer development. Pouchitis is ileal pouch inflammation, mainly caused after colectomy or anastomosis of the ileal pouch. The irregular immune system, genetic and environmental factors, oxidative stress, and intestinal dysbiosis are reported as important features of IBD commencement [64].

Crohn's disease generally affects the digestive system or any organ of the digestive tract and lead to ulcer development and irritation. The Bacterial strains that are involved in Crohn's disease include *Campylobacter jejuni*, *Salmonella*, *Mycoplasma*, *Clostridium difficile*, and Adenovirus. Probiotics may exert therapeutic effects on IBD conditions by producing antimicrobic substances, affecting the configuration of microbial ecology and competing for adhesion sites and nutrients via cell-to-cell communication, and. In addition, probiotics interact with the immune system of the host via bacterial products, DNA, or cell wall components [39].

Kelesidis et al. [43] studied the beneficial effect of probiotics *Saccharomyces boulardii, Lactobacillus casei,* and *Bifidobacterium bifidum* for ulcerative colitis treatment. Asto et al. [6] also support this study and suggested the use of probiotic bacteria *Bifidobacterium bifidum, Bifidobacterium breve,* and *Lactobacillus acidophilus* in the fermented milk for IBD treatment. Mechanisms involved in the therapeutic role of probiotics in Crohn's disease treatment may be competition with bacteria, maintaining the immune system and defensive functions of the intestinal mucosa and destruction of pathogenic microorganisms [105]. The use of VSL3, a formulation of eight probiotic strains plus *Escherichia coli* Nissle 1917, for the treatment of pouchitis was suggested by Gionchetti et al. [29]; and Lichtenstein et al. [53]. There is no adverse or antagonistic effect reported in patients when probiotics are used for the treatment of IBDs. However, Sivamaruthi [90] reported mild dry cough in a patient suffering from ulcerative colitis when *Bifidobacterium longum* 536 supplementations were used for treatment.

9.5.5 IRRITABLE BOWEL SYNDROME (IBS)

This syndrome is characterized by bloating, constipation, intestinal pain, and looseness of the bowels [8] that occur primarily during childhood. The important factors that lead to irritable bowel syndrome (IBS) are stress, gut mobility and sensitivity, and emotional status [63]. There is no precise treatment for this condition. However, some studies suggested a reduction in the severity of abdominal pain in patients by the use of probiotics. Kim et al. [45]; and Shin et al. [89] found a reduction in diarrhea and other symptoms of IBS when using the probiotic *Lactobacillus gasseri* BNR17 derived from breast milk. Multi-strain probiotic supplementation is more effective for reducing the severity of IBS symptoms as compared to a single strain [18].

9.5.6 MANAGING LACTOSE INTOLERANCE

People all over the world suffer from an inherited deficiency of the enzyme β-galactosidase. Due to the lack of this enzyme, hydrolyzation of lactose into the monosaccharides does not occurs. This unabsorbed lactose is then metabolized by gut microflora [70] and the resultant by-products causes diarrhea, intestinal discomfort, nausea, bloating, and flatulence after milk or dairy products consumption [96]. This symptom may reduce by using probiotics.

There are two possible mechanisms of probiotics for use in treatment or management of lactose intolerance conditions, i.e., fermentation that reduces lactose in dairy products and release of lactase by probiotics in the GI tract. Vonk et al. [109] found easy digestion of milk in lactose-intolerant individuals when milk is incorporated with probiotic cultures *Bifidobacterium longum* and *Lactobacillus acidophilus*. Kechagia et al. [42] suggested the use of probiotics *L. delbrueckii* subsp. *bulgaricus* and *S. thermophilus* in yogurt for alleviating lactose intolerance. Gingold-Belfer et al. [28] found a decrement in lactose intolerance cases and symptoms after the use of probiotics for 6 weeks.

9.5.7 HEPATIC ENCEPHALOPATHY

In healthy individuals, detoxification of ammonia produced by the action of gut microflora, is done by the liver. But, when liver failure occurs, the concentration of ammonia in the blood may reach up to toxic levels. This condition is called hepatic encephalopathy. It is a neurological disorder. Several studies suggested that by using probiotics, gut microflora count changes and this strategy may use for the handling of hepatic encephalopathy [57].

Dalal et al. [17] also supported this strategy. In previous studies, patients were treated with both probiotic *Lactobacillus acidophilus* and neomycin. Patients treated with probiotics + neomycin showed a higher reduction in the activity of fecal urease as compared to the patients treated with neomycin alone. Reduction in fecal urease activities is directly related to lowering of serum ammonia levels and enhancements in the quantifiable status of patients.

9.5.8 REDUCTION OF CHOLESTEROL

Cholesterol is a significant fat-like and biological waxy compound related to lipoproteins. Hypercholesteremia is a metabolic disorder facilitated by cellular cholesterol levels and abnormal serum [49]. High serum cholesterol levels may lead to diabetes type II, hypertension, coronary heart diseases (CHDs) and atherosclerosis [13].

Recently, probiotic specially *Bifidobacteria* and *Lactobacilli* are regarded as innovative therapeutic approaches against hypercholesteremia [76]. Bile salt hydrolysis (BSH) activity is the main mechanism of probiotics for a reduction in cholesterol levels. Kim et al. [46] by his *in-vivo* study suggested the use of mixer of probiotic cultures in which 3 strains of *Bifidobacterium* and 2 strains of *Lactobacillus* was used for reduction of cholesterol in mice model.

A similar *in vivo* study was done by Wang et al. [110] for BSH activity of probiotics *Lactobacillus casei*, *Lactobacillus plantarum* and *Lactobacillus casei*. Use of these probiotics may reduce body weight, total serum cholesterol, atherogenic index and low-density lipoprotein cholesterol levels.

9.5.9 PREVENTION OF RESPIRATORY TRACT INFECTION (RTI)

Respiratory tract infection (RTI) is a common infection occurred in respiratory part of human body. Recent research suggested that probiotics might reduce the risk and the period of symptoms of this infection. Rautava et al. [80] reported a reduction in RTI by using the probiotic *Bifidobacterium animalis* subsp. *Lactis* with combination of *Lactobacillus rhamnosus* GG.

Laursen & Hojsak [50] also observed reduction in RTI by use of probiotics *Lactobacillus rhamnosus* GG but positive results were not observed with *Bifidobacterium animalis* subsp. *Lactis*. In a study done by Pu et al. [78] on reduction in RTI was recorded with the administration of probiotic yogurt supplemented with *Lactobacillus paracasei* N1115 by increasing T cells (CD3+). The above supplementation modulates the native immune system. Though, the exact mechanisms of probiotics are yet to be ascertained [52].

9.5.10 PREVENTION OF URINARY TRACT INFECTIONS (UTIS)

Urinary tract infections (UTIs) are caused by intestinal uro-pathogens viz., *Proteus*, *Escherichia coli*, *Staphylococcus* spp. and *Klebsiella*. This type of infection is most common in women as compared to men. Falagas et al. [23] suggested the importance of probiotics in the deterrence of UTI and reported positive effects of *Lactobacillus acidophilus* in urogenital health. A reduction in UTI with probiotic *Lactobacillus crispatus* was also reported by Stapleton et al. [92].

Sadeghi-Bojd et al. [83] found *E. coli* and *Klebsiella pneumoniae* responsible for urinary tract contagion in children and the risk of urinary tract contagion reduction was recorded with mixture of probiotic cultures *Lactobacillus rhamnosus*, *Lactobacillus acidophilus*, *Bifidobacterium Lactis* and *Bifidobacterium bifidum* when administered orally.

9.5.11 TREATMENT OF BACTERIAL VAGINOSIS

Normal microflora of healthy women's vagina consists of *Lactobacilli* strains that protects vagina from infections but due to unstable microflora, bacterial vaginosis is caused. Women with bacterial vaginosis have an excessive danger of human immunodeficiency syndrome. The pathological severity of bacterial vaginosis is allied with eminent risks of endometritis, pelvic inflammation, preterm delivery, abrupt abortion. *Gardnerella vaginalis* is the foremost causative agent of bacterial vaginosis. Some studies recommended the use of probiotics for urogenital health [113]. The preventive mechanism of probiotics involves retard sialidase enzyme activity, amelioration of inflammatory cytokines and reduced *Gardnerella vaginalis* binding ability to epithelial cells of vagina.

Liu et al. [56] reported reduction in spread of sexual immunodeficiency virus contamination by use of the probiotic strain *Lactobacillus reuteri* RC-14, as this probiotic strain gave strong protection to the vagina from the virus that caused immunodeficiency infection. In an *in-vivo* study, Sabbatini et al. [82] also found inhibition of *Gardnerella vaginalis* by the use of probiotic *Saccharomyces cerevisiae*.

9.5.12 PREVENTION OF OSTEOPOROSIS

Bone tissues, bone mass and bone porosity deteriorated in osteoporosis and this leads loss in strength and density of bone [9]. Imbalance in gut microflora causes osteoporosis. The major cause of this imbalance includes oxidative stress, inflammatory bowel disorder, systematic inflammation, IBS, diabetes mellitus, hyperthyroidism, abnormality in somatostatin profile, celiac disorder and non-alcoholic liver fatty disease [68]. Some of the drugs such as heparin, cyclosporine, rifampicin, protein kinase inhibitors, phenobarbital, glucocorticoids, thiazolidinedione, cinacalcet, doxorubicin induce Ca, P, and vitamin D malabsorption and thus negatively associated with osteoporosis [1]. The symbiotic effect of probiotics and lactose-based prebiotics for the treatment of osteoporosis was demonstrated by Nath et al. [68].

Liu et al. [55] also emphasized the importance of *Lactobacillus rhamnoses* GG for the prevention of tenofovir-induced mandibular bone loss in the *in-vivo* study. The possible mechanism was inhibition of mandible-derived mesenchymal stem cells and osteogenesis by use of probiotic *Lactobacillus rhamnoses* GG. A comparative *in-vivo* study done by Montazeri-Najafabady et al. [65] concluded that the efficacy of probiotics is mainly strained specific and recorded high bone mineral density, bone marrow concentration and bone area by use of probiotics *Lactobacillus acidophilus* and *Lactobacillus casei*.

9.5.13 ANTICARCINOGENIC EFFECT OF PROBIOTICS

Cancer is a chronic, non-communicable disease (NCD) that causes loss of the natural apoptotic cell cycle. This loss may be due to mutagens accumulation during the DNA replication process [98]. For treatment of cancer chemotherapy is done followed by immunotherapy and radiotherapy, but due to this therapy, cancer patients face severe health side effect. There are several studies carried on assessing the effectiveness of probiotic for managing cancer by improvement of gut microflora [108]. Mechanisms of anticancer activity of probiotics involve strengthening of the GIT mucosal barrier, stabilization of gut microflora, enhancing the anti-inflammatory response and metastasis development and delaying proliferation of cancer cells. Lee et al. [51]; and Uccello et al. [101] studied the positive effect of probiotic *Propionibacterium freudenreichii* on intestinal and colon cancer cells in humans by producing short-chain fatty acids.

Ranjbar et al. [79] conducted a human trial and suggested probiotics for breast cancer treatment. Shamekhi et al. [88] reported prevention of cancer growth by the anti-inflammatory potential of probiotics. The anti-inflammatory potential is due to nuclear factor kappa B and interferon-γ modulation, increment in Ig-A secretion and cyclooxygenase-2 (COX-2) regulation. The exact mechanism for the anticarcinogenic activity of probiotics is not clear, thus there is a need for extensive investigations and clinical trials before probiotics might be considered treatment of cancer.

9.5.14 PREVENTION OF TYPE 2 DIABETES MELLITUS

Type 2 diabetes mellitus is a non-transmissible disease that may later results in obesity, retinopathy, CHDs, kidney failure, atherosclerosis, and sexual weakness [94]. The most common symptoms of type 2 diabetes mellitus are high urination, frequent weight loss, poor foot lesion healing power and thirst. The role of probiotics in the secretion's regulation of insulin and intestinal hormones are studied by Will et al. [112].

Vallianou et al. [103] suggested the importance of gut microflora in type 2 *diabetes mellitus* management. Improvement in function of gut barrier, increment in insulin sensitivity, reduction in serum lipopolysaccharides levels and oxidative and endoplasmic reticulum stress amelioration are therapeutic potential of probiotics in type 2 diabetes mellitus prevention. Probiotics also delay and reverse the complications of type 2 diabetes, in turn, reducing the incidence as reported by Sun et al. [95].

Yan et al. [116] found *Lactobacillus acidophilus* as antidiabetic probiotic. In an *in-vitro* study, the probiotic showed inhibitory activity against dipeptidyl peptidase IV, retarded insulin resistance, increased glucose level in blood and oxidative stress. Limitations of the use of probiotics in the human model for type 2 *diabetes mellitus* treatment are preparation of supplements, inconsistency in measurements, sample size, monitoring of intervention, questionnaire biases and sampling strategies.

In order to comprehend the relationship between the role of gut microflora, glycemic control and collaboration of probiotics with the host further, well-designed studies are need of the hour.

9.6 SUMMARY

Probiotics have significant practical characteristics that might accomplish our maximum basic dietary and quantifiable necessities. These microorganisms showed optimistic responses in dealing of several different diseases and disorders. The importance of probiotics in prevention of diarrhea, intestinal infections, diabetes, UTI, obesity, RTI, osteoporosis, IBD, cancer, IBS, etc., is an exhilarating and rapidly proceeding research area. In both dairy and non-dairy products, probiotics are found. *Lactobacillus* and *Bifidobacteria* are the most common and prominent members of the gut microflora categorized under probiotics and generally recognized as safe (GRAS). In revealing some unusual purposes of specific probiotic strains now a days several clinical and nutritional assessments is successfully done. However, still there is a need of well-designed and well-controlled clinical trials for use of probiotics as therapeutic agents for human.

KEYWORDS

- **anticarcinogenic effect**
- **antimicrobials**
- **diabetes mellitus**
- **gut microflora**
- **immunity**
- **intestinal health**
- **probiotics**
- **therapeutic agents**

REFERENCES

1. Aggarwal, A., Sharma, M., Maisnam, I., Ghosh, S., Aggarwal, S., Bhattacharya, S., & Dutta, D., (2019). Drug-induced bone disorders: A systematic review. *Indian Journal of Rheumatology, 14*(Suppl. S1), 44–51.
2. Akbari, O., DeKruyff, R. H., & Umetsu, D. T., (2001). Pulmonary dendritic cells producing IL-10 mediate tolerance induced by respiratory exposure to antigen. *Nature Immunology, 2*(8), 725–731.
3. Alakomi, H. L., Skytta, E., Saarela, M., Mattila-Sandholm, T., Latva-Kala, K., & Helander, I. M., (2000). Lactic acid permeabilizes gram-negative bacteria by disrupting the outer membrane. *Applied and Environmental Microbiology, 66*(5), 2001–2005.
4. Alberda, C., Marcushamer, S., Hewer, T., Journault, N., & Kutsogiannis, D., (2018). Feasibility of a *Lactobacillus casei* drink in the intensive care unit for prevention of antibiotic associated diarrhea and *Clostridium difficile*. *Nutrients, 10*(5), 539–545.
5. Alvi, S., Javeed, A., Akhtar, B., Sharif, A., & Akhtar, M. F., (2017). Probiotics for cure of *Helicobacter pylori* infection: A review. *International Journal of Food Properties, 20*(10), 2215–2222.
6. Asto, E., Mendez, I., Audivert, S., FarranCodina, A., & Espadaler, J., (2019). The efficacy of probiotics, prebiotic inulin-type fructans, and synbiotics in human ulcerative colitis: A systematic review and meta-analysis. *Nutrients, 11*(2), 293–298.
7. Bae, J. M., (2018). Prophylactic efficacy of probiotics on travelers' diarrhea: An adaptive meta-analysis of randomized controlled trials. *Epidemiology and Health, 40,* E-article ID 2018043.
8. Barbara, G., Cremon, C., Annese, V., Basilisco, G., Bazzoli, F., Bellini, M., Benedetti, A., et al., (2016). Randomized controlled trial of mesalazine in IBS. *Gut, 65*(1), 82–90.
9. Behera, J., Ison, J., Tyagi, S. C., & Tyagi, N., (2020). The role of gut microbiota in bone homeostasis. *Bone, 135*(4), E-article ID 115317.
10. Brown, A. C., & Valiere, A., (2004). Probiotics and medical nutrition therapy. *Nutrition in Clinical Care, 7*(2), 56–68.

11. Canducci, F., Cremonini, F., Armuzzi, A., Di Caro, S., Gabrielli, M., Santarelli, L., Nista, E., et al., (2002). Probiotics and *Helicobacter pylori* eradication. *Digestive and Liver Disease, 34*(Suppl 2), S81–S83.
12. Chenoll, E., Casinos, B., Bataller, E., Astals, P., Echevarría, J., Iglesias, J. R., Balbarie, P., et al., (2011). Novel probiotic *Bifidobacterium bifidum* CECT 7366 strain active against the pathogenic bacterium *Helicobacter pylori*. *Applied and Environmental Microbiology, 77*(4), 1335–1343.
13. Chien, Y. L., Wu, L. Y., Lee, T. C., & Hwang, L. S., (2010). Cholesterol-lowering effect of phytosterol-containing lactic-fermented milk powder in hamsters. *Food Chemistry, 119*(3), 1121–1126.
14. Chu, H., Kang, S., Ha, S., Cho, K., Park, S. M., Han, K. H., Kang, S. K., et al., (2005). *Lactobacillus acidophilus* expressing recombinant K99 adhesive fimbriae has an inhibitory effect on adhesion of enterotoxigenic *Escherichia coli*. *Microbiology and Immunology, 49*(11), 941–948.
15. Chuayana, E. L., Ponce, V. C., Rivera, M. R. B., & Cabrera, E. C., (2003). Antimicrobial activity of probiotics from milk products. *Philippine Journal of Microbiology and Infectious Diseases, 32*(2), 71–74.
16. Collado, M. C., Gueimonde, M., Hernández, M., Sanz, Y., & Salminen, S., (2005). Adhesion of selected *Bifidobacterium* strains to human intestinal mucus and the role of adhesion in enteropathogen exclusion. *Journal of Food Protection, 68*(12), 2672–2678.
17. Dalal, R., McGee, R. G., Riordan, S. M., & Webster, A. C., (2017). Probiotics for people with hepatic encephalopathy. *Cochrane Database of Systematic Reviews, 2*(2), E-article ID 008716.
18. Dale, H. F., Rasmussen, S. H., Asiller, Ö. Ö., & Lied, G. A., (2019). Probiotics in irritable bowel syndrome: An up-to-date systematic review. *Nutrients, 11*(9), E-article ID 2048.
19. De Keersmaecker, S. C., Verhoeven, T. L., Desair, J., Marchal, K., Vanderleyden, J., & Nagy, I., (2006). Strong antimicrobial activity of *Lactobacillus rhamnosus* GG against *Salmonella typhimurium* is due to accumulation of lactic acid. *FEMS Microbiology Letters, 259*(1), 89–96.
20. Derwa, Y., Gracie, D. J., Hamlin, P. J., & Ford, A. C., (2017). Systematic review with meta-analysis: The efficacy of probiotics in inflammatory bowel disease. *Alimentary Pharmacology Therapeutics, 46*(4), 389–400.
21. Eiseman, B., Silen, W., Bascom, G. S., & Kauvar, A. J., (1958). Fecal enema as an adjunct in the treatment of pseudo membranous *enterocolitis*. *Surgery, 44*(5), 854–859.
22. Eslami, M., Yousefi, B., Kokhaei, P., Hemati, M., Nejad, Z. R., Arabkari, V., & Namdar, A., (2019). Importance of probiotics in the prevention and treatment of colorectal cancer. *Journal of Cellular Physiology, 234*(10), 17127–17143.
23. Falagas, M. E., Betsi, G. I., Tokas, T., & Athanasiou, S., (2006). Probiotics for prevention of recurrent urinary tract infections in women: A review of the evidence from microbiological and clinical studies. *Drugs, 66*(9), 1253–1261.
24. Felley, C., & Michetti, P., (2003). Probiotics and *Helicobacter pylori*. *Best Practice & Research: Clinical Gastroenterology, 17*(5), 785–791.
25. Food and Agriculture Organization of the United Nations/World Health Organization FAO/WHO, (2001). *Health and Nutritional Properties of Probiotics in Food including Powder Milk with Live Lactic Acid Bacteria*. https://www.fao.org/3/a0512e/a0512e.pdf (accessed on 28 November 2023).

26. Furrie, E., Macfarlane, S., Kennedy, A., Cummings, J. H., Walsh, S. V., O'neil, D. A., & Macfarlane, G. T., (2005). Synbiotic therapy (*Bifidobacterium longum* /Synergy 1) initiates resolution of inflammation in patients with active ulcerative colitis: A randomized controlled pilot trial. *Gut, 54*(2), 242–249.
27. Giddings, S. L., Stevens, A. M., & Leung, D. T., (2016). Traveler's diarrhea. *Medical Clinics of North America, 100*(2), 317–330.
28. Gingold-Belfer, R., Levy, S., Layfer, O., Pakanaev, L., Niv, Y., Dickman, R., & Perets, T. T., (2020). Use of a novel probiotic formulation to alleviate lactose intolerance symptoms-a pilot study. *Probiotics and Antimicrobial Proteins, 12*(1), 112–118.
29. Gionchetti, P., Rizzello, F., Helwig, U., Venturi, A., Lammers, K. M., Brigidi, P., Vitali, B., et al., (2003). Prophylaxis of pouchitis onset with probiotic therapy: A double-blind, placebo-controlled trial. *Gastroenterology, 124*(5), 1202–1209.
30. González-Rodríguez, I., Sánchez, B., Ruiz, L., Turroni, F., Ventura, M., Ruas-Madiedo, P., Gueimonde, M., & Margolles, A., (2012). Role of extracellular trans aldolase from *Bifidobacterium bifidum* in mucin adhesion and aggregation. *Applied and Environmental Microbiology, 78*(11), 3992–3998.
31. Haller, D., Colbus, H., Ganzle, M. G., Scherenbacher, P., Bode, C., & Hammes, W. P., (2001). Metabolic and functional properties of lactic acid bacteria in the gastrointestinal ecosystem: A comparative *in vitro* study between bacteria of intestinal and fermented food origin. *Systematic and Applied Microbiology, 24*(2), 218–226.
32. Hamilton-Miller, J. M., (2003). The role of probiotics in the treatment and prevention of *Helicobacter pylori* infection. *International Journal of Antimicrobial Agents, 22*(4), 360–366.
33. Hassan, M., Kjos, M., Nes, I. F., Diep, D. B., & Lotfipour, F., (2012). Natural antimicrobial peptides from bacteria: Characteristics and potential applications to fight against antibiotic resistance. *Journal of Applied Microbiology, 113*(4), 723–736.
34. Hill, C., Guarner, F., Reid, G., Gibson, G. R., Merenstein, D. J., Pot, B., Morelli, L., et al., (2014). The international scientific association for probiotics and prebiotics consensus on the scope and appropriate use of the term probiotic. *Nature Reviews Gastroenterology & Hepatology, 11*(8), 506–514.
35. Hill, D. R., & Ryan, E. T., (2008). Management of travelers' diarrhea. *BMJ* (*Clinical Research Edition*), *337*, E-Article ID 1746.
36. Hirano, J., Yoshida, T., Sugiyama, T., Koide, N., Mori, I., & Yokochi, T., (2003). The effect of *Lactobacillus rhamnosus* on enterohemorrhagic *Escherichia coli* infection of human intestinal cells *in vitro*. *Microbiology and Immunology, 47*(6), 405–409.
37. Homan, M., & Orel, R., (2015). Are probiotics useful in *Helicobacter pylori* eradication? *World Journal of Gastroenterology, 21*(37), 10644–10653.
38. Ianiro, G., Masucci, L., Quaranta, G., Simonelli, C., Lopetuso, L. R., Sanguinetti, M., Gasbarrini, A., & Cammarota, G., (2018). Randomized clinical trial: Fecal microbiota transplantation by colonoscopy plus vancomycin for the treatment of severe refractory *Clostridium difficile* infection-single versus multiple infusions. *Alimentary Pharmacology & Therapeutics, 48*(2), 152–159.
39. Jonkers, D., Penders, J., Masclee, A., & Pierik, M., (2012). Probiotics in the management of inflammatory bowel disease. *Drugs, 72*(6), 803–823.
40. Juntunen, M., Kirjavainen, P. V., Ouwehand, A. C., Salminen, S. J., & Isolauri, E., (2001). Adherence of probiotic bacteria to human intestinal mucus in healthy infants

and during rotavirus infection. *Clinical* and *Diagnostic Laboratory Immunology, 8*(2), 293–296.

41. Kalliomaki, M., Salminen, S., Arvilommi, H., Kero, P., Koskinen, P., & Isolauri, E., (2001). Probiotics in primary prevention of atopic disease: A randomized placebo-controlled trial. *The Lancet, 357*(9262), 1076–1079.
42. Kechagia, M., Basoulis, D., Konstantopoulou, S., Dimitriadi, D., Gyftopoulou, K., Skarmoutsou, N., & Fakiri, E. M., (2013). Health benefits of probiotics: A review. *ISRN Nutrition, 2013*(5), E-article ID 481651.
43. Kelesidis, T., & Pothoulakis, C., (2012). Efficacy and safety of the probiotic *Saccharomyces boulardii* for the prevention and therapy of gastrointestinal disorders. *Therapeutic Advances in Gastroenterology, 5*(2), 111–125.
44. Kelly, C. P., (2013). Fecal microbiota transplantation – an old therapy comes of age. *The New England Journal of Medicine, 368*(5), 474–475.
45. Kim, J. Y., Park, Y. J., Lee, H. J., Park, M. Y., & Kwon, O., (2018). Effect of *Lactobacillus gasseri* BNR17 on irritable bowel syndrome: A randomized, double-blind, placebo-controlled, dose-finding trial. *Food Science and Biotechnology, 27*(3), 853–857.
46. Kim, S. J., Park, S. H., Sin, H. S., Jang, S. H., Lee, S. W., Kim, S. Y., Kwon, B., et al., (2017). Hypocholesterolemia effects of probiotic mixture on diet-induced hypercholesterolemic rats. *Nutrients, 9*(3), 293–298.
47. Kim, Y. S., & Ho, S. B., (2010). Intestinal goblet cells and mucins in health and disease: Recent insights and progress. *Current Gastroenterology Reports, 12*(5), 319–330.
48. Kroger, M., Kurmann, J. A., & Rasic, J. L., (1992). Fermented milks – Past, present, and future; Chapter 7. In: National Research Council (US) Panel on the Applications of Biotechnology to Traditional Fermented Foods, (ed.), *Applications of Biotechnology to Fermented Foods: Report of an ad Hoc panel of the Board on Science and Technology for International Development* (pp. 61–67). Washington (DC) – US: National Academic Press.
49. Kumar, M., Nagpal, R., Kumar, R., Hemalatha, R., Verma, V., Kumar, A., Chakraborty, C., et al., (2012). Cholesterol-lowering probiotics as potential biotherapeutics for metabolic diseases. *Experimental Diabetes Research, 2012*, E-article ID 902917.
50. Laursen, R. P., & Hojsak, I., (2018). Probiotics for respiratory tract infections in children attending day-care centers: A systematic review. *The European Journal of Pediatrics, 177*(7), 979–994.
51. Lee, S. Y., Jeon, H. M., Ju, M. K., Kim, C. H., Yoon, G., Han, S. I., Park, H. G., & Kang, H. S., (2012). WNT/ Snail signaling regulates cytochrome C oxidase and glucose metabolism. *Cancer Research, 72*(14), 3607–3617.
52. Lehtoranta, L., Pitkaranta, A., & Korpela, R., (2014). Probiotics in respiratory virus infections. *European Journal of Clinical Microbiology & Infectious Diseases, 33*(8), 1289–1302.
53. Lichtenstein, L., Avni-Biron, I., & Ben-Bassat, O., (2016). The current place of probiotics and prebiotics in the treatment of pouchitis. *Best Practice & Research: Clinical Gastroenterology, 30*(1), 73–80.
54. Lilly, D. M., & Stillwell, R. H., (1965). Probiotics: Growth-promoting factors produced by microorganisms. *Science, 147*(3659), 747–748.
55. Liu, H., Gu, R., Li, W., Xue, J., Cong, Z., Wei, Q., & Zhou, Y., (2020). Probiotics protect against tenofovir-induced mandibular bone loss in mice by rescuing mandible-derived

mesenchymal stem cell proliferation and osteogenic differentiation. *Journal of Oral Rehabilitation, 47*(Suppl 1), 83–90.

56. Liu, J. J., Reid, G., Jiang, Y., Turner, M. S., & Tsai, C. C., (2006). Activity of HIV entry and fusion inhibitors expressed by the human vaginal colonizing probiotic *Lactobacillus reuteri* RC-14. *Cellular Microbiology, 9*(1), 120–130.
57. Liu, Q., Duan, Z. P., Da Ha, K., Bengmark, S., Kurtovic, J., & Riordan, S. M., (2004). Synbiotic modulation of gut flora: Effect on minimal hepatic encephalopathy in patients with cirrhosis. *Hepatology, 39*(5), 1441–1448.
58. Makras, L., Triantafyllou, V., Fayol-Messaoudi, D., Adriany, T., Zoumpopoulou, G., Tsakalidou, E., Servin, A., & DeVuyst, L., (2006). Kinetic analysis of the antibacterial activity of probiotic *lactobacilli* towards *Salmonella enterica serovar typhimurium* reveals a role for lactic acid and other inhibitory compounds. *Research in Microbiology, 157*(3), 241–247.
59. Marteau, P., Lémann, M., Seksik, P., Laharie, D., Colombel, J. F., Bouhnik, Y., Cadiot, G., et al., (2006). Ineffectiveness of *Lactobacillus johnsonii* LA1 for prophylaxis of postoperative recurrence in Crohn's disease: A randomized, double blind, placebo controlled GETAID trial. *Gut, 55*(6), 842–847.
60. Martin, F. P., Wang, Y., Sprenger, N., Yap, I. K., Lundstedt, T., Lek, P., Rezzi, S., et al., (2008). Probiotic modulation of symbiotic gut microbial-host metabolic interactions in a humanized microbiome mouse model. *Molecular Systems Biology, 4*, 157–163.
61. McFarland, L. V., (2015). From yaks to yogurt: History, development, and current use of probiotics. *Clinical Infectious Diseases, 60*(suppl 2), S85–S90.
62. McFarland, L. V., (2006). Meta-analysis of probiotics for the prevention of antibiotic-associated diarrhea and the treatment of *Clostridium difficile* disease. *The American Journal of Gastroenterology, 101*(4), 812–820.
63. Lacy, B. E., Mearin, F., Chang, L., Chey, W. D., Lembo, A. J., Simren, M., & Spiller, R., (2016). Bowel disorders. *Gastroenterology, 150*(6), 1393–1407.
64. Moeinian, M., Ghasemi-Niri, S., Mozaffari, S., & Abdollahi, M., (2013). Synergistic effect of probiotics, butyrate, and l-carnitine in treatment of IBD. *Journal of Medical Hypotheses and Ideas, 7*(2), 50–53.
65. Montazeri-Najafabady, N., Ghasemi, Y., Dabbaghmanesh, M. H., Talezadeh, P., Koohpeyma, F., & Gholami, A., (2019). Supportive role of probiotic strains in protecting rats from ovariectomy-induced cortical bone loss. *Probiotics* and *Antimicrobial Proteins, 11*(4), 1145–1154.
66. Muñoz, J. A., Chenoll, E., Casinos, B., Bataller, E., Ramón, D., Genovés, S., Montava, R., et al., (2011). Novel probiotic *Bifidobacterium longum* subsp. *infantis* CECT 7210 strain active against rotavirus infections. *Applied* and *Environmental Microbiology, 77*(24), 8775–8783.
67. Nakamura, S., Kuda, T., An, C., Kanno, T., Takahashi, H., Kimura, B., (2012). Inhibitory effects of *Leuconostoc mesenteroides* 1RM3 isolated from narezushi, a fermented fish with rice, on *Listeria monocytogenes* infection to Caco-2 cells and A/J mice. *Anaerobe, 18*(1), 19–24.
68. Nath, A., Molnár, M. A., Csighy, A., Kőszegi, K., Galambos, I., Huszár, K. P., Koris, A., & Vatai, G., (2018). Biological activities of lactose-based prebiotics and symbiosis with probiotics on controlling osteoporosis, blood-lipid, and glucose levels. *Medicina (Kaunas, Lithuania), 54*(6), 98–103.

69. Neurath, M. F., (2007). IL-23: Master regulator in Crohn disease. *Nature Medicine, 13*(1), 26–28.
70. Oak, S. J., & Jha, R., (2019). The effects of probiotics in lactose intolerance: A systematic review. *Critical Reviews in Food Science and Nutrition, 59*(11), 1675–1683.
71. Ohland, C. L., & Macnaughton, W. K., (2010). Probiotic bacteria and intestinal epithelial barrier function. *American Journal of Physiology – Gastrointestinal and Liver Physiology, 298*(6), G807–G819.
72. Ouwehand, A. C., Salminen, S., Tolkko, S., Roberts, P., Ovaska, J., & Salminen, E., (2002). Resected human colonic tissue: New model for characterizing adhesion of lactic acid bacteria. *Clinical* and *Diagnostic Laboratory Immunology, 9*(1), 184–186.
73. Ozen, M., & Dinleyici, E. C., (2015). The history of probiotics: The untold story. *Beneficial Microbes, 6*(2), 159–165.
74. Patel, A., Shah, N., & Prajapati, J. B., (2014). Clinical application of probiotics in the treatment of *Helicobacter pylori* infection: A brief review. *Journal of Microbiology, Immunology, and Infection, 47*(5), 429–437.
75. Pathmakanthan, S., Li, C. K., Cowie, J., & Hawkey, C. J., (2004). Lactobacillus plantarum 299: Beneficial *in vitro* immunomodulation in cells extracted from inflamed human colon. *Journal of Gastroenterology and Hepatology, 19*(2), 166–173.
76. Pavlovic, N., Stankov, K., & Mikov, M., (2012). Probiotics – interactions with bile acids and impact on cholesterol metabolism. *Applied Biochemistry* and *Biotechnology, 168*(7), 1880–1895.
77. Perdigon, G., Maldonado-Galdeano, C., Valdez, J. C., & Medici, M., (2002). Interaction of lactic acid bacteria with the gut immune system. *European Journal of Clinical Nutrition, 56*(Suppl 4), S21–S26.
78. Pu, F., Guo, Y., Li, M., Zhu, H., Wang, S., Shen, X., He, M., et al., (2017). Yogurt supplemented with probiotics can protect the healthy elderly from respiratory infections: A randomized controlled open-label trial. *Clinical Interventions in Aging, 12*, 1223–1231.
79. Ranjbar, S., Seyednejad, S. A., Azimi, H., Rezaeizadeh, H., & Rahimi, R., (2019). Emerging roles of probiotics in prevention and treatment of breast cancer: A comprehensive review of their therapeutic potential. *Nutrition and Cancer, 71*(1), 1–12.
80. Rautava, S., Salminen, S., & Isolauri, E., (2008). Specific probiotics in reducing the risk of acute infections in infancy– A randomized, double-blind, placebo-controlled study. *The British Journal of Nutrition, 101*(11), 1722–1726.
81. Roškar, I., Švigelj, K., Štempelj, M., Volfand, J., Štabuc, B., Malovrh, Š., & Rogeli, I., (2017). Effects of a probiotic product containing *Bifidobacterium animalis* subsp. animalis IM386 and *Lactobacillus plantarum* MP2026 in lactose intolerant individuals: Randomized, placebo-controlled clinical trial. *Journal of Functional Foods, 35,* 1–8.
82. Sabbatini, S., Monari, C., Ballet, N., Mosci, P., Decherf, A. C., Pélerin, F., Perito, S., et al., (2018). Saccharomyces cerevisiae-based probiotic as novel anti-microbial agent for therapy of bacterial vaginosis. *Virulence, 9*(1), 954–966.
83. Sadeghi-Bojd, S., Naghshizadian, R., Mazaheri, M., Ghane, S. F., & Assadi, F., (2020). Efficacy of probiotic prophylaxis after the first febrile urinary tract infection in children with normal urinary tracts. *Journal* of the *Pediatric Infectious Diseases Society, 9*(3), 305–310.

84. Sakaguchi, T., Köhler, H., Gu, X., McCormick, B. A., & Reinecker, H. C., (2002). *Shigella flexneri* regulates tight junction-associated proteins in human intestinal epithelial cells. *Cellular Microbiology, 4*(6), 367–381.
85. Sanchez, B., Delgado, S., Blanco-Miguez, A., Lourenco, A., Gueimonde, M., & Margolles, A., (2017). Probiotics, gut microbiota, and their influence on host health and disease. *Molecular Nutrition & Food Research, 61*(1), E-article ID 1600240.
86. Schiffrin, E. J., & Blum, S., (2002). Interactions between the microbiota and the intestinal mucosa. *The European Journal of Clinical Nutrition, 56*(Suppl 3), S60–S64.
87. Sgouras, D., Maragkoudakis, P., Petraki, K., Martinez-Gonzalez, B., Eriotou, E., Michopoulos, S., Kalantzopoulos, G., et al., (2004). *In vitro* and *in vivo* inhibition of *Helicobacter pylori* by *Lactobacillus casei* strain Shirota. *Applied* and *Environmental Microbiology, 70*(1), 518–526.
88. Shamekhi, S., Lotfi, H., Abdolalizadeh, J., Bonabi, E., & Zarghami, N., (2020). An overview of yeast probiotics as cancer biotherapeutics: Possible clinical application in colorectal cancer. *Clinical and Translational Oncology, 22*(8), 1227–1239.
89. Shin, S. P., Choi, Y. M., Kim, W. H., Hong, S. P., Park, J. M., Kim, J., Kwon, O., et al., (2018). A double blind, placebo-controlled, randomized clinical trial that breast milk derived – *Lactobacillus gasseri* BNR17 mitigated diarrhea-dominant irritable bowel syndrome. *Journal of Clinical Biochemistry and Nutrition, 62*(2), 179–186.
90. Sivamaruthi, B., (2018). A comprehensive review on clinical outcome of probiotic and synbiotic therapy for inflammatory bowel diseases. *Asian Pacific Journal of Tropical Biomedicine, 8*(3), 179–186.
91. Sniffen, J. C., McFarland, L. V., Evans, C. T., & Goldstein, E. J. C., (2018). Choosing an appropriate probiotic product for your patient: An evidence-based practical guide. *PLoS One, 13*(12), E-article ID 0209205.
92. Stapleton, A. E., Au-Yeung, M., Hooton, T. M., Fredricks, D. N., Roberts, P. L., Czaja, C. A., Yarova-Yarovaya, Y., et al., (2011). Randomized, placebo-controlled phase 2 trial of a *Lactobacillus crispatus* probiotic given intravaginally for prevention of recurrent urinary tract infection. *Clinical Infectious Diseases: An Official Publication of the Infectious Diseases Society of America, 52*(10), 1212–1217.
93. Stecher, B., & Hardt, W. D., (2008). The role of microbiota in infectious disease. *Trends in Microbiology, 16*(3), 107–114.
94. Sun, J., Xu, J., Yang, B., Chen, K., Kong, Y., Fang, N., Gong, T., et al., (2020). Effect of *Clostridium butyricum* against microglia-mediated neuroinflammation in Alzheimer's disease via regulating gut microbiota and metabolites butyrate. *Molecular Nutrition & Food Research, 64*(2), E-article ID 1900636.
95. Sun, Z., Sun, X., Li, J., Li, Z., Hu, Q., Li, L., Hao, X., Song, M., & Li, C., (2020). Using probiotics for type 2 diabetes mellitus intervention: Advances, questions, and potential. *Critical Reviews in Food Science and Nutrition, 60*(4), 670–683.
96. Syngai, G. G., Gopi, R., Bharali, R., Dey, S., Lakshmanan, G. A., & Ahmed, G., (2016). Probiotics-the versatile functional food ingredients. *The Journal of Food Science and Technology, 53*(2), 921–933.
97. Todoriki, K., Mukai, T., Sato, S., & Toba, T., (2001). Inhibition of adhesion of food-borne pathogens to Caco-2 cells by *Lactobacillus* strains. *Journal of Applied Microbiology, 91*(1), 154–159.
98. Tomasetti, C., & Vogelstein, B., (2015). Variation in cancer risk among tissues can be explained by the number of stem cell divisions. *Science, 347*(6217), 78–81.

99. Tsai, C. C., Lin, P. P., & Hsieh, Y. M., (2008). Three *Lactobacillus* strains from healthy infant stool inhibit enterotoxigenic *Escherichia coli* grown *in vitro*. *Anaerobe, 14*(2), 61–67.
100. Tubelius, P., Stan, V., & Zachrisson, A., (2005). Increasing work place healthiness with the probiotic *Lactobacillus reuteri*: A randomized, double-blind placebo-controlled study. *Environmental Health, 4*(1), 25–30.
101. Uccello, M., Malaguarnera, G., Basile, F., D'agata, V., Malaguarnera, M., Bertino, G., Vacante, M., et al., (2012). Potential role of probiotics on colorectal cancer prevention. *BMC Surgery, 12*(Suppl 1), S35–S38.
102. Ulisse, S., Gionchetti, P., D'Alò, S., Russo, F. P., Pesce, I., Ricci, G., Rizzello, F., et al., (2001). Expression of cytokines, inducible nitric oxide synthase, and matrix metalloproteinases in pouchitis: Effects of probiotic treatment. *The American Journal of Gastroenterology, 96*(9), 2691–2699.
103. Vallianou, N. G., Stratigou, T., & Tsagarakis, S., (2018). Microbiome and diabetes: Where are we now? *Diabetes Research and Clinical Practices, 146*, 111–118.
104. Van-Gossum, A., Dewit, O., Louis, E., De Hertogh, G., Baert, F., Fontaine, F., DeVos, M., et al., (2007). Multicenter randomized-controlled clinical trial of probiotics (*Lactobacillus johnsonii*, LA1) on early endoscopic recurrence of Crohn's disease after ileocecal resection. *Inflammatory Bowel Diseases, 13*(2), 135–142.
105. Van-Immerseel, F., Ducatelle, R., De Vos, M., Boon, N., Van De Wiele, T., Verbeke, K., Rutgeerts, P., et al., (2010). Butyric acid-producing anaerobic bacteria as a novel probiotic treatment approach for inflammatory bowel disease. *Journal of Medical Microbiology, 59*(Pt 2), 141–143.
106. Van-Tassell, M. L., & Miller, M. J., (2011). *Lactobacillus* adhesion to mucus. *Nutrients, 3*(5), 613–636.
107. Veerappan, G. R., Betteridge, J., & Young, P. E., (2012). Probiotics for the treatment of inflammatory bowel disease. *Current Gastroenterology Reports, 14*(4), 324–333.
108. Vivarelli, S., Salemi, R., Candido, S., Falzone, L., Santagati, M., Stefani, S., Torino, F., et al., (2019). Gut microbiota and cancer: From pathogenesis to therapy. *Cancers, 11*(1), 38–43.
109. Vonk, R. J., Reckman, G. A., Harmsen, H. J., & Priebe, M. G., (2012). Probiotics and lactose intolerance; Chapter 7. In: Rigobelo, E. C., (ed.), *Probiotics* (pp. 149–160). London – UK: IntechOpen Ltd.
110. Wang, G., Huang, W., Xia, Y., Xiong, Z., & Ai, L., (2019). Cholesterol-lowering potentials of *Lactobacillus* strain overexpression of bile salt hydrolase on high cholesterol diet-induced hypercholesterolemic mice. *Food & Function, 10*(3), 1684–1695.
111. Watts, T., Berti, I., Sapone, A., Gerarduzzi, T., Not, T., Zielke, R., & Fasano, A., (2005). Role of the intestinal tight junction modulator zonulin in the pathogenesis of type I diabetes in BB diabetic-prone rats. *Proceedings of the National Academy of Sciences of the United States of America, 102*(8), 2916–2921.
112. Will, S., Hornigold, D. C., Baker, D. J., Coghlan, M. P., Mesquita, M., Trevaskis, J. L., & Naylor, J., (2017). Gut check on diabesity: Leveraging gut mechanisms for the treatment of type 2 diabetes and obesity. *Current Opinion in Pharmacology, 37*, 10–15.
113. Williams, A. B., Yu, C., Tashima, K., Burgess, J., & Danvers, K., (2001). Evaluation of two self-care treatments for prevention of vaginal candidiasis in women with HIV. *Journal of the Association of Nurses in AIDS Care, 12*(4), 51–57.

114. Williamson, E., Bilsborough, J. M., & Viney, J. L., (2002). Regulation of mucosal dendritic cell function by receptor activator of NF-kappa B (RANK)/ RANK ligand interactions: Impact on tolerance induction. *Journal of Immunology, 169*(7), 3606–3612.
115. Working Group on Drafting Guidelines for the Evaluation of Probiotics in Food, (2002). *Guidelines for the Evaluation of Probiotics in Food by Joint FAO/WHO* (p. 63). Ontario – Canada: Food and Agriculture Organization of the United Nations (FAO)/ World Health Organization (WHO).
116. Yan, F., Li, N., Yue, Y., Wang, C., Zhao, L., Evivie, S. E., Li, B., & Huo, G., (2020). Screening for Potential Novel Probiotics with Dipeptidyl Peptidase IV-Inhibiting Activity for Type 2 Diabetes Attenuation *in vitro* and *in vivo*. *Frontiers in Microbiology, 10*, E-article ID 2855.
117. Zhang, Z., Hinrichs, D. J., Lu, H., Chen, H., Zhong, W., & Kolls, J. K., (2007). After interleukin – 12p40, are interleukin – 23 and interleukin – 17 the next therapeutic targets for inflammatory bowel disease? *International Immunopharmacology, 7*(4), 409–416.

CHAPTER 10

Probiotics, Prebiotics, and Synbiotics as Dietary Supplements: A Healthy Approach

Manali Chakraborty, Savita Budhwar, and Suneel Kumar

ABSTRACT

The intestinal tract is colonized by various bacterial species to facilitate different metabolic pathways and maintain healthy gut environment. The complex community of these microorganisms undergo co-evolution along with their host. However, due to alteration in lifestyle, food habits, and environment, the population of the health-beneficial microbiota gets disturbed. Therefore, recovery and restoration of the gut environment and adaptation of the flora with the alteration is crucial to prevent health damage, infections, metabolic disorders and other fatal health conditions. Inclusion of prebiotics like carbohydrates, fibers, etc., in diet can enrich the environment for the healthy growth of the probiotics. Moreover, according to ongoing studies, the byproduct of probiotic fermentation, also known as postbiotics, also support the biological functions. Hence, presence of probiotic, prebiotics, and synbiotics (combination of probiotic-prebiotic in a form of synergism) in daily diet can enrich health status. Immune modulation, cancer prevention, inflammatory condition, cardiovascular diseases (CVDs), bowel problems, metabolic disorders can be managed through the dietary supplements consisting of prebiotic-probiotic properties. Consumption of certain raw vegetables, fruits, fermented products, dairy based products can be helpful. Besides such products, use of certain plant-based byproducts viz., rice bran, wheat bran, fruit peels in food formulation can provide favorable medium to probiotics for sustainability. Besides retaining nutrient properties

The Functional Foods: Nutrient and Health Benefits.
Megh R. Goyal, Junaid Ahmad Malik, and Anu Kumari (Eds.)

of the functional foods possessing probiotic/prebiotic properties, it is also necessary to maintain the shelf-life of such products. Various processing technologies to increase the storage stability of the products. Application of these processing techniques like drying technique, dry-freezing, blanching, irradiation technology, chemical treatment sometimes effects the nutrient availability and food properties of the products. In this present book chapter, the potential role of prebiotics, probiotics, synbiotic as dietary supplement, probable product formulation, processing, preservation of the products have been discussed. The mechanistic relationship among such functional food and health status will be established to help in management of personalized gut microbiota.

10.1 INTRODUCTION

Now-a-days, awareness regarding the health and diet is rising among consumers along with the demand of healthier products with better nutritional properties to prevent the risk of fatal diseased conditions. Hippocrates' aphorism "*Let food be your medicine and medicine be your food*" [45] is becoming a trend in society leading to novel opportunities for nutrient-dense products to meet consumer demand resulting in the increasing popularity of the functional food market worldwide. These functional food products possess several bioactive compounds, majorly obtained from plant sources. These bioactive compounds include probiotics, paraprobiotics, prebiotics, synbiotics along with polyunsaturated fatty acids, minerals, and essential vitamins. Hippocrates stated that "All diseases begin in the gut." Scientists evaluating the human microbiome, also propose that inclusion of fermented foods in daily diet can enhance live microorganisms in our gut temporarily [45].

Human gut consists of different bacterial phyla along with *Bacteroidetes, Firmicutes,* and *Actinobacteria.* "Microbiota," "microflora," "normal flora" terms describe these microbes [28, 35]. According to the available data, the gastrointestinal tract (GIT) contains the largest microbial ecosystem in the body [29, 47]. It comprises the digestive system, in which colon is the most colonized organ. Due to exposure to different food particles and other foreign particles (Antigen of microbial origin), changes in gut microbial flora can cause illness. Alteration in gut leads to diseased conditions such as, colon cancer, hypercholesterolemia, bowel diseases, metabolic disorders. Studies show that the lactic acid bacteria (LAB) (Genus *Lactobacillus* and

Bifidobacterium) exhibit positive effect to maintain the gut health. These bacteria, also termed as 'probiotics,' are being utilized for the treatment and management of health issues.

According to FAO/WHO, probiotics are the living microbes, which are beneficial to health [13, 14, 16]. "Probiotic" came from Greek word, which means "for life" [46]. A selective fermented ingredient or medium suitable for the growth of these probiotics conferring health benefits, is known as 'prebiotic.' These prebiotics were described as non-viable and non-digestible food elements which modify the microflora resulting in improved health condition [14]. The synergistically active prebiotic and probiotic together is established as synbiotics. The International Scientific Association for Probiotics and Prebiotics (ISAPP) along with the International Probiotic Association are investigating the beneficial properties of several microbes. Introduction of these beneficial bacteria through ingestion of certain strains along with the consumption of prebiotic such as soluble fibers (growth substance for the probiotics) can boost the gut health. These probiotics and prebiotics are available in commercial forms which includes food products, dietary supplements.

From the ancient times, the awareness of health benefits of the probiotics like LAB were popular among the population. Fermented milk and sour milk have been in knowledge of Ancient Romans and Greek people. Various fermented milk drinks used to be prepared in ancient times. "Leben," a sour milk drink, was popular in Egypt. "Jahurt" was famous in Balkans, "arjan" in Central Russia and "tarho" in Hungary in 14th century were well-known drinks [21].

In the beginning of the 20th Century, research studies on lactic acid fermentation became more prominent due to immense passion of a Russian Immunologist, Ilia Miecznikow, who stated that "*Studies on Optimism – with various foods undergoing lactic acid fermentation and consumed raw (sour milk, kefir, sauerkraut, pickles) humans introduced huge amounts of proliferating lactic acid bacteria to their alimentary tracts*" [33].

In fermentation, complex molecules are converted into their simpler form due to action of microbes present in the medium. Yeast enzymes convert sugars into alcohol and proteins to peptides/amino acids. It is a natural phenomenon to increase bioavailability of nutrients viz., vitamins, minerals, sensory qualities of food, viz., texture, flavors, and aroma along with reducing antinutrient activity, present in the food ingredients. Hence, activities of microorganisms through fermentation exhibit a potential role to improve chemical and physical qualities of foods. Fermented foods possess longer shelf-life in comparison with other foods. Cooking time for needed for the fermented food preparation is relatively lesser. These food products possess higher antioxidant activity. Due to release of certain biopeptides and proteolysis of proteins (such as: α-casein, α-lactalbumin, and β-lactoglobulin), fermented products (yogurt) show higher antioxidant activities than the milk. These foods are influenced by the content of the substrates in use as well as the fermenting microorganisms. Furthermore, the handling of food and the period of fermentation during processing have an impact on food fermentation. LAB is the dominating microbe in all fermented foods and beverages that have been detected and presence of this microorganism in fermented foods and beverages exerts potential benefits [42].

This chapter focuses on various applications of probiotics, prebiotics, synbiotics, and their sources; prebiotics from plant byproducts, and benefits of inactivated probiotics (Parabiotics).

10.2 PROBIOTICS, PREBIOTICS, AND SYNBIOTICS: MICROORGANISMS AND APPLICATIONS

Human probiotics mainly belong to *Lactobacillus, Bifidobacterium, Enterococcus, Streptococcus, Lactococcus* (Table 10.1). Majorly used probiotics utilized for food formulation are mainly of the genus *Bacillus* and *Saccharomyces* [31]. These probiotics, used for food purpose, are marked safe by different general food law bodies in different countries. GRAS (Generally Regarded as Safe) status, regulated by FDA (Food and Drug Administration) is being assigned in USA. QPS (qualified presumption of safety) by EFSA (European Food Safety Authority) assigns the safety mark to the probiotics in Europe [31].

In case of prebiotics, they stimulate the growth of probiotics in definite environment, pH. Edible plant parts, fruits (Banana, berries), vegetables (Tomato, garlic, onion, green vegetables), cereals, legumes rich in carbohydrates, fiber mainly exhibits prebiotic property. Synthetic prebiotics include fructo-oligosaccharides (FOS), cyclodextrins, lactulose, lacto-saccharose, and many more [9, 25]. Prebiotics are mainly non-digestible portion. Fermentation of these prebiotics results in enhanced production of short-chain fatty acids (SCFA) leading to increase stool mass resulting in reduced colon pH, fecal enzymes along with improvement of immune system [10]. They act as support system for the probiotics.

TABLE 10.1 Utilization of Probiotics for Pharmaceutical and Food Purposes [31]

Pharmaceutical Use	Food Additives
*L. acidophilus**	*L. amylovorus**
*L. casei**	*L. casei**
*L. gasseri**	*L. johnsonii**
*L. helveticus**	*L. pentosus**
*L. reuteri**	*L. plantarum**
*L. rhamnosus**	*L. rhamnosus**
B. adolescentis	*B. breve*
*B. animalis**	*Lactococcus lactis**
B. bifidum	
B. infantis	
*B. longum**	
Enterococcus faecium	
*Streptococcus thermophiles**	
*Bacillus clausii**	
Escherichia coli Nissle 1917	
*Saccharomyces cerevisiae (boulardi)**	

Note: *QPS: (Qualified presumption of safety) microorganisms.

Prebiotics are classified based on their properties, which are non/partially-digestible, poor absorption in small intestine, fermentation by bacteria/pathogens. Products in which prebiotic elements favor the growth and survival of probiotic microorganisms are established as synbiotics. *Lactobacilli, S. boulardii, Bifidobacteria* spp., and *B. coagulans* are major probiotic strains used for the preparation of synbiotics along with the potential prebiotics, viz., GOS, xylose-oligosaccharide, FOS, and inulin (Figure 10.1). According to the studies, synbiotics exert positive effects on the growth of the gut microflora, improvement in liver cirrhosis, immune system [50].

The researchers developed buttermilk, rich in synbiotics, by using inulin (which acted as prebiotics) and *Lactobacillus acidophilus and Bifidobacterium* spp. (which were the probiotics) [30]. For the preparation of control buttermilk, commercially available mesophilic starter, CHN22 was used, which contained different species of *Lactococcus lactis*.

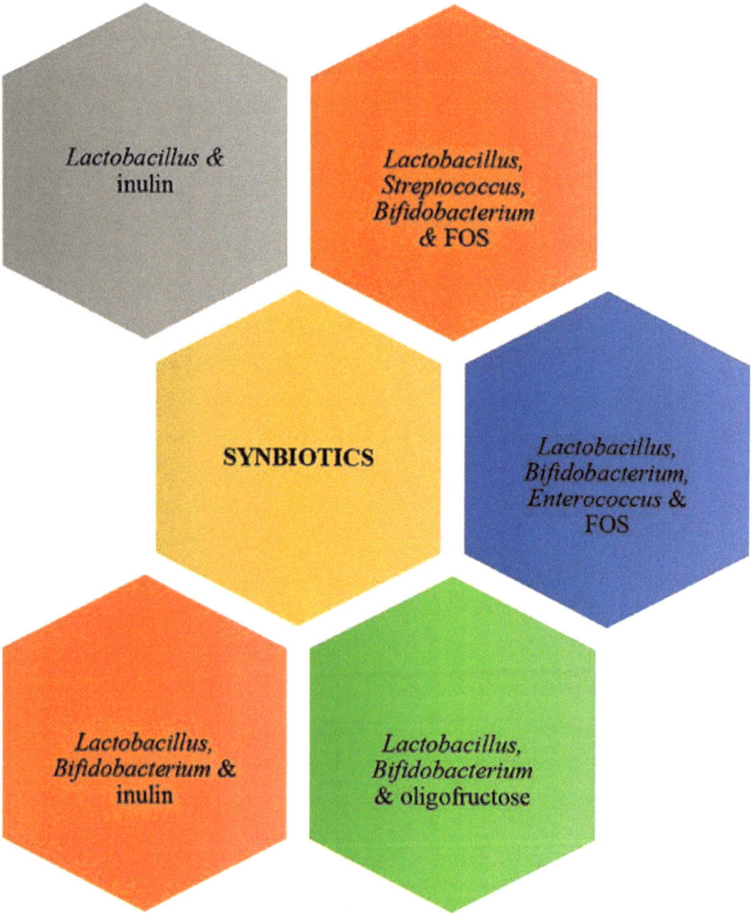

FIGURE 10.1 Examples of synbiotics in human nutrition fructo-oligosaccharides [50].

The chemical composition, survival of the probiotics, growth of mold, yeast, and coliform in synbiotics buttermilk was evaluated. The buttermilk was stored at 4°C for 12 weeks, pH, titratable acidity, and proteolysis were all assessed as well. The control and synbiotic buttermilk contained protein 3.29 and 3.30%, fat 3.28 and 3.26%, carbohydrate 4.55 and 5.16%, total solids 11.81 and 12.43%, and ash 0.69 and 0.70%, respectively. *Lactobacillus acidophilus* and *Bifidobacterium* spp. populations were at first higher than 107 cfu/ml. The growth count was higher than 106 cfu/ml during the experiment of 12 weeks, although there was no mold or yeast identified. Significant differences in pH and titratable acidity were observed between the control and synbiotics buttermilk ($p<0.05$). Proteolysis did not differ significantly between the two samples. The survival of probiotics was substantially higher in synbiotic buttermilk than in other fermented foods, indicating that it may be regarded a functional food [51].

Linear oligosaccharides consisting of β-1,4-linked glucopyranose units are known as cello-oligosaccharides (COS) that are collection of oligosaccharides that are gaining popularity as possible prebiotic sources in the food, chemical, pharmaceutical industries. COS derived from lignocellulosic biomass, notably agro-industrial leftovers and forestry sector byproducts are considered as novel biomass resources with prebiotic property. For the manufacture of COS, two main techniques have been used: acid and enzyme-mediated cellulose hydrolysis. Because of the utilization of milder reaction conditions and less monomer synthesis, the latter has been deemed more appealing. Although COS possesses potential prebiotic with various applications, more study on large-scale manufacturing is required for its sustainable usage in various field [4].

10.3 PREBIOTIC PROPERTY OF PLANT-BASED BYPRODUCTS: A BOON TO CIRCULAR ECONOMY

Due to processing of fruits and vegetables, huge mass of food byproducts is generated with less commercial use causing economic as well as environmental damage. In spite of possessing a significant amount of dietary fiber and prebiotics activities such as gut regulation, maintaining glycemic load, and improved food texture, these byproducts are not properly utilized. The international scientific community has considered the use of the byproduct extracts or their flour to fortify food products along with the preservation of the products to fulfill the demand of today's consumers to develop safer and health-promoting foods [38].

Properties of chickpea was studied to analyze its influence as an antioxidant and prebiotic property enriched thickener on the probiotic viability and physicochemical qualities of formulated bio-yogurt. *Bifidobacterium bifidum* and a combination of *Bifidobacterium animalis* subsp. *lactis* and *Lactobacillus acidophilus*) were mixed with chickpea flour at various percentages to make stirred bio-yogurts (1, 2, and 3%). During storage, the stirred bio-physicochemical, yogurt's antioxidant, rheological, microbiological, and sensory properties were assessed. Presence of the chickpea flour in formulation, resulted in enhanced growth of probiotics. With decreasing percentages of chickpea flour, the bacterial counts in formulated bio-yogurt were also decreasing, i.e., for 3% chickpea flour the count was log 8.28, for 2% flour the count was 8.12, for 1% flour it was 8.04, and the control (0% flour) showed 7.32 $CFUg^{-1}$, counts, respectively.

Antioxidant capacity was high in bio-yogurt incorporated with chickpea flour, although with decreasing flour percentage the antioxidant activity was decreasing. At concentrations of 3%, 2%, and 1% chickpea flour, the antioxidant activities were 91%, 83%, and 76.6%, respectively. In control the antioxidant activity was 31.12%. Besides antioxidant activity, Chickpea addition also enhanced the viscosity of bio-yogurt. Panelists favored bio-yogurt with 1 and 2% chickpea flour in texture and flavor attributes, according to the sensory evaluation. According to the findings, chickpea flour might be used instead of skimmed milk powder to improve the quality of bio-yogurt in cost-effective way [22].

Due to the high production of the organic content along with its toxicity, proper management of olive byproducts is a major concern. Municipal wastes (TPOMW) from Olive mill and the olive pruning residues (OLPR) were utilized as substrates to culture mushrooms, *Ganoderma lucidum* and *Pleurotus ostreatus*, respectively. Grown mushrooms were analyzed for the antioxidant property, phenol level, nutrient composition along with their chemical composition, glucans, total phenolic content (TPC), and antioxidant activity, as well as their prebiotic potential, by observing their impact on the growth of gut bacteria.

P. ostreatus mushroom production was improved by olive byproducts. When compared to beech sawdust, increased percentage of OLPR to wheat-straw led to 42% increase in crude protein level in *P. ostreatus*. However, *G. lucidum* mushrooms cultured in OLPR-based substrates showed increase in α-glucan up to three-fold, along with significant increment of β-glucan in comparison with the control, i.e., beech sawdust. The qualitative/quantitative distinction identified by routine assays was corroborated by the mushrooms' FTIR spectra. In terms of prebiotic characteristics, mushroom flours showed positive effect on the growth of *Lactobacillus acidophilus* and *Lactobacillus gasseri* after 24/48 hours of incubation. The growth of these probiotics was accelerated. Moreover, in growth of *Bifidobacteria*, a specific pattern was noticed. Growth of *Bifidobacterium bifidum* was hindered by mushrooms, however growth for *B. longum* was not affected by the mushrooms [27].

In another study, 13 plants and their components were collected in southern Thailand, and their polysaccharide content and prebiotic characteristics were analyzed. At ambient and boiling temperatures, ground materials were extracted with ethanol and water. After freeze-drying the extract was digested using acidic buffer and amylase to assess the amount of indigestible polysaccharide. Around 10 combined samples of skin, flesh, pericarp, and seeds of different fruits (such as: palm fruit (*Borrassus flabellifer* L.), jackfruit (*Artocarpus heterophyllus* Lam.), rambutan (*Nephelium lappaceum* L.), jampadah (*Artocarpus integer* Merr.), and young coconut (*Cocos nucifera* Linn.), okra pod (*Hibiscus esculentus* Linn.) were identified as prospective prebiotic sources based on the higher level of indigestible polysaccharide content as well as the extract yield [48].

The foliar steviol glycosides of *Stevia rebaudiana* (Bertoni) have been extensively researched. Its roots have lately yielded fructan-type polysaccharides. Fructans are type of carbohydrates that has number of health benefits as well as potential food applications. Prebiotic potential of FOSs obtained from *S. rebaudiana* roots and *in vitro* adventitious root cultures was assessed. A roller bottle system was used to obtain *in vitro* adventitious root cultures. Chemical investigations of FOSs taken from various sources (off-line electrospray ionization-mass spectrometry, gas chromatography–mass spectrometry, and 1H nuclear magnetic resonance (NMR)) demonstrated similar chemical characteristics. FOSs extracted from *S. rebaudiana* roots had potential prebiotic effects with improved fermentation capabilities leading to increased growth of both *Bifidobacteria* and *lactobacilli* [30].

Probiotics are living, nonpathogenic bacteria that dwell in the gut and provide health benefits. They work by inhibiting pathogen bacteria adherence, enhancing innate immunity, reducing inflammation caused by pathogen. They also promote survival of the intestinal epithelial, barrier function, as well as the protective responses through molecular and cellular processes. Secreted substances obtained from probiotics known as "post-biotic," are thought to be responsible for some of these positive benefits. Although several studies have been reported for different probiotic strains, majority of the work focuses on *Lactobacilli* derived postbiotic. The known mechanistic findings on the studied activities of numerous secretory products from distinct *Lactobacillus* species. The soluble components that mediate probiotics' therapeutic effects could lead to not only a better understanding of the mechanisms, but also the development of efficient pharmaceutical techniques involving the activity of live bacteria treatments [7].

Crude polysaccharides of Bamboo shoot (BSCP) isolated from *Gigantochloa levis* shoots yielded 3.27–0.18% (dry basis) and negligible amount of protein (0.02–0.01%). By gel chromatography, the molecular weight of BSCP was determined 7.49 Da. The spectrum of BSCP obtained by using FTIR demonstrated probable presence of glucan in extract, useful for the pharma and food industries which indicated use of BSCP as a potential prebiotic [5].

The 90% of target component was obtained in the refined form after purification of pectic oligosaccharides (POS) using membrane filtration. POS was extracted from orange peel wastes (OPW) after the hydrothermal treatment. After the *in vitro* fermentation, human *fecal inocula* was used to examine the concentrate's prebiotic potential. Similar studies were carried out utilizing pectin obtained from orange and commercially available FOS to analyze the prebiotic property as well as for the comparison of the degree of fermentation. Fluorescent *in situ* hybridization was used to examine the activity of chosen population of the microorganisms. Gas production, pH, and short-chain fatty acids (SCFA) were measured to understand the post phase of fermentation, in which the counts of microbes were increased. The counts of both the genera and the overall cell population were altered. The counts were increased from 17% in the inocula to 27% after fermentation which depicted that presence of POS enhanced the growth of Bifidobacteria and lactobacilli. Although SCFA production was equivalent to the concentration of FOS, yet butyrate production was lower due to pectin fermentation [18].

The use of prebiotics to formulate functional food products is being recognized in recent years. Although prebiotics have been studied for their stability, the application of the same to improve the storage stability of food is rare. Biologically stable five prebiotics (FOS, inulin, galacto-oligosaccharides (GOS), resistant starch, and polydextrose (PDX)) were studied in a variety of food matrices under various experimental conditions. The storage stability of various food products like baked products, snack bar, breakfast cereal, and drink containing these prebiotics were tested using assay to check the prebiotic activity. This assay measures a prebiotic's ability to support probiotic strain development in comparison to enteric strain growth and growth on a non-prebiotic substrate. Although exposure to the mild heating did not affect the stability of FOS and inulin, yet acidic environment along with moderate heating destabilize both. Irrespective any treatment, i.e., the mild temperature and acidic environment, GOS was stable. Due to the complex food matrices of the food products, the prebiotic assay was not sensitive enough to analyze the activity [34].

Commercially available food products generally lack dietary fiber and contain excess of calories. This is the reason food manufacturers are keen to formulate novel products by using fiber rich flour such as chickpea flour. Even after the processing of chickpea, the outer part of the legume is discarded in spite of presence of high dietary fiber. Gluten free pasta was made using prebiotic property rich chickpea flour. The formulated pasta showed enhanced nutrient composition along with satisfactory sensory qualities [8].

The combination culture (synbiotics) of probiotic microorganisms and prebiotics, has numerous health benefits. The cereal-based synbiotics products could be a beneficial approach to provide a nutritious alternative [41]. Despite the fact that many fruit byproducts are nutrient-dense, less is known about the prebiotic potential they possess. The prebiotic property of the byproducts obtained from pineapple on probiotics along with the antioxidant and antimutagenic peptides in yogurt was analyzed. Peel and pomace powders were ground separately and tested for prebiotic activity after being oven and freeze dried. The granules were also added to milk separately while making yogurt with or without probiotics. The pineapple powder incorporated yogurts showed increased probiotic counts (by 0.3–1.4 log cycle). *In vitro* studies also revealed that the peptide extract obtained from the formulated yogurts exerts excellent antioxidant activity, specifically scavenging activity of 1,1-diphenyl-2-picrylhydrazyl and hydroxyl radicals. Moreover, according to the Salmonella mutagenicity test, potential antimutagenic activity as a sodium azide inhibitor implied pineapple byproducts as significant prebiotic source to formulate novel functional food [40].

Germ and bran from wheat milling contain higher amount of nutrient composition (viz., dietary fiber, vitamins, protein, minerals) along with potential antioxidant activity. In a study, after treatment with xylanase, wheat milling byproducts were utilized to formulate cake at various percentages, ranging from 0% to 30%. Xylanase treatment was carried out to improve the nutrient availability of the byproducts. Proximate composition along with mineral content enhanced with the increasing percentages of the byproducts incorporated in cake preparation. Wheat germ-based cake showed the highest mineral contents (Mn, Zn) along with improved protein level. The addition of xylanase increased the volume of the cakes while decreasing the hardness and color brightness. Xylanase treated wheat byproducts improved the quality and nutritional content of cakes, according to this study. The findings of this study paved the way for xylanase to be used in the production of enriched cakes [6].

Lactic acid-producing bacterial growth was also evaluated to study the potential of prebiotics to boost the growth of probiotics in rice drink under high pressure situation [24]. Glucan (BG), extracted from oat bran and treated with xylanase to obtain β-glucan modification. The prebiotic effects of BG and MBG on *B. bifidum* in the human colon were investigated. The addition of BG and MBG to the fermentation broths resulted in a lower pH. In the culture, lactic acid and acetic acid production increased. BG was found to increase *B. bifidum* proliferation [39].

Steam treated Rice bran inactivate the enzyme before the fermentation with *Rhizopus oryzae* and *Aspergillus oryzae. In vitro* human fecal microbiota fermentation was used to evaluate the prebiotic potentials of rice bran, fermented using fungus. This fermented rice bran enhanced proximate composition and total phenol content as compared to the naturally fermented rice bran. Furthermore, due to the prebiotic nature of the fungal fermented rice bran, short-chain fatty acids were produced in the medium along with the improved population growth of the beneficial bacteria. In comparison to the individual effect of *Rhizopus oryzae* and *Aspergillus oryzae*, combined effect of *Rhizopus oryzae* and *Aspergillus oryzae* upon rice bran fermentation showed the higher growth of *Bifidobacterium* and Bacteroidaceae in the fermented residues. These findings suggested that rice bran fermented by fungus has higher prebiotic potential. Hence, steam treated, and fungal fermented rice bran can be a novel ingredient for food formulation [49].

The ISAPP presented the most recent and revised definition of prebiotics in an expert consensus statement. Instead of only non-digestible oligosaccharides of dietary origin, prebiotics are now described as a substrate, selectively utilized by bacteria resulting into health beneficial effects [17] rather than just non-digestible oligosaccharides of dietary origin. The potential of the probiotics (such as: *Lactobacillus plantarum* ATCC 8014 and *Pediococcus acidilactici* ATCC 8042) was studied to find out the *in-vitro* prebiotic role of the husk of green gram (*Vigna radiata*) and black gram (*Vigna mungo*). The study indicated that the husks are resistant to *in-vitro* gastrointestinal (GI) digestion, indicating that they possess non-digestible parts and supporting the proliferation of beneficial microbes found in the GI tract. In comparison to inulin, *Pediococcus* sp. grew up to 80% faster in the husks. The study also shows that husks have antioxidant activity, which might be beneficial to health. As a result, the husks could be considered prebiotic possibilities [44].

In a study, oat bran based grape flavored non-dairy beverage was formulated through fermentation using probiotic strain *Bifidobacterium lactis*. Physicochemical properties, probiotic bacteria viability, and sensory differentiation were studied in beverage after 3 weeks of storage in refrigerator. The pH of the beverage got increased with the increasing amount of oat bran extract. Increasing added amount of carbon dioxide (CO_2) and oat bran also enhanced the bacteria's vitality. When the storage time was increased to 21 days, the viability drastically dropped. Oat bran in a grape-flavored probiotic drink also boost the nutrient content [2].

Production of underutilized byproducts obtained from fruits like apple is higher due to consumer demand for apple juice or cider. The insoluble dietary fraction (IDF) is the most abundant, pectins from the soluble dietary fraction (SDF). This includes homogalacturonan and rhamnogalacturonan type I branched with arabinans, galactans, and arabinogalactans. *In vitro Bifidobacterium bifdus* used these polysaccharides to produce acetate, short chain fatty acid. In an *in vivo* study, butyrate was enhanced to 3-folds, demonstrating that butyrate bacteria may utilize apple byproduct. As a result, it demonstrated a potential bifidogenic and butyrogenic action, which is being investigated further for novel prebiotics. Furthermore, an apple byproduct incorporated diet boosted HDL along with decreasing trygliceridemia, indicating a potential hypolipidemic effect due to the bile acids binding effect [32].

The expanding circular economy trend has inspired the development of creative ways for valorizing byproducts. Winery byproduct like *Cava lees* made up of *Saccharomyces cerevisiae*, which also possesses oligosaccharides, can be employed as potential stimulant for microbial growth. The most effective amount for increasing microbial counts was determined to be 5% of cava lees. Majority of these strains showed a bigger and faster drop in pH. The study suggested possible value addition to the winery byproduct as synbiotics culture to enhance the nutritional quality of fermented products resulting in improved health status [19].

Being the most consumed nuts, Hazelnuts generates a lot of waste, particularly shells, which may be valorized for better purpose. Recent research has discovered that this nut shells possess xylo-oligosaccharide (XOS) and arabino-xylo-oligosaccharide ((A)XOS) precursors, previously characterized as developing prebiotics with significant health benefits for humans. The manufacturing of these compounds could have significant impact on the functional food sector. Studies have revealed the use of hazelnut wastes as a circular economy approach through the development of novel functional products [15].

XOS are non-digestible food elements having prebiotic qualities that promote the growth of probiotics selectively, leading to potential applications in the food and pharma industries. In a research study, XOS was derived from a variety of agro-wastes, which were sugarcane straw, Endo-xylanase, L-arabino-furanosidase, and coffee husk. These byproducts were subjected in an optimal enzymatic reaction. XOS derived from these agricultural residues exhibited prebiotic property along with potential antioxidant activity, making it suitable for use as a functional component in food [3]. In another study, artichoke byproduct showed prebiotic potential. The use of important bio-resources, such as artichoke byproduct, is a beneficial approach towards waste management and the reduction of agro-industrial byproducts [20].

Consumer demand for cereal and legume-based products with rich nutrient composition and health-beneficial characteristics is increasing. Application of sourdough technology has shown enhanced nutrients of grain flours. Sourdough hydrolyzes dietary fiber, lowers lipid rancidity, improves carbohydrate and protein digestibility, and increases vitamin and mineral bioavailability. Furthermore, during fermentation, beneficial bioactive compounds are generated along with the reduction of the components that affect the biological activities. Sourdough fermented items have been found to play a role in gut microbiota control. Sourdough stands out among other fermentation processes because of its health-promoting potential, allowing for the development of novel products [12].

After the preparation of tofu and soymilk from soybean seeds, an aqueous extract is obtained. This left-over extract known as Okara has rich proximate composition, nutrients (viz.: soluble and insoluble fibers, minerals), bioactive compounds (viz.: isoflavones and soyasaponins). Its rich nutrient makes okara a potent ingredient to use in the food industry. Okara can also utilized as dietary supplement to manage various diseased conditions like metabolic disorders, cardiovascular diseases (CVDs). Moreover, it showed beneficial property to enhance the growth of intestinal microbes and the production of microbe-derived metabolites (xeno-metabolites), since gut dysbiosis (imballed microbiota) has been linked to a variety of diseases. According to the studies, soybean residue (okara) showed significant prebiotic property depicting probable use of the extract as functional diet on gut issues, liver, and kidney malfunctions [43].

According to recent research studies, diets comprising fermented foods improve microflora environment in our gut, resulting in increasing demand of fermented foods among consumers. Hippocrates also stated that "*Let food be your medicine and medicine be your food* – and *all diseases originate in the gut.*" Individuals allergic to milk proteins are unable to ingest dairy products. As a result, there is a need to investigate different non-dairy matrices as probiotic carriers for lactose intolerant people to provide an alternative to dairy products. Several sources for the formulation of fermented beverages (viz.: cereal, pseudo cereal, legumes, fruits) can be enriched with beneficial bioactive compounds with prebiotic properties [45].

10.4 INACTIVATED PROBIOTICS

Probiotics are commonly used orally now-a-days. The safety profile of live probiotics, on the other hand, is currently being debated. Cases of systemic infections as a result of translocation, especially in immune-compromised individuals and children, leading to presence of antibiotic resistance genes, neonatal gut colonization are the main hazards. To avoid these issues, non-viable heat-killed probiotic bacteria are becoming more popular. These treated probiotics showed positive benefits, including immunomodulation and immune response similar to that of active probiotics. The products incorporated with inactivated probiotic strains exhibited positive effect on GI issues such as bloating, infantile coli, and diarrhea. According to studies, heat-treated probiotics may potentially be useful in the treatment of allergic disorders of the skin and lungs [37].

The ability of beneficial bacteria (probiotics) to deliver health benefits is dependent on their viability. This belief was challenged after the term "paraprobiotics" was coined, implying that non-viable microorganisms may provide health advantages comparable to those provided by living probiotics. Giving live probiotics to those who have a weak immune system can be risky at times. In such circumstances, taking paraprobiotics may be a viable alternative. Due to their long shelf-life, safety, and positive benefits, paraprobiotics have gotten a lot of attention. These characteristics indicated probable use of paraprobiotics for health benefits, although the exact mechanisms are still unknown [1].

"*They are live microorganisms (that when taken in proper numbers and frequencies) can deliver health advantages to the host,*" according to the traditional definition of probiotics. The effect of probiotics on the interaction of gut microbiota and immune system results into improved health status. Non-viable probiotics have been referred to as "ghost probiotics," "postbiotics," and "inactivated probiotics" or "paraprobiotics."

Paraprobiotics have been linked to a variety of biological impacts, implying that they could be an ideal way to improve one's health and well-being. Although paraprobiotics have been related to health benefits, the majority of research shows that these effects are only seen when they are consumed directly. As a result, using foods as paraprobiotic carriers appears to be a field worth exploring, with both benefits and obstacles [11, 23].

LAB (probiotic containing therapeutic drug) can assist astronauts in preserving their health. One of the most pressing concerns in space crew medical care is infection prevention. Although astronauts' susceptibility to infection during the stay in space has not been confirmed, studies showed decline in LAB (*lactobacilli* and *Bifidobacteria*) indicating need for antibiotics. LAB prevents microflora multiplication from increasing virulence. Use of LAB for astronauts in space could help them cope with extremely stressful space circumstances including microgravity and low doses of cosmic radiation. Exposure to cosmic rays, on the other hand, may inactivate or modify LAB, affecting its positive benefits for humans. As a result, before the formulation of such probiotic rich drug for astronauts, radiation-resistant LAB strains were identified in some studies by assessing their capacity to withstand ionizing radiation. According to the obtained data, *L. acidophilus* showed the most resistant to gamma irradiation depicting probable use of the strain for the preparation of probiotics containing drugs for astronauts to maintain their health in space [23].

In another study, non-irradiated and irradiated *Lactobacillus paracasei* was used in combination with glycerol and starch along with gelatin to prepare edible coating to study the effect of the coating on the storage stability of tomato stored in low temperature. At a dose of 3.5 kGy radiation, the antimicrobial effect of the strain was the highest. The study revealed that irradiated strain incorporated edible coating exerted lower bacterial load along with retained sensory and nutritional quality of tomato indicating improved storage stability [26].

10.5 SUMMARY

Probiotics are necessary for the modulation of the human gut microflora to be maintained. Several scientific studies have confirmed their significant effect in enhancement of health. Metabolic disorders, liver issues, CVD are among conditions where probiotic bacteria are thought to have a significant therapeutic potential. Bowel problems, bacterial infections, and a variety of GI illnesses and diarrhea all appear to benefit from probiotics. Lactose intolerance can be relieved with probiotic microbes, and atopic dermatitis can be treated with them as well. Probiotics also showed potential effect on definite neoplastic disorders and anti-cancer therapy side effects. Prebiotics can be used as a replacement for probiotics or as a supplement to them. It turns out that creating bio-therapeutic formulae with proper combination of probiotic and prebiotics can enrich the environment of small intestine and colon. More research on probiotics and prebiotics in combination, as well as the creation of synbiotics, appears to be on the horizon. Future research could reveal the mechanisms of action of such components, which could have a beneficial impact on health.

KEYWORDS

- **dietary supplement**
- **nutrient bioavailability**
- **prebiotics**
- **probiotics**
- **shelf-life**
- **synbiotics**

REFERENCES

1. Akter, S., Park, J. H., & Jung, H. K., (2020). Potential health-promoting benefits of paraprobiotics, inactivated probiotic cells. *Journal of Microbiology and Biotechnology, 30*(4), 477–481.
2. Asadzadeh, A., Jalali, H., Azizi, M. H., & Mohammadi-Nafchi, A., (2021). Production of oat bran functional probiotic beverage using *Bifidobacterium lactis*. *Journal of Food Measurement and Characterization, 15*(2), 1301–1309.
3. Ávila, P. F., Martins, M., De Almeida, C. F. A., & Goldbeck, R., (2020). Xylooligosaccharides production by commercial enzyme mixture from agricultural wastes and their prebiotic and antioxidant potential. *Bioactive Carbohydrates and Dietary Fiber, 24*(2), E-Article ID: 100234.
4. Ávila, P. F., Silva, M. F., Martins, M., & Goldbeck, R., (2021). Cello-oligosaccharides production from lignocellulosic biomass and their emerging prebiotic applications. *World Journal of Microbiology and Biotechnology, 37*(5), 1–11.
5. Azmi, A. F. M. N., Mustafa, S., Hashim, D. M., & Manap, Y. A., (2012). Prebiotic activity of polysaccharides extracted from *Gigantochloa levis* shoots. *Molecules, 17*(2), 1635–1651.
6. Çakır, N., Bilgiçli, N., & Yaver, E., (2021). Impact of xylanase-treated wheat milling by-products on the physical and chemical properties of cakes. *Journal of the Science of Food and Agriculture, 101*(15), 6331–6337.
7. Cicenia, A., Scirocco, A., Carabotti, M., Pallotta, L., Marignani, M., & Severi, C., (2014). Postbiotic activities of lactobacilli-derived factors. *Journal of Clinical Gastroenterology, 48*(Suppl 1), S18–S22.
8. Costantini, M., Summo, C., Faccia, M., Caponio, F., & Pasqualone, A., (2021). Kabuli and Apulian black chickpea milling byproducts as innovative ingredients to provide high levels of dietary fiber and bioactive compounds in gluten-free fresh pasta. *Molecules, 26*(15), E-Article ID 4442.

9. Crittenden, R., & Playne, M. J., (2008). Nutrition News. Facts and functions of prebiotics, probiotics and synbiotics; Chapter 4. In: Lee, Y. K., & Salminen, S., (eds.), *Handbook of Probiotics and Prebiotics* (2nd edn., pp. 562–582). Hoboken – USA: John Wiley & Sons.
10. Crittenden, R., & Playne, M. J., (2008). Prebiotics; Chapter 7. In: Lee, Y. K., & Salminen, S., (eds.), *Handbook of Probiotics and Prebiotics* (2nd edn., pp. 535–561). Hoboken – New Jersey: A John Wiley & Sons.
11. De Almada, C. N., Almada, C. N., Martinez, R. C., & Santa-Ana, A. S., (2016). Paraprobiotics: Evidences on their ability to modify biological responses, inactivation methods and perspectives on their application in foods. *Trends in Food Science & Technology, 58*(C), 96–114.
12. Fernández-Peláez, J., Paesani, C., & Gómez, M., (2020). Sourdough technology as a tool for the development of healthier grain-based products: An update. *Agronomy, 10*(12), 1962–1970.
13. Food and Agriculture Organization of the United Nations/World Health Organization FAO/WHO, (2001). *Health and Nutritional Properties of Probiotics in Food Including Powder Milk with Live Lactic Acid Bacteria.* https://www.fao.org/3/a0512e/a0512e.pdf (accessed on 2 November 2023).
14. Food and Agriculture Organization, (2007). *FAO Technical Meeting Report on Prebiotics: Food– Quality and Standards Service (AGNS)* (p. 12). Rome- Italy: Food and Agriculture Organization of the United Nations (FAO).
15. Fuso, A., Risso, D., Rosso, G., Rosso, F., Manini, F., Manera, I., & Caligiani, A., (2021). Potential valorization of hazelnut shells through extraction, purification, and structural characterization of prebiotic compounds: A critical review. *Foods, 10*(6), E-Article ID 1197.
16. Gibson, G. R., Scott, K. P., Rastall, R. A., Tuohy, K. M., Hotchkiss, A., Dubert-Ferrandon, A., & Buddington, R., (2010). Dietary prebiotics: Current status, and new definition. *Food Science & Technology Bulletin Functional Foods, 7*(1), 1–19.
17. Gibson, G. R., Hutkins, R., Sanders, M. E., Prescott, S. L., Reimer, R. A., Salminen, S. J., Scott, K., et al., (2017). The international scientific association for probiotics and prebiotics (ISAPP) consensus statement on the definition and scope of prebiotics. *Nature Reviews Gastroenterology & Hepatology, 14*(8), 491–502.
18. Gómez, B., Gullon, B., Remoroza, C., Schols, H. A., Parajo, J. C., & Alonso, J. L., (2014). Purification, characterization, and prebiotic properties of pectic oligosaccharides from orange peel wastes. *Journal of Agricultural and Food Chemistry, 62*(40), 9769–9782.
19. Hernández-Macias, S., Comas-Basté, O., Jofré, A., Bover-Cid, S., Latorre-Moratalla, M. L., & Vidal-Carou, M. C., (2021). Growth-promoting effect of cava lees on lactic acid bacteria strains: A potential revalorization strategy of a winery byproduct. *Foods, 10*(7), E-Article ID 1636.
20. Holgado, F., Campos-Monfort, G., De Las Heras, C., & Rupérez, P., (2021). *In vitro* fermentability of globe artichoke byproduct by *Lactobacillus acidophilus* and *Bifidobacterium bifidum*. *Bioactive Carbohydrates and Dietary Fiber, 26*, E-Article ID 100286.
21. Hosono, A., (1992). Fermented milk in the orient. In: Nakazawa, Y., & Hosono, A., (eds.), *Functions of Fermented Milk: Challengers for the Health Sciences* (pp. 61–78). London- UK: Elsevier Applied Science.

22. Hussein, H., Awad, S., El-Sayed, I., & Ibrahim, A., (2020). Impact of chickpea as prebiotic, antioxidant, and thickener agent of stirred bio-yoghurt. *Annals of Agricultural Sciences, 65*(1), 49–58.
23. Hwang, E. N., Kang, S. M., Kim, J. K., Lee, J. W., & Park, J. H., (2013). Screening of radio-resistant lactic acid bacteria. *Food Science of Animal Resources, 33*(3), 335–340.
24. Jaichakan, P., Nakphaichit, M., Rungchang, S., Weerawatanakorn, M., Phongthai, S., & Klangpetch, W., (2021). Two-stage processing for xylo-oligosaccharide recovery from rice byproducts and evaluation of products: Promotion of lactic acid-producing bacterial growth and food application in a high-pressure process. *Food Research International, 147*, E-Article ID 110529.
25. Jakubczyk, E., & Kosikowska, M., (2000). New generation of fermented milk products with probiotics and prebiotics, synbiotic products. *Prz Mlezna, 12*, 397–400.
26. Khalil, O. A., Mounir, A. M., & Hassanien, R. A., (2020). Effect of gamma irradiated *Lactobacillus* bacteria as an edible coating on enhancing the storage of tomato under cold storage conditions. *Journal of Radiation Research and Applied Sciences, 13*(1), 318–330.
27. Koutrotsios, G., Patsou, M., Mitsou, E. K., Bekiaris, G., Kotsou, M., Tarantilis, P. A., Pletsa, V., et al., (2019). Valorization of olive byproducts as substrates for the cultivation of *Ganoderma lucidum* and *Pleurotus ostreatus* mushrooms with enhanced functional and prebiotic properties. *Catalysts, 9*(6), 537–545.
28. Kunz, C., Kuntz, S., & Rudloff, S., (2009). Intestinal flora. *Advances in Experimental Medicine and Biology, 639*, 67–79.
29. Ley, R. E., Peterson, D. A., & Gordon, J. I., (2006). Ecological and evolutionary forces shaping microbial diversity in the human intestine. *Cell, 124*(4), 837–848.
30. Lopes, S. M. S., Francisco, M. G., Higashi, B., De Almeida, R. T. R., Krausová, G., Pilau, E. J., Goncalves, J. A., et al., (2016). Chemical characterization and prebiotic activity of fructo-oligosaccharides from *Stevia rebaudiana* (Bertoni) roots and *in vitro* adventitious root cultures. *Carbohydrate Polymers, 152*, 718–725.
31. Markowiak, P., & Śliżewska, K., (2017). Effects of probiotics, prebiotics, and synbiotics on human health. *Nutrients, 9*(9), E-Article ID 1021.
32. Mateos-Aparicio, I., De La Peña, A. R., Pérez-Cózar, M. L., Rupérez, P., Redondo-Cuenca, A., & Villanueva-Suárez, M. J., (2020). Apple byproduct dietary fiber exhibits potential prebiotic and hypolipidemic effects in high-fat fed Wistar rats. *Bioactive Carbohydrates and Dietary Fiber, 23*, E-Article ID 100219.
33. Markowiac, P., & Slizewska, K., (2018). The role of probiotics, prebiotics and synbiotics in animal nutrition. *Gut Pathogen, 10*, 21–25.
34. Moore, K. E., (2011). *Biological Analysis of Prebiotics in Various Processed Food Matrices* (p. 96). Master's degree thesis, University of Nebraska, Lincoln, Nebraska.
35. Morelli, L., (2008). Postnatal development of intestinal microflora as influenced by infant nutrition. *The Journal of Nutrition, 138*(9), 1791S–1795S.
36. Neish, A. S., (2009). Microbes in gastrointestinal health and disease. *Gastroenterology, 136*(1), 65–80.
37. Piqué, N., Berlanga, M., & Miñana-Galbis, D., (2019). Health benefits of heat-killed (Tyndallized) probiotics: An overview. *International Journal of Molecular Sciences, 20*(10), E-Article ID 2534.
38. Pop, C., Suharoschi, R., & Pop, O. L., (2021). Dietary fiber and prebiotic compounds in fruits and vegetables food waste. *Sustainability, 13*(13), E-Article ID 7219.

39. Qiu, X. J., Zheng, W. X., Zhang, L., Shi, Y. L., Hu, J. H., Li, Y. L., Liu, Z. Y., & Zhu, M. D., (2020). Prebiotic effects of xylanase modification of β-glucan from oat bran on *Bifidobacterium bifidum. Italian Journal of Food Science, 32*(1), 4–15.
40. Sah, B. N. P., Vasiljevic, T., McKechnie, S., & Donkor, O. N., (2016). Effect of pineapple waste powder on probiotic growth, antioxidant and antimutagenic activities of yogurt. *Journal of Food Science and Technology, 53*(3), 1698–1708.
41. Saman, P., Fuciños, P., Vázquez, J. A., & Pandiella, S. S., (2019). Byproducts of the rice processing obtained by controlled de-branning as substrates for the production of probiotic bacteria. *Innovative Food Science & Emerging Technologies, 51*, 167–176.
42. Sharma, R., Garg, P., Kumar, P., Bhatia, S. K., & Kulshrestha, S., (2020). Microbial fermentation and its role in quality improvement of fermented foods. *Fermentation, 6*(4), 106–110.
43. Swallah, M. S., Fan, H., Wang, S., Yu, H., & Piao, C., (2021). Prebiotic impacts of soybean residue (okara) on eubiosis/ dysbiosis condition of the gut and the possible effects on liver and kidney functions. *Molecules, 26*(2), 326–331.
44. Tsafrakidou, P., Michaelidou, A. M., & Biliaderis, C., (2020). Fermented cereal-based products: Nutritional aspects, possible impact on gut microbiota and health implications. *Foods, 9*(6), 734–739.
45. Valero-Cases, E., Cerdá-Bernad, D., Pastor, J. J., & Frutos, M. J., (2020). Non-dairy fermented beverages as potential carriers to ensure probiotics, prebiotics, and bioactive compounds arrival to the gut and their health benefits. *Nutrients, 12*(6), E-Article ID 1666.
46. Vergin, F., (1954). Antibiotics and probiotics. *Hipokrates, 25*(4), 116–119.
47. Whitman, W. B., Coleman, D. C., & Wiebe, W. J., (1998). Prokaryotes: The unseen majority. *Proceedings of the National Academy of Sciences, 95*(12), 6578–6583.
48. Wichienchot, S., Thammarutwasik, P., Jongjareonrak, A., Chansuwan, W., Hmadhlu, P., Hongpattarakere, T., Itharat, A., & Ooraikul, B., (2011). Extraction and analysis of prebiotics from selected plants from southern Thailand. *Songklanakarin Journal of Science & Technology, 33*(5), 517–523.
49. Yang, M., Ashraf, J., Tong, L., Wang, L., Zhang, X., Li, N., Zhou, S., & Liu, L., (2021). Effects of *Rhizopus oryzae* and *Aspergillus oryzae* on prebiotic potentials of rice bran pretreated with superheated steam in an *in vitro* fermentation system. *LWT-Food Science & Technology, 139*, E-Article ID 110482.
50. Zhang, M. M., Cheng, J. Q., Lu, Y. R., Yi, Z. H., Yang, P., Wu, X. T., (2010). Use of pre-, pro-and symbiotics in patients with acute pancreatitis: A meta-analysis. *World Journal of Gastroenterology, 16*(31), 3970–3978.
51. Zhang, D., (2015). *Chemical Composition, Probiotic Survivability and Shelf-Life Studies of Symbiotic Buttermilk* (p. 57). Master's degree thesis, University of Vermont, Burlington, US.

CHAPTER 11

Postbiotics: Types, Mechanism of Action, and Applications

Areeba Tariq and Sebnem Ozturkoglu-Budak

ABSTRACT

Postbiotics are cell-free suspensions that are enriched with bacterial by-products. These are obtained from various beneficial bacteria like probiotics and useful yeasts. These probiotic bacteria are naturally enriched with organic acids, SCFAs, peptides, bacteriocins, exopolysaccharides, and oligosaccharides. Due to certain limitations of incorporating live bacteria in humans (especially immunocompromised populations), these postbiotics present a valuable alternative, as they exhibit similar properties as that of their bacterial source. The postbiotics also show positive immunomodulatory response, anti-microbial, anticancer, antiproliferative, anti-diabetic, and anti-inflammatory roles in our body. Because of their functional characteristics, these are potentially used as bio-preservatives in the food industry. Postbiotics are obtained by implying various methods of bacterial cell disruption, such as: treatment with high temperature, high pressure, sonication, tyndallization, radiations or by using chemical agents. The methods of bacterial cell inactivation results in cell free supernatant (CFS) or cell-free extracts (CFE), which are mostly lyophilized before supplementing into products.

11.1 INTRODUCTION

Man has always been directly or indirectly in contact with the microbes. Microorganisms can either be beneficial to human health like LAB, some

The Functional Foods: Nutrient and Health Benefits.
Megh R. Goyal, Junaid Ahmad Malik, and Anu Kumari (Eds.)

yeasts, etc.; or could be harmful enough to even become a reason for epidemics or pandemics.

Human bodies are complex structures that are made up proteins, fats, various organic and inorganic matter. In addition, the human gut consists of plethora of microorganism that reside inside the stomach, small intestines, and large intestines.

Continuous efforts are employed to improve the quality of products and made them available for human consumption as much as possible. The concept of probiotics and prebiotics is well discussed in literature. Probiotics are healthy bacteria, which are added to food to introduce healthy and beneficial bacteria in the gut. There are numerous ways of adding probiotics in dairy productsm, such as: yogurt, kefir, fermented drinks including fermented milk. Sometimes the prebiotics are also added to provide the food for probiotics. However, the consumption of live microorganisms sometimes might impose some negative effects on consumers. Due to this, the concept of parabiotics and postbiotics came into being. Being an innovative term in the food industry, a proper definition of postbiotics has not yet been proposed. While on research basis, it is defined as by-products or the left behind metabolites of live bacteria that are suspended in supernatant that does not contain any live bacterial cell matter.

Postbiotics commonly cover wide range of metabolites and by-products that are released by the probiotics. This also includes the substances released after bacterial cell lysis. In addition, the term includes substances that are released by bacteria when exposed to harmful chemicals or microbes. These substances might be helpful to hinder the negative effects of those chemicals.

Postbiotics (also known as cell free supernatant (CFS) or cell free extract (CFE)) display wide functionality both in laboratory studies and in animal studies. The common properties delivered by postbiotics are anti-microbial, anti-inflammatory, anti-diabetic, anti-cancer, antioxidant, and antihypertensive. Once exposed to target microbe of environment, postbiotics act in the same way as expected from their healthy bacterial source.

Metabolites, produced by probiotic bacteria, exhibit same or similar properties as their source. Hence, they are believed to mimic the activity of probiotic bacteria against pathogens while providing same beneficial effects. The utilization of postbiotics in food still needs to be studied as the mode of action (MoA) of some postbiotic metabolites is still unknown.

In the dairy industry, postbiotics are naturally produced in many fermented products including yogurts, probiotics yogurts, cheese, etc. Keeping in mind the whitish color of milk and other products, the use of MRS broth (used for postbiotic preparation) might restrict postbiotic usage in these products. Hence for large scale production, alternatives (such as: milk media) are under consideration. In dairy sector, the use of mixture of postbiotics is preferred over metabolites from a single source. In addition, spray drying of postbiotics as a biopreservative over hard cheeses is considered as a natural strategy to counter microbial pathogens and other spoilage agents [18].

This chapter provides an in-depth review of postbiotics including its types, methods of preparation and their functionality or utilization. Also, their role of action at the target site is explored.

11.2 DEFINITION OF POST-BIOTICS

Postbiotics are yet to be formally defined by FAO. However, according to various sources, postbiotics are metabolites or by-products produced by many probiotics and non-probiotic bacteria. These could be either chemical compounds that are produced as a result of bacteria's normal activity (*in-vitro* or *in-vivo*) or the substances after the bacterial cells have been disrupted. In some cases, these also include cellular fragments produced after the bacterial cell is ruptured.

The term postbiotics is made up of two words, i.e., "Post" is used as a prefix means "after," while "Biotic" means "resulting from living organisms." Hence, the word "Postbiotic" means "after life," which describes the non-living factors that are obtained after the organism death. As the most common way of obtaining postbiotics is via deactivating/killing the bacterial cell and suspending the remaining substances in supernatant; therefore, the supernatant does not contain any live bacteria and it is considered as a CFS or CFE.

According to Salminen et al. [27], "*Postbiotics are inanimate form of microorganisms as components or metabolites to provide specific health benefits to the host.*"

11.2.1 COMMON SOURCES OF OBTAINING POST-BIOTICS

Postbiotics could either be obtained by disrupting the bacterial cell membrane and obtaining its fragments or extracting intracellular matter as a result of cell lysis [7]. Disruption of bacterial cell membrane yields many cell surface components, such as: exopolysaccharides, lipoteichoic acids, glycolipids, etc. Cell lysis gives not only membrane components but also the intracellular components of the bacterial cells.

11.3 TYPES OF POST-BIOTICS

There are variety of substances, chemical compounds, and molecules (both having lower molecular weight as well as larger molecular weight) that could be characterized under vast umbrella of post-biotics. This includes many peptides, organic acids, bacteriocins, etc. Hence, it is more convenient to categorize them. Postbiotics are divided according to their structural composition (includes many lipids, organic acids, peptides, some carbohydrates, vitamins especially vitamin B group, minerals, etc.) or the function they perform in the body (antioxidant activity, anti-microbial, anti-hypertensive, etc.) as shown in Figure 11.1.

11.3.1 TYPES OF POST-BIOTICS: ACCORDING TO STRUCTURE

Structurally, postbiotics are described either according to the type of chemical bond or structure present and/or by the presence of certain functional groups in the structure. According to structure, postbiotics are differentiated under this section.

11.3.1.1 SHORT CHAIN FATTY ACIDS (SCFAS)

The short chain fatty acids (SCFs) commonly include fatty acids with carbon number ranging from C_2 to C_6. These include four carbons contain butyric acid, three carbon containing propionic acid or two carbon containing acetic acid. Among these, butyric acids provide maximum anti-inflammatory properties. In addition, all these common fatty acids play significant role in the colon health; and have antineoplastic characteristics [22].

Among SCFA, butyrate is significantly important in clinical studies because of its multiple benefits. It helps in GI tract infection, supports intestinal health, has anti-inflammatory properties, and acts as immunomodulatory agent. Moreover, butyrate has also gained importance in protecting against various food allergies. Homayouni et al. [10] highlighted on the functionality of butyrate in the protection against food allergens.

Butyrate attaches to G-protein receptors and stimulates macrophages to produce IL-10. They also develop Treg in lymphatic system, which suppress the activity of allergens, hence can inhibit allergic reactions.

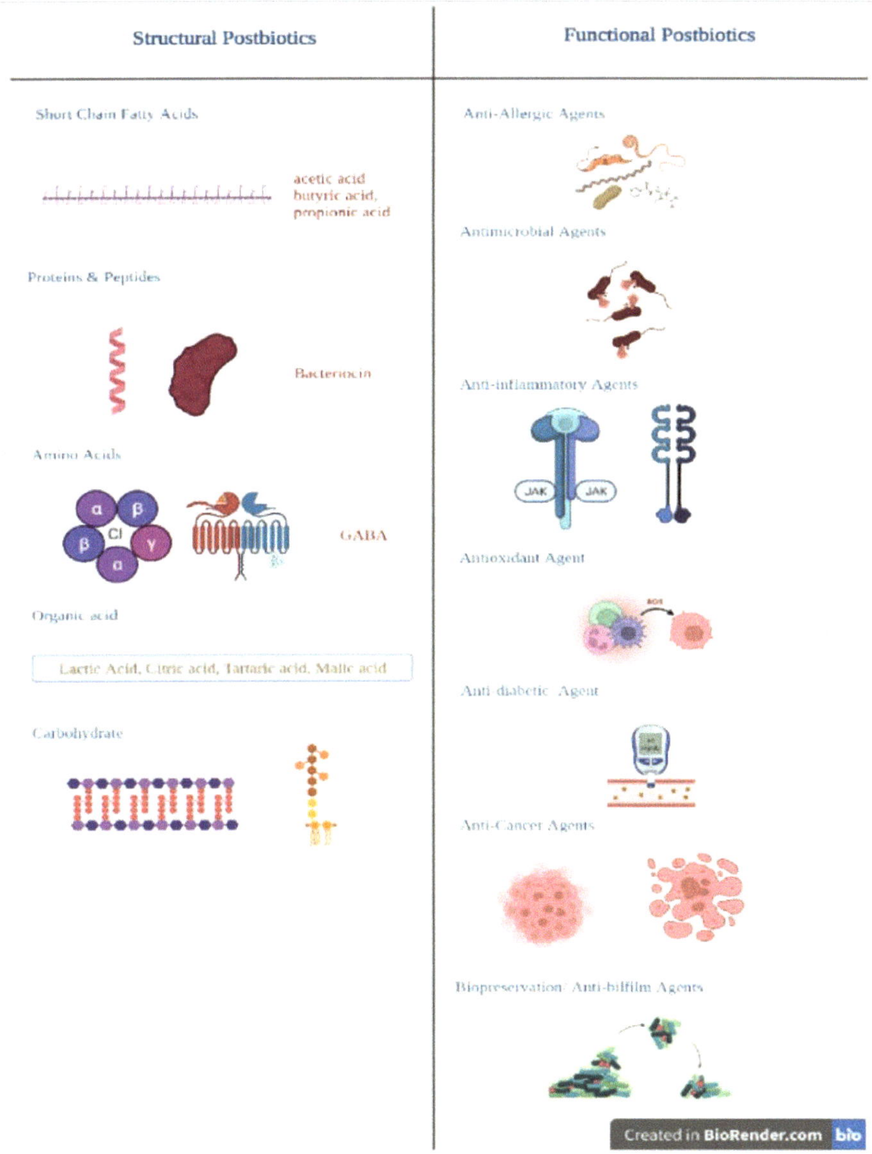

FIGURE 11.1 Types of post-biotics.

Source: Created by authors with BioRender.com.

11.3.1.2 PROTEINS AND PEPTIDES

One of the major classes of peptides are bacteriocins that constitute significant role as postbiotics as they are released from live bacteria. These are structurally protein in nature and are effective antimicrobials.

Bacteriocins obtained from *Enterococcus hirae* ST57ACC strain and *Pediococcus pentosaceus* ST65ACC strain have excellent activity against *L. monocytogenes* and *L. innocua*. These bacteriocins have shown high stability at high temperatures and wide pH ranges and are considered as potential biopreservatives [4]. Stability and activity of bacteriocins obtained from described bacterial strains are described in Table 11.1.

TABLE 11.1 Stability Limits of Postbiotics Extracted from Cheese Made with Raw Milk

Source	Stability	Limitations and Results
Enterococcus hirae and *Pediococcus pentosaceus:* Extracted from cheeses made with raw milk.	Temperature	Up to 100°C
	Sterilization	At 121°C for 15 min.
	pH	From 2.0–10.0 for 2 h.
	Proteases	Not stable (activity lost).
	α-amylase and lipase enzyme.	Stable
	Skim milk, Tween 80 and NaCl.	Stable
Enterococcus hirae: Inhibitory activity	*L. monocytogenes* 211	100% inhibition after 12 h.
	L. monocytogenes 422	100% inhibition after 12 h.

Abo-Amer [2] indicated similar biopreservative role of postbiotics using 11 various strains of *L. acidophilus*. These strains represented antimicrobial activity due to the presence of highly stable bacteriocins. Bacteriocins were active against wide range of pathogens and retained their bioactivity for up to 70 days. Noteworthy antibacterial properties were observed by AA105 strain of *L. acidophilus* against *Staphylococcus aureus*, *B. cereus* and *B. subtilis*. Moreover, bacteriocins appeared to retard the development of many bacteria especially Gram-negative bacteria.

11.3.1.3 AMINO ACIDS

These are basic building blocks on which the foundation of proteins is built. Some amino acids are also found in non-protein sources. A classic example of non-protein amino acid is ɣ-Aminobutyric acid (also known as GABA). GABA are considered as GRAS and are helpful against anxiety and depression when delivered to body. They also help in regulating the hormonal secretions in body.

Various LAB species are tested for the ability to produce GABA. Wu & Shah [31] showed that various strains of *Lactobacillus brevis* produce GABA by decarboxylation of glutamate.

11.3.1.4 ORGANIC ACIDS

Organic acids constitute major segment of postbiotics. These include acetic acids, propionic acid, butyric acid, lactic acid, etc. The main role of organic acids is to lower the pH in both bacterial cells and/or within the bacterial membrane. As a result of changing of pH, the membrane becomes weak, and it is disrupted. A mixture of organic acids is preferred as compared to the use of individual organic acid for antimicrobial activity [11]. The same study showed the high antimicrobial activity of mixing various acids (including lactic acid, malic acid, tartaric acid, and citric acid). These acids were obtained from different Chinese foods (Milk thistle, yogurt, urum, kumis). Their functionality was observed for *salmonella*, *Staphylococcus aureus* and *E. coli* strains.

In addition to above-mentioned benefits, organic acids have also lipophilic properties against pathogenic cell-membrane. According to Rathod et al. [26], various organic acids can interact with lipids present in outer membrane of microbes, thus solubilizing them and leading to their entry in the microbial cytoplasm. Once in the cytoplasm, these acids can hinder the enzyme activity of pathogenic bacterial cells to reduce their proliferation.

11.3.1.5 CARBOHYDRATES

Postbiotics in the form of carbohydrates include soluble polysaccharides, such as: exopolysaccharides (both linear and branched), oligosaccharides, endo-polysaccharides, and GOS, etc. Carbohydrates are important in the dairy industry to contribute towards texture, mouthfeel, and taste of the products. In terms of dairy products, these are helpful in delivering water-binding properties, viscosity, and stability to commodities.

In the pharmaceutical sector, polysaccharides could be utilized in the efficient drug delivery due to their destructive nature at low pH. In this way, they could be molded into dosage forms of drugs. They are non-toxic, gel-forming, and stable in GI tract. These features make them better option for the targeted drug delivery. They also have antioxidant and antitumor properties and can lower the blood cholesterol levels.

Exo-polysaccharides postbiotics constitute major class of carbohydrates. They could be isolated from LAB bacteria, such as: *L. fermentum*, *L. helveticus*, and *S. thermophilus*, etc. Pullulan, for example, is an exopolysaccharide, which is used as a starch replacer and in the form of protective glaze over fruits. Due to its high impermeability towards oxygen and good mechanical properties, pullulan is widely used as a biopreservative [15].

Postbiotics in the form of various oligosaccharides are important growth factors in infants. A clinical study carried was carried out to show that infant formulas supplemented with a mixture of oligosaccharides (such as: 2′-Fucosyllactose, 3′-Galactosyllactose, fructo-oligosaccharides (FOS), and GOS) could effectively mimic the characteristics of human milk. A 17-week study on infant was carried out. Infant formula containing mixture of specific percentages of described oligosaccharides and milk fat was given. The observed results indicated weight gain and adequate infant growth without any adverse effects on the infant immunity and their gastrointestinal (GI) microflora [29].

Soluble polysaccharides from *Lactobacillus cesei* ATCC606 were tested for their anti-cancer properties. Choi et al. [5] showed that soluble polysaccharides caused DNA fragmentation in HT-29 cancer cells and caused apoptosis of these cells. Moreover, these polysaccharides presented high antiproliferative properties in cancer cells.

11.3.1.6 VITAMINS

Postbiotics in the form of vitamins (especially B-vitamins) are essential in accelerating various metabolic pathways in food commodities and in human bodies. Some bacterial species including LAB are able to yield lower to moderate quantities of some B-vitamins.

Zago et al. [33] determined the ability of some strains of *L. plantarum* to produce folate (vitamin B_9) and riboflavin (B_2) that are commonly consumed by bacteria during their growth; and it was observed via their production in relative folate free (FACM) and riboflavin free media (RAM). It was found that all tested strains of *L. plantarum* (Lp790, Lp813, and Lp998) delivered lower quantities of vitamin B_9 ranging from 2 to 6 ng/ml, while only single strain (Lp998) produced lower concentration of riboflavin (detected quantity was 6.36 ng/ml). Riboflavin and folate are important vitamins, which play vital role in the growth and development of bacteria.

11.3.2 TYPES OF POST-BIOTICS: ACCORDING TO FUNCTION

11.3.2.1 ANTIMICROBIAL AGENTS

Antimicrobial properties are mostly due to presence of postbiotics in the form of bacteriocins, SCFAs, peptides, and other small metabolites. These metabolites inhibit growth of pathogen microorganisms. Antimicrobial agents are normally categorized based on their molecular weight [19]: lower molecular weight compounds, such as, H_2O_2 or CO_2; or higher molecular weight compounds, such as: bacteriocins.

The antimicrobial activity of metabolites produced from LAB showed that four LAB strains including *L. delbrueckii* indicated maximum antimicrobial activity against multidrug resistance and pathogenic bacteria [12]. The increased activity of these postbiotic metabolites showed that the metabolic acids may cause increased antimicrobial activity.

Moradi et al. [19] indicated that the addition of *L. salivarius* lyophilized CFE in pasteurized milk (10–45 mg/ml) showed excellent activity against *Listeria monocytogenes*; while it also helps in removing its biofilm.

Yordshahi et al. [32] showed the activity of postbiotics from *Lactobacillus plantarum* against *Listeria monocytogenes*. Their study suggested that the heterofermentative bacteria *L. plantarum* excretes some major classes of postbiotics, such as, bacteriocins, lactic acid and other antimicrobial metabolites that could be added in active packaging. These postbiotics are released slowly in the environment and can increase the shelf-life of product for up to 9 days in case of meat products. The diffusion of postbiotics is dependent on the type of product that is packaged. If the packed product has higher hydration levels, then the probability and rate of diffusion is more. The study also suggests that the activity of postbiotics from *L. plantarum* is heat resistant (at elevated temperature of 100°C) and are stable against wide range of pH.

11.3.2.2 ANTI-INFLAMMATORY AGENTS

The metabolites of cell lysates from treating various LAB resulted in formation of anti-inflammatory metabolites. Their exposure to pathogen bacteria results in formation of many proinflammatory and inflammatory mediators. It is shown that two different forms of postbiotics obtained *Lactobacillus rhamnosus* (i.e., secreted media from *L. rhamnosus*) and its cell lysates produce distinctive effects against *Prohormonal gingivalis*, which is a biofilm forming pathogen present on gingival epithelial cells in the mouth.

It was observed that co-culturing cell lysates with pathogen increased the transcription of IL-1, TNF-α, and IL-6, in the infected cells. Furthermore, the presence of both types of postbiotics resulted in an enhanced receptor response of TLR-2 and TLR-4. Other noteworthy outcomes were [28]: the increased ability of suppressing *P. gingivalis* to adhere to the gingival epithelial cells; improved cell recognition; and increase of proinflammatory cytokines.

A mixture of postbiotics obtained from a variety of probiotic sources is considered to be more effective in treating inflammatory bowel disease (IBD) and other GI tract inflammations. De Marco et al. [9]; and Das et al. [8] observed the effect of postbiotics from *L. acidophilus*, *L. lactis*, *L. casei*, and *Saccharomyces boulardii* strains, against pro-inflammatory cytokines. The study revealed the highest capacity of *L. lactis* to reduce production of IL-8 and IL-6. In addition, the experiments also described the low production of PGE-2, which is an important factor causing inflammations and progressions of colon-rectal cancers. The same study observed the downregulatory role of postbiotics from *L. acidophilus* and *L. casei*, for TNF-α secretion.

11.3.2.3 ANTIOXIDANTS

The ability to scavenge free radicals in the target environment is an important feature of postbiotics metabolites produced by various LAB species. It is even believed that postbiotics show greater antioxidant properties when compared to live bacterial cells. These activities are mainly proposed by some ionic compounds, such as: Mg^{2+}, peptides, and polysaccharides, etc. While synthetic antioxidants have wide market in the form of BHA, BHT, and TBHQ (tert-butyl hydro quinone), yet similar antioxidant benefits could also be obtained from postbiotics. Various studies show high scavenging properties of postbiotics resulting in reduced chances of reactive oxygen accumulation.

Postbiotics of *Lactobacillus plantarum* DM5 strain, produced by sonication, were presented by Das & Goyal [8]. The results showed that maximum antioxidant enzymes are obtained via cell disruption methods. These enzymes included NADH oxidase, ascorbate peroxidase and superoxide dismutase, etc. The presence of ascorbate peroxidase was proved by increased inhibition of autooxidation of ascorbate; hence, lower free radical formation. Also, the cell components of *L. plantarum* DM5 resulted in lower chances of lipid peroxidation due to its ability of scavenging hydroxyl radicals. The other activities of two enzymes seems to be utilized against superoxide anions as shown in same study.

11.3.2.4 ANTI-DIABETIC AGENTS

Diabetes is characterized by insulin resistance and inability of body organs to utilize insulin for glucose breakdown as a source of energy. Postbiotics are large family of metabolites including peptides, enzymes, SCFA, organic acids and GABA compounds, etc., that could be potentially consumed in diabetic patients.

GABA is added as a food additive to increase its therapeutic properties. According to Abdelazez et al. [1], two strains of *L. brevis* including $KLDS_{1.0727}$ and $KLDS_{1.0373}$ (having an ability to produce high amounts of GABA) could be exploited as an anti-diabetic agent. Both these strains of bacteria have large quantities of GAD enzyme, which is linked to high production of GABA. Their study showed that post-prandial levels of glucose were reduced after 4 weeks of inducing postbiotics mixed with camel milk. Also, the hypolipidemic conditions were reduced in all treated organisms.

11.3.2.5 IMMUNOREGULATORY COMPOUNDS

Postbiotics or the CFEs help to stimulate immune responses [14]. The study showed that the CFEs of *Bifidobacterium* sp. BGN4 helped to enhance phagocytosis in experimental tests; while at the same time its cellular fractions were released in the supernatant to help in the release of cytokine, which is an anti-inflammatory agent.

Ashraf et al. [3] showed that cytokine production (both anti-inflammatory and pro-inflammatory) is considerably increased when postbiotics from a variety of probiotics and LAB are produced. The study was carried on the PBMC (peripheral blood mononuclear cells) cells. These cells were taken *in-vitro* from buffy coats. The results showed that strain of LAB produced high amounts TGF-ß, interleukin-10 and TNF-α. In addition to this, other cytokines including IL-12, IFN-ɣ and IL-2 were also produced during 24 h of stimulation. The highest activity was observed at 48 h and lasted up to 72 h except for IL-10 and IL-2, which showed maximum activity after 96 h. Hence, the utilization of postbiotics or the cell components could be effective for stimulating immune responses in body.

Vinderola et al. [30] evaluated the postbiotic exopolysaccharides and peptides for their immunomodulatory characteristics in the intestinal mucosa. The ingested postbiotics over a period of 5–7 days increased IL-6 production along with IgA in the small intestine. Therefore, the overall immunoregulatory properties could be enhanced against pathogen bacteria in the intestine.

11.3.2.6 ANTICANCER AGENTS

Postbiotics in the form of soluble polysaccharides help in the apoptosis of colon-cancer cells. These postbiotic fractions were obtained by treating the *L. acidophilus* 606 strain at 95C for one hour while filtering the soluble polysaccharides with ethanol. The polysaccharides helped in the reduction of proliferative colon-cancer cells [5].

Lactobacillus reuteri is another LAB bacteria, whose CFEs and high temperature treated with combined sonication (HK-sonicated) fractions were tested against colon-rectal cancer-causing cells. Various concentrations of both supernatant and high temperature and sonicated fractions showed high apoptotic activity in cancer inducing HT-29-ShE cells. Matrix metalloproteinases, which are the main facilitators of metastasis of cancer cells, could be significantly reduced by treating with more than 100 kDa of HK-sonication fraction of *L. reuteri*. Another mediator known as cyclooxygenase-2 (COX-2) present in cancer cells can help in invasion of hepatocellular carcinoma and can maintain the stemness of cancer stem cells. To counter their activity, more than 50 kDa of CFS was enough to significantly suppress the proliferation of COX-2. Maghsood et al. [16] indicated antiproliferative, antitumor, antimetastatic, and apoptotic activities of postbiotics from *L. reuteri*.

11.3.2.7 BIOPRESERVATIVE AGENTS

Postbiotics have significant antimicrobial, anti-biofilm, and antibacterial properties. Hence, postbiotics are recognized as main source of biopreservatives. Bacteriocins are highly stable against wide temperature ranges and pH extremes; hence, it could be used to enhance shelf-life of many food commodities. Jiang et al. [13] showed that bacteriocins from *Lactobacillus paracasei* strains disrupted the cell membrane of *S. aureus* cells and decreased their metabolic activity. Disruption of cell membrane causes cytoplasmic matter to come out of the cell and thus causing the deformation in cell-structure.

11.4 PREPARATION OF POSTBIOTICS

Postbiotics are the products of probiotics that are produced in the media called supernatant. Around 99% of postbiotics are produced from healthy bacteria (probiotics), while only a small percentage of postbiotics is obtained from other bacterial or fungal sources (Figure 11.2).

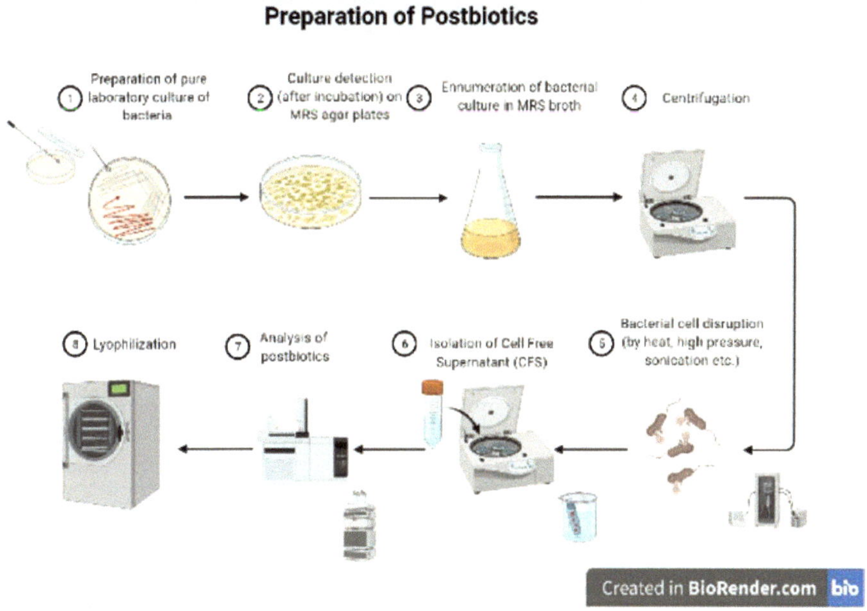

FIGURE 11.2 Step by step preparation of postbiotics.

Source: Created by authors with BioRender.com.

Probiotics include large class of LAB (*Lactobacillus acidophilus* spp.), bifidobacterium and yeasts, etc. LAB are heterogenous, gram-positive, and catalase negative bacteria while Bifidobacterium are also considered as prominent probiotics. LAB contains many species that could potentially be utilized for the production of postbiotics, such as: *Streptococcus* spp., *Lactococcus* spp., *Lactobacillus* spp. or Leuconostic bacteria. Among these, *Lactobacillus* spp. are widely used for the production of postbiotics. They include *Lactobacillus casei*, *Lactobacillus acidophilus*, etc.

While most postbiotics come from probiotic bacterial species, yet some prebiotics like FOS or gluco-oligosaccharides (GOS) are also added to the medium to maintain the appropriate production of postbiotics in supernatant.

For the laboratory production of postbiotics major procedures followed are (Figure 11.2):

- Preparation of pure laboratory culture of bacteria;
- Enumeration of isolated bacterial culture;
- Bacterial cell disruption; and
- Isolation of cell free supernatant (CFS) or postbiotics.

11.4.1 PREPARATION OF PURE LABORATORY CULTURE OF BACTERIA

In the preparation of postbiotics, firstly the bacteria from which the postbiotics are to be obtained are selected. Nutrient agar or the appropriate growth media is selected and us prepared according to the pre-escribed instructions. After the media is prepared, the dilution of bacterial colonies/bacterial is streaked onto the plates. Two methods for streaking are described as below:

1. **Pour Plate Method:** In this method, petri plate is inoculated with bacterial colonies and appropriated media is poured over it. The petri plate is moved in a swirling motion to mix the contents. The plates are then placed in incubator at 37C till the bacterial growth is observed.
2. **Spread Plate Method:** In this method, the appropriate media is poured onto the plates and enough time is given to solidify it. The bacterial inoculum is then streaked onto the media and the plates are kept in the incubator till the growth is obtained.

11.4.2 ENUMERATION OF BACTERIAL CULTURE

For this purpose, the diluted media is prepared in tubes. A single colony is randomly picked from the petri plate and is streaked into the media containing tubes. These are again placed in incubator at 37°C for enough time till maximum growth is observed.

11.4.3 ISOLATION OF BACTERIAL CELLS

After an appropriate number of probiotic bacteria are obtained, the obtained bacterial cells are isolated via centrifugation. The centrifugation conditions vary according to the type of probiotic bacterial cells. For example: *Lactobacillus acidophilus*, the broth containing bacterial cells is centrifuged at 3,500 g for 20 minutes. The normal conditions vary from 4,000 g to 12,000 g for 10 min to 30 min. The centrifuge temperature is normally kept at 4°C. After centrifugation, the obtained bacterial cells are washed three times with PBS solution.

11.4.4 BACTERIAL CELL DISRUPTION

An important step in the preparation of postbiotics CFS is the implication of bacterial cell death or bacterial lysis. Various methods are employed to kill the live bacterial cells obtained after centrifugation, such as:

- High pressure treatment;
- High temperature treatment;
- Inactivation by using chemical agents like formalin;
- Radiations (ultraviolet rays UV and ionization);
- Sonication;
- Tyndallization.

Among the these-described methods, the high temperature treatment and sonication are the most utilized one as described in subsections.

11.4.4.1 HIGH TEMPERATURE TREATMENT

In this process, the postbiotic metabolites are produced as a result of heat treatment at elevated temperatures. Postbiotic components are generally heat resistant while the bacteria producing those postbiotics could survive at high temperature. A temperature ranging from 70°C to 100°C is enough to kill maximum cont. Normally, the temperature given is high enough to ensure that no live bacteria would remain in the end.

Nielsen et al. [21] compared the peptides of normal fermented yogurt to yogurt prepared with high temperature treatment done in the form of pasteurization after fermentation. The temperature was adjusted to 95°C for 30s. The high temperature treatment showed an increased number of peptides (482 peptides) as compared to yogurt without heat treatment. This showed that thermal denaturation occurring at high temperatures could increase the proteolytic activity in the yogurts.

11.4.4.2 TYNDALLIZATION

The concept of tyndallization is based on the sterilization. This process is basically used for killing bacterial pathogens along with their spores of vegetative forms. In terms of postbiotics, it is made sure that even the inactive or spore form of bacterial is eliminated from the supernatant.

In this process, the probiotic bacteria are killed with the application of heat in intervals. The bacterial colonies are first killed by giving heat (temperature ranging from 80°C to above 100°C). After heat treatment, the supernatant is kept in the incubator at 37°C for at least 1-day. The stay time is long enough to ensure that if any bacteria survive it can proliferate. In terms of spore forming bacteria, the stay time is adjusted according to spores to germinate. Once the growth is observed, the supernatant is heated again under same conditions. This process is time consuming and usually takes up to 3-days. The main advantage of this process is the certainty that complete cell death or cell lysis is achieved [23].

11.4.4.3 SONICATION

It is a method of killing bacterial cells by giving sound waves to the media containing bacterial cells. In this process, the probiotic cells are cultured and centrifuged as in this chapter. The bacterial cells obtained after centrifugation are washed and suspended in the PBS solution (20 ml). The tube containing bacterial cells is placed in ice-water bath. The sound waves at the frequency of 20 kHz (kilohertz) is generally applied at 30% intensity. The frequencies are applied for approximately 25 minutes. The sonication process disrupts the bacterial cell-wall and causes damage to its contents resulting in cell death. After sonication process, the residue dead cells are removed from the supernatant by filtering with double filters (with 0.4 and 0.2 μm pore size). The obtained cell-free supernatant (CFS) is stored and used for further analysis.

Methods for sonication are manually adjusted according to the type and number of bacterial colonies. Ashraf et al. [3] showed that different conditions are applied for the bacterial cells obtained from a variety of probiotics, rather than a single bacterial strain. The described method started with placing cell suspension over ice bath and continued with five cycles at 100 W each. Each sonication cycle lasted for 15 minutes with 2-minute break between cycles. Sonication done in five cycles ensured maximum cell death of all types of probiotic bacteria and the clear supernatant was obtained by centrifugation at 10,000 g for 10 minutes [25].

11.4.5 ISOLATION OF CELL FREE SUPERNATANT (CFS) OR POSTBIOTICS

After the destruction of bacterial cells via above-mentioned methods, the next step involves the isolation or in other words, cleaning of the dead cell debris from supernatant. Isolation involves the extraction of pure CFS. For this purpose, cleaning could be done by carrying out either 2nd centrifugation, dialysis or column purification.

11.4.6 CHECKING THE PRESENCE OF LIVE BACTERIA

Cell free supernatant (CFS) or postbiotics are checked to make sure that no live bacteria remain in the supernatant. This step is necessary as it involves the assurance of basic principle of postbiotics because presence of any viable bacterial cell renders the medium to be called Postbiotic. Culture Dependent and Culture Independent methods are adopted to check the viable bacterial count in CFS.

11.4.6.1 CULTURE DEPENDENT METHOD

This method is commonly recognized as Plate Count Method. In case of postbiotics, this method involves preparation of media according to the source of postbiotics (for examples, if postbiotics are obtained from LAB, then MRS media is prepared and are solidified in petri plates). Prepared postbiotics are streaked onto the petri plate and incubation time is given. Plates are evaluated for the presence of any probable bacterial growth. Advantages of culture dependent method are Formation of visible colonies if present; and Easy and Low cost. Limitation of culture dependent method are: Requires specific time and temperature; Works only under specific conditions; and does not give information about unculturable and metabolically active bacteria.

11.4.6.2 CULTURE INDEPENDENT METHOD

These methods do not require cultures for the evaluation of viable counts. Mostly the principle involves the presence of nucleic acids in bacterial cells. Despite sensitive nature of these techniques, these are used mostly for evaluation of cellular components and compounds rather than for testing viable counts. Commonly involved techniques are: Real Time-PCR, rt-PCR, and/or Flow cytometry. Advantages of culture independent methods are: Gives highly precise results; Rapid and less time consuming; and it is a direct method.

11.4.6.3 STORAGE OF POSTBIOTICS

Once the postbiotics are tested for the absence of any live bacteria, these are stored in extreme low temperatures. Their storage temperature ranges from –20°C to –80°C. For most lyophilized postbiotics, the supernatant in semi-solid freeze form is suspended in PBS buffer solution before use.

11.4.7 ANALYSIS OF POSTBIOTICS

The analysis of postbiotic metabolites in CFS is necessary to identify their future use [20]. Various techniques and instruments are used for the quantitative and qualitative analysis of postbiotics from probiotics sources. Some common metabolites identified or quantified through various techniques are given in Table 11.2.

TABLE 11.2 Analysis Techniques for Identifying Postbiotics

Analysis Technique	Identified Components/Application
Fourier transform infrared (FTIR) spectroscopy.	Metabolite identification, characterization of biosurfactants.
Gas chromatography–mass spectrometry (GC-MS).	Organic acids quantification, EPS composition, etc.
Gas chromatography (GC)	Free fatty acids (e.g., 2-butanone), organic acids and other volatile compounds.
Gas chromatography–flame ionization detector (GC-FID)	SCFAs
Liquid chromatography (LC)/high-performance liquid chromatography (HPLC)	Organic acids (acetic acid, lactic acid, malic acid, etc.)
NMR (^{1}H NMR and ^{13}C NMR)	Phenolic components, lipid components, identify antimicrobial and antifungal compounds.
Thin layer chromatography (TLC)	Identify polysaccharides, and lipid contents, structural determination of biosurfactants.
Ultra-performance liquid chromatography (UPLC)	Intracellular components (enzymes and proteins), antifungal components, etc.

11.5 METHODS FOR INCORPORATION OF POSTBIOTICS

11.5.1 ADDITION OF LYOPHILIZED POSTBIOTICS

Lyophilization is a common way to keep postbiotics and preserve its functionality. Lyophilization includes drying of CFS at extremely low temperatures. After bacterial cell disruption and subsequent separation of killed cells through centrifugation, the supernatant is filtered with an appropriate pore size filter paper. The filtered supernatant/postbiotics are placed at the freezing temperature of –40°C. The pump pressure is maintained at 100 mTorr while the temperature of shelf is reduced to –60°C. This procedure is considered as the freeze drying of postbiotics and lyophilized postbiotics are obtained as a result [32]. Spray drying, although uncommon, yields better results against pathogens and spoilage.

11.5.2 CO-CULTURING

Co-culturing involves the growth of culture having beneficial properties with the culture of pathogenic bacteria. After given the proper inoculation time, the activity of both cultures is observed. In case of negative growth of pathogen bacteria occurs, the metabolites or the postbiotics that were involved in the process are analyzed.

Co-culturing is mostly done at laboratory levels to exploit the functional properties of postbiotics against the pathogen bacteria. A relevant example of co-culturing includes disc diffusion method or agar diffusion method:

1. **Disc Diffusion Method:** In this method, disc paper of 5 mm diameter is first immersed in supernatant and then in the media containing *S. typhi* pathogen. This disc paper is then kept in incubator at the temperature of 37°C for 24 h. After a specific time, diameter is estimated and is calculated for inhibition zones, with the caliper [24].
2. **Agar Diffusion Method:** In this method, petri plates are prepared with nutrient agar. Upon solidification, wells are made in solidified agar. Postbiotic-treated pathogens are introduced into the wells and incubated. After appropriate time, the inhibition zone around wells is observed.

11.5.3 MICROENCAPSULATION

Some postbiotics especially organic acids are sensitive to light and oxygen. In a similar way, a few proteins (including single cell proteins) might be denatured by protease enzymes in intestine before they reach the target area. To overcome these difficulties, an innovative approach recognizes the microencapsulation of postbiotics so that they could reach the target area without losing their integrity. Mora et al. [17] showed the successful microencapsulation of oleic acid in dairy beverages with additional partial incorporation of single-cell protein.

11.6 MODE OF ACTION (MOA) OF POSTBIOTICS

Huge number of postbiotics are produced and isolated from bacteria including organic acids, proteins, amino acids, polysaccharides, etc. Each of these exhibits its own distinctive characteristics and activity against pathogens while delivering bacterial benefits at the same time. The activity varies with the type of postbiotic used in the product. The mechanism of some structurally described postbiotics is explained in this section.

11.6.1 MOA OF ORGANIC ACIDS

As organic acids have lower pH values, therefore these acts by destroying the outer membrane of Gram-negative bacteria. Organic acids interact with the lipids present in these membranes and solubilize them. This results in the breakdown of outer bacterial core and the leakage of internal cytoplasmic constituents into the environment. There are several factors that could affect the activity of organic acids, such as:

- Acid dissociation constant (pK_a);
- pH of the medium;
- Polarity of molecules;
- Resistance of targeted pathogens towards acidic media;
- Size of molecules.

11.6.2 MOA OF FATTY ACIDS

Fatty acids act via mechanisms, such as:

- Causes bacterial cell lysis;
- Disrupts electron transport chain occurs;
- Disrupts enzyme activity of pathogen;
- Increase membrane permeability.

11.6.3 MOA OF PROTEINS AND PEPTIDES

Peptides, in the form of bacteriocins, have low molecular weights (LMWs). Their main role is to act as antimicrobial agents and inhibition growth of pathogens. They bind with cell surface receptors of pathogens or might gain direct entry to the host cells. Once they are in contact with the pathogenic bacteria, they exert their activity by pathways, such as:

- Formation on pores in cell's outer membrane;
- Degrade DNA of pathogenic bacteria;
- Inhibit the formation of peptidoglycan.

Two major types of proteins (P40 and P75) which was obtained from *Lactobacillus rhamnosus*; shown positive effect in regulating homeostasis of intestinal epithelial cells. Their mode of action (MoA) depends on signal transduction. In the GI tract, P40 proteins activate EGF factor that promotes intestine health and helps in building damaged epithelium cells. This also plays role in Akt activity, which is the major beneficial component in inducing proapoptotic signals along with antiapoptotic signals [6].

11.6.4 MOA OF CARBOHYDRATES

Carbohydrates have been used in food industry for texture development, emulsification, and water-binding properties. In case of Postbiotics, carbohydrates are mostly present in the outer membranes of bacterial cells as exopolysaccharides. These are obtained by cell disruption methods. The mechanism of action is based on the increased immunomodulatory properties and production of anti-proliferative agents. Some major activities of these are:

- Activation of Dectin-1 by ß-glucans;
- Antioxidant properties due to iron binding potential and increasing bioavailability of carotenoids;
- Inhibition of cholesterol absorption;
- Interaction with dendritic cells to increase proliferation of T-lymphocytes;
- Secretion of nitrogen oxide (NO), resulting in phagocytosis of macrophages;
- Stabilize blood glucose levels.

11.7 SUMMARY

The use of postbiotics is an innovative term in food and dairy technology. While some forms of supplements obtained from bacteria exist in market, but their role is rarely utilized in food industry. For future aspects, the preparation of postbiotics from high temperature treatment and sonication is an effective approach on an industrial scale. For heat-resistant bacteria, a combination of two treatments could be applied to achieve high quality postbiotics. Single bacterial source delivers variety of structural postbiotics with numerous roles, such as: antibiotic, anti-inflammatory, anti-microbial, and anti-diabetic agents in the body. Hence, a variety of probiotic bacterial species with unique properties and byproducts could be used for developing innovative postbiotics.

KEYWORDS

- **cell free extract**
- **cell free supernatant**
- **mode of action**
- **postbiotics**
- **probiotics**
- **tyndallization**

REFERENCES

1. Abdelazez, A., Alshehry, G., Algarni, E., Al-jumayi, H., Abdelmotaal, H. M., & Meng, X., (2022). Postbiotic gamma-aminobutyric acid and camel milk intervention as innovative trends against hyperglycemia and hyperlipidemia in streptozotocin-induced C57BL/6J diabetic mice. *Frontiers in Microbiology Journal, 13*, E-article ID 943930.
2. Abo-Amer, A. E., (2013). Inhibition of foodborne pathogens by a bacteriocin-like substance produced by a novel strain of *Lactobacillus acidophilus* isolated from camel milk. *Applied Biochemistry and Microbiology, 49*(3), 270–279.
3. Ashraf, R., Vasiljevic, T., Smith, S. C., & Donkor, O. N., (2014). Effect of cell-surface components and metabolites of lactic acid bacteria and probiotic organisms on cytokine production and induction of CD25 expression in human peripheral mononuclear cells. *Journal of Dairy Science, 97*(5), 2542–2558.
4. Cavicchioli, V. Q., Camargo, A. C., Todorov, S. D., & Nero, L. A., (2017). Novel bacteriocin genic *Enterococcus hirae* and *Pediococcus pentosaceus* strains with

anti-listerial activity isolated from Brazilian artisanal cheese. *Journal of Dairy Science, 100*(4), 2526–2535.
5. Choi, S. S., Kim, Y., Han, K. S., You, S., Oh, S., & Kim, S. H., (2006). Effects of *Lactobacillus* strains on cancer cell proliferation and oxidative stress *in vitro*. *Letters in Applied Microbiology, 42*(5), 452–458.
6. Cicenia, A., Scirocco, A., Cuevas-González, P. F., Liceaga, A. M., & Aguilar-Toalá, J. E., (2020). Postbiotic activities of Lactobacilli-derived factors. *Journal of Clinical Gastroenterology, 48*(Suppl 1), S18–S22.
7. Cuevas-González, P. F., Liceaga, A. M., & Aguilar-Toalá, J. E., (2020). Postbiotics and paraprobiotics: From concepts to applications. *Food Research International, 136,* E-article ID 109502.
8. Das, D., & Goyal, A., (2015). Antioxidant activity and γ-aminobutyric acid (GABA) producing ability of probiotic *Lactobacillus plantarum* DM5 isolated from Marcha of Sikkim. *LWT-Food Science and Technology, 61*(1), 263–268.
9. De Marco, S., Sichetti, M., Muradyan, D., Piccioni, M., Traina, G., Pagiotti, R., & Pierrella, D., (2018). Probiotic cell-free supernatants exhibited anti-inflammatory and antioxidant activity on human gut epithelial cells and macrophages stimulated with LPS. *Evidence-Based Complementary and Alternative Medicine, 2018,* E-article ID 1756308.
10. Homayouni, R. A., Aghebati, M. L., Samadi, K. H., & Abbasi, A., (2021). Postbiotics: A novel strategy in food allergy treatment. *Critical Reviews in Food Science and Nutrition, 61*(3), 492–499.
11. Hu, C. H., Ren, L. Q., Zhou, Y., & Ye, B., (2019). Characterization of antimicrobial activity of three *Lactobacillus plantarum* strains isolated from Chinese traditional dairy food. *Food Science & Nutrition, 7*(6), 1997–2005.
12. Hussain, N., Tariq, M., Saris, P. E. J., & Zaidi, A., (2021). Evaluation of the probiotic and postbiotic potential of lactic acid bacteria from artisanal dairy products against pathogens. *The Journal of Infection in Developing Countries, 15*(1), 102–112.
13. Jiang, Y. H., Xin, W. G., Yang, L. Y., Ying, J. P., Zhao, Z. S., Lin, L. B., Li, X. Z., & Zhang, Q. L., (2022). A novel bacteriocin against *Staphylococcus aureus* from *Lactobacillus paracasei* isolated from Yunnan traditional fermented yogurt: Purification, antibacterial characterization, and antibiofilm activity. *Journal of Dairy Science, 105*(3), 2094–2107.
14. Kim, N. J., & Ji, G. E., (2006). Modulatory activity of *Bifidobacterium sp.* BGN4 cell fractions on immune cells. *Journal of Microbiology and Biotechnology, 16*(4), 584–589.
15. Krithika, R., & Balasasirekha, R., (2021). FTIR spectrum and XRD of postbiotics – exopolysaccharides zinc oxide nanoparticles. *Journal of Advanced Scientific Research, 12*(3), 292–300.
16. Maghsood, F., Johari, B., Rohani, M., Madanchi, H., Saltanatpour, Z., & Kadivar, M., (2020). Anti-proliferative and anti-metastatic potential of high molecular weight secretory molecules from probiotic *Lactobacillus reuteri* cell-free supernatant against human colon cancer stem-like cells (HT29-ShE). *International Journal of Peptide Research and Therapeutics, 26*(4), 2619–2631.
17. Mora, H. M. B., Pineros, M. A., Moreno, D. E., Restrepo, S. R., Jaramillo, J. E. C. C., Solano, O. A. A., Fernandez-Nino, M., & Barrios, A. F. G., (2019). Multiscale design of a dairy beverage model composed of *Candida utilis* single cell protein supplemented with oleic acid. *Journal of Dairy Science, 102*(11), 9749–9762.
18. Moradi, M., Kousheh, S. A., Almasi, H., Alizadeh, A., Guimaraes, J. T., Yilmaz, N., & Lotfi, A., (2020). Postbiotics produced by lactic acid bacteria: The next frontier in food safety. *Comprehensive Reviews in Food Science and Food Safety, 19*(6), 3390–3415.

19. Moradi, M., Mardani, K., & Tajik, H., (2019). Characterization and application of postbiotics of *Lactobacillus* spp. on *Listeria monocytogenes in vitro* and in food models. *LWT, 111*(4), 457–464.
20. Moradi, M., Molaei, R., & Guimarães, J. T., (2021). A review on preparation and chemical analysis of postbiotics from lactic acid bacteria. *Enzyme and Microbial Technology, 143*, E-article ID 109722.
21. Nielsen, S. D., Jakobsen, L. M. A., Geiker, N. R. W., & Bertram, H. C., (2022). Chemically acidified, live and heat-inactivated fermented dairy yoghurt show distinct bioactive peptides, free amino acids and small compounds profiles. *Food Chemistry, 376*, E-article ID 131919.
22. Ou, J., Carbonero, F., Zoetendal, E. G., DeLany, J. P., Wang, M., Newton, K., Gaskins, H. R., & O'Keefe, S. J. D., (2013). Diet, microbiota, and microbial metabolites in colon cancer risk in rural Africans. *The American J. of Clinical Nutrition, 98*(1), 111–120.
23. Piqué, N., Berlanga, M., & Miñana-Galbis, D., (2019). Health benefits of heat-killed (tyndallized) probiotics: An overview. *International Journal of Molecular Sciences, 20*(10), E-article ID 2534.
24. Pratiwi, S. T., Aditya, M. B., & Abbas, E. H., (2021). *Temperature Optimization and Inhibition Test of Lactobacillus Acidophilus Bacteriocin Against Salmonella Typhi Bacteria.* Paper Presentation at 12th Annual Scientific Meeting: International Symposium on Emergency Preparedness and Disaster Response during COVID-19; Medical Faculty, Universitas Jenderal Achmad Yani.
25. Rad, A. H., Aghebati-Maleki, L., Kafil, H. S., Gilani, N., Abbasi, A., & Khani, N., (2021). Postbiotics, as dynamic biomolecules, and their promising role in promoting food safety. *Biointerface Research Applied Chemistry, 11*(6), 14529–14544.
26. Rathod, N. B., Phadke, G. G., Tabanelli, G., Mane, A., Ranveer, R. C., Pagarkar, A., & Ozogul, F., (2021). Recent advances in bio-preservatives impacts of lactic acid bacteria and their metabolites on aquatic food products. *Food Bioscience, 44*, E-article ID 101440.
27. Salminen, S., Collado, M. C., Endo, A., Hill, C., Lebeer, S., Quigley, E. M., Sanders, M. E., et al., (2021). The international scientific association of probiotics and prebiotics (ISAPP) consensus statement on the definition and scope of postbiotics. *Nature Reviews Gastroenterology & Hepatology, 18*(9), 649–667.
28. Vale, G. C., & Mayer, M. P. A., (2021). Effect of probiotic *Lactobacillus rhamnosus* by-products on gingival epithelial cells challenged with *Porphyromonas gingivalis*. *Archives of Oral Biology, 128*, E-article ID 105174.
29. Vandenplas, Y., De Halleux, V., Arciszewska, M., Lach, P., Pokhylko, V., Schoen, S., Abrahamse-Berkeveld, M., et al., (2020). A partly fermented infant formula with postbiotics including 3′-GL, specific oligosaccharides, 2′-FL, and milk fat supports adequate growth, is safe and well-tolerated in healthy term infants: A double-blind, randomized, controlled, multi-country trial. *Nutrients, 12*(11), 3560–3567.
30. Vinderola, G., Matar, C., Palacios, J., & Perdigon, G., (2007). Mucosal immunomodulation by the non-bacterial fraction of milk fermented by *Lactobacillus helveticus* R389. *International Journal of Food Microbiology, 115*(2), 180–186.
31. Wu, Q., & Shah, N. P., (2015). Gas release-based prescreening combined with reversed-phase HPLC quantitation for efficient selection of high-γ-aminobutyric acid (GABA)-producing lactic acid bacteria (LAB). *Journal of Dairy Science, 98*(2), 790–797.

32. Yordshahi, A. S., Moradi, M., Tajik, H., & Molaei, R., (2020). Design and preparation of antimicrobial meat wrapping nano-paper with bacterial cellulose and postbiotics of lactic acid bacteria. *International Journal of Food Microbiology, 321*, E-article ID 108561.
33. Zago, M., Scaltriti, E., Bonvini, B., Fornasari, M. E., Penna, G., Massimiliano, L., Carminati, D., et al., (2017). Genomic diversity and immunomodulatory activity of *Lactobacillus plantarum* isolated from dairy products. *Beneficial Microbes, 8*(4), 597–604.

CHAPTER 12

Extraction Methods for Essential Oils from Citrus Peels

Ramesh Sharma

ABSTRACT

The extraction of necessary oils from citrus skin is an important option to reduce the operational cost in fruit and vegetable processing industries. The flavedo part of citrus peel is rich of essential oil (EO) and it is very much sensitive to heat and light, which is the biggest challenge for the oil recovery from citrus peel. Different conventional procedures of oil extraction including cold pressing (CP) distillation, and solvent extraction (SE) are commonly used to extract EO from the citrus peel, but the main disadvantages of traditional method are long processing time, solvent residue, lower yield and loss of volatile compounds from the oil. In this chapter, the emerging oil extraction techniques (such as: supercritical fluid (SCF) extraction, ultrasound assisted extraction (UAE), microwave assisted extraction (MAE), high pressure extraction (HPE) and pulse electric field (PEF) extraction) have been discussed and reviewed. The emerging techniques are alternatives for EO extraction with reducing extraction time, increasing the yield, free from solvent residues with giving higher recovery of the volatile compounds in EO.

12.1 INTRODUCTION

Essential oil (EO) is found maximum in citrus fruits that belongs to *Rutaceae* family. The term citrus is the genus belonging to flowering tree, which includes lemon (*citrus Limon*), grapes (*citrus paradise*), mandarin (*citrus*

The Functional Foods: Nutrient and Health Benefits.
Megh R. Goyal, Junaid Ahmad Malik, and Anu Kumari (Eds.)

reticulate) and orange (*citrus sinensis*). It consists of 25–30% of peel. The EO is rich in functional ingredients, such as: vitamin C, flavanones (hesperidin and neo-hesperidin), and naringin, which shows the antioxidant properties; thus, these functional ingredients have multiple uses in the food and pharmacy industry. These compounds are highly volatile in nature and are critical to heat, light, and extraction time.

The peel parts are separated between two parts: (i) the inner part called albedo whose color is white and is rich in pectin and phenol; and (ii) exterior fragment, i.e., flavedo that is enriched with EOs with green color. The gland is present to hold the oil, which is called oil sac. Many industries are processing the citrus fruits, but the peels are wasted, which are potential source of functional ingredients [56]. The citrus fruits are rich in EO, and the oil is used for food flavor [12].

Different techniques (such as: CP, HD, CFE, SE, ME, and SD methods) are used to extract the EO from citrus peels. Many researchers are trying to develop the innovative techniques for the oil extraction from fruits and vegetables [9]. Conventional oil extraction techniques (such as: HD [17], SD, CP, and SE method [78]) are used to extract the EO. The limitations of the conventional method of oil extraction is related to low yield, and solvent residue affected by processing pressure and temperature of oil chemical components [6]. The use of microwave steam distillation (MSD) for oil extraction is in contrast with normal SD and it was concluded that MSD gave better output with regarding to time, energy saving and retention of nutritional quality [66].

The use of eight oil extraction techniques with old techniques like SD and HD in comparison with modern techniques like UAE, microwave assisted extraction (MAE) including MHD, MSD, microwave-generated HD and gravity were performed to extract oil from lavandin flower [60]. The excellent outcome was given by gravity and microwave hydro-diffusion (MHD) with maximum yield of 5.4% at 15 min time interval with power consumption of 1.3 kWh. The maximum time and power consumption of steam distillation (SD) was reported with 120 min and 8.06 kWh. The HD and solvent free microwave (SFM) techniques were used to extract oil from Algerian Myrtle leaf [49]. The oil extracted from SFM technique had greater amount of oxygenated compound. The oil extracted from hydro-distillation (HD) deteriorated easily due to volatile nature of terpenes [4]. Supercritical fluid (SCF) extraction method is considered as a safe method for EO extraction, and it is operated at ambient temperature without using any solvent [39].

This chapter aims to study the emerging oil extraction techniques from citrus peels and to compare these with conventional techniques with respect to extraction duration, efficiency, and biocomponents of EO.

12.2 COMPOSITION OF ESSENTIAL OIL (EO) IN CITRUS FRUITS

The oil percentage for different citrus fruits has been mentioned in literature [40]. It is observed that maximum oil content was in *Citrus sinensis* (0.24–1.07%) followed by *Citrus reticulate* (0.30–0.50%) and *Citrus paradisii* (0.20–0.40%). The citrus oil composition is mainly classified in four groups, such as: oxygenated monoterpenes, hydrocarbon monoterpenes, oxygenated sesquiterpenes and hydrocarbon sesquiterpenes.

The EO is composed of lower molecular weight volatile compounds, such as: monoterpenes, sesquiterpenes hydrocarbons, terpenoids, aromatic, and aliphatic components and non-volatile components [29]. The non-volatile compounds include paraffin and waxes, fatty acid, sterols, carotenoids, and esters, etc. The EO components are also classified between two groups: (i) terpenes; and (ii) oxygenated compounds (such as: alcohol, aldehydes, ester, ketones, phenols, and oxides).

The volatile compound in oil gets vaporized at 100°C during the distillation process. Therefore, it is the challenging task for the researcher to retain the volatile compound. The terpenes hydrocarbons give aroma and flavor to food products, and they are characterized with insoluble in alcohol solution and their oxidation gives rise to off-flavor of EO. Limonene and linalool are the important components of EO. Limonene composition is highest in terpenes group in peel oil of orange (90–95%) and lemon (75%) and 95% are hydrocarbon monoterpenes among these [48].

The GC-MS were used in identifying the composition of lime (*C. aurantifolia*) oil and total 22 components were identified, among which limonene was a major compound with 28.27% [2]. The biocomponents in citrus oil vary with location and methods of its extraction [34]. This limonene contributes to small amounts for the aroma; and also, there was presence of sesquiterpenes and some oxygenated compounds in small amounts that contribute majorly to the aroma development. EO is composed of both polar and non-polar group of compounds [53]. Terpenes contributes aroma and flavor in the EO. The organoleptic characteristics of EO are also due to oxygenated terpenes and oxygenated compound derivatives, i.e., aldehydes, ketones, etc.

Citrus peel is rich in hesperidin and neo-hesperidin (a flavanone glycosidic compound), and it helps in reducing the blood pressure and it improves blood circulation [73]. The hesperidin and neo-hesperidin have the potential application as natural sweeteners in food industry [73]. The citrus also has the anti-inflammatory function in human body [76]. The hesperidin controls the LDL oxidation from the body [17].

Naringin is a glycosidic flavonoid extracted from citrus peel and is bitter in taste. The importance of this flavonoid is known due to its antioxidant qualities, and it is used as an oil stabilizer [30].

12.3 USE OF CITRUS OIL AS PRESERVATIVES

The EO from citrus is rich in volatile compound, therefore there is a need of advanced techniques to add such preservatives to improve packaging techniques, such as: vacuum packaging and canning during beverage production. Addition of EO as preservatives in combination with chemical preservatives reduces the cost and it is also a health-conscious step with reduction of doses of chemical preservatives. The EO can provide aroma and flavor to soft drinks, ice cream and cake, etc. [33]. The EO has the ability to protect foods from pathogens and spoilage organisms [31].

FDA has recognized citrus oil as GRAS item among the food products [29]. The citrus peel oil has been used in food products, such as: soft drinks, candy, and ice cream for food preservation [71]. The EO components (like thymol and carvacrol) can inhibit the microbial and chemical deterioration in human body [12].

The anti-oxidant and anti-microbial characteristics of the phenolic compounds in oil were increased while combining with chemical preservatives [65]. Terpenes and phenolic compounds act as anti-oxidants for preventing the oxidation process in EO [53]. The lemon oil was more effective in destruction of *E. coli* pathogen [31].

The antibacterial effect of orange oil in salmonella species was observed in skimmed milk at 1,000 μl concentration in one liter of milk; and quality of milk was acceptable [18]. Here, EO was extracted from citrus peel using HD method [10]. The extracted chemical components of EO were analyzed using GC-MS and the main components identified were β-myrcene, 4-terpineol β-pinen, α-pinene, sabinene, d-limonene, m-cymene, linalool; and d-limonene was the major component with 66–93% of extracted oil. The test with antimicrobial effect of EO was performed against selected microorganisms (such as:

S. aureus, E. coli, S. typhimurium B. cereus, E. faecalis and *L. monocytogenes*) using a disc-diffusion method. The oil extracted from blood orange (*Citrus sinensis*) had shown the maximum activity against these bacteria.

12.4 PREPARATION OF PEEL SAMPLES BEFORE EXTRACTION

Sample preparation of citrus peels involves washing, peeling, cutting, drying, grinding, and storing the powder. The unit operation like dehydration and grinding caused the breakdown of cellular molecules and improved the yield of extraction of phenolic compounds [66]. During the sample preparation, dehydration, and grinding unit operations had significant effect on functional ingredients of citrus oil. Drying is essential for removing the free-water molecules from plant materials, but it also effects the properties of phenols and antioxidants of citrus peel. The various methods are available to extract EO from biological samples (Table 12.1).

TABLE 12.1 Oil Extraction Yield by Different Extraction Methods from Different Plant Materials

Plant Material	Method of Extraction	Extraction Time (min)	Yield (%)	References
Dryopteris fragans	SFME	34.00	0.340	[47]
Dryopteris fragans	HD	300.00	0.290	[47]
Lemon	SE	240.00	0.950	[48]
Lemon	HPHT	30.00	2.970	[48]
Peels from lemon	SFME	10.00	2.00	[5]
Peels from orange	MSD	12.00	1.540	[25]
Peels from orange	CSD	40.00	1.510	[25]
Peels from orange	SFME	10.00	3.70	[5]
Plant of rosemary	SFME	30.00	0.540	[28]
Plant of rosemary	HD	90.00	0.570	[28]

12.5 OIL EXTRACTION METHODS FROM CITRUS PEEL

12.5.1 TRADITIONAL METHODS

12.5.1.1 STEAM DISTILLATION (SD)

Distillations is the process in which the materials are steamed or boiled in water; as a result, EO evaporates with water and get separated through the

Florentine flask [21, 37]. This technique is commonly used to separate EO from heat sensitive food materials. Overall, oil extraction is around 93% with SD method [52]. The oil extracted by distillation process deteriorates faster and develops off-flavor due to instability of the terpene compound [4]. Though distillation process is an economical process for EO extraction in comparison with cold press method [26]. Distillation process give the recovery of 0.21% in comparison with 0.05% by cold pressing [27].

In the distillation process, washed citrus peel is pre-heated at 50°C for 30 minutes before SD procedure. The preheated citrus peel is transferred into the round base flask and steam is supplied to bottom of a flask. The switch of the condenser is on and heated up to 100°C for 2 h and distillate is collected and transfer to Florentine flask for separation of oil from the water, which takes place due to the density difference in the flask and the presence of orange oil is confirmed by presence of cloudy appearance on top of distillate.

The separation of citrus oil was performed from orange skin using diverse extraction procedures including SE, SD, and HD method and the efficiency of the extraction by different method was compared [34]. The efficiency of 4.40% for SD, 3.47% for HD and 2.536% for solvent extraction (SE) were obtained. Similarly, the extraction of citrus oil from orange peel was carried out using SD and the oil composition of the EO was calculated by GC-MS method [15], with which 15 compounds were identified and out of which D-limonene was the major component in oil.

12.5.1.2 HYDRO-DISTILLATION (HD)

Hydro-distillation (HD) is an ancient method for the extraction of EO from citrus peels [69]. In this method, initially citrus peel is weighed and dipped in buffer solution at 4.5 pH maintained at 33°C for 3 h. Then, it was crushed to make into a pulp with the help of a mixer and was used for distillation process. The desirable oil is separated by heating the mixture of water and pulp at atmospheric pressure. Azeotropic mixture containing water is separated in the Florentine flask as per their immiscible properties and based on the difference in densities [47].

The extraction of EO from grape (*Citrus paradise*) fruits using HD procedure was performed [58]. The composition of extracted EO was measured using GC-MS. The observations indicated the composition with D-limonene (75.05% highest amount) followed by beta-myrene (7.25%), alpha-pinene (2.11%), caryophyllene (1.88%); and other components in trace amount was identified.

12.5.1.3 STEAM EXPLOSION

The EO extraction by the steam explosion is an emerging method, which involves self-evaporation that resulted to displacement of volatile molecules from the plant matrix [54]. Mainly, the separation occurs due to volatile nature of compound with exposure of raw sample to saturated vapor for short duration [36]. The experiment for oil extraction from orange peel was performed using steam explosion and HD method and these were compared for the oil composition by GC-MS (Tables 12.2 and 12.3). Tables 12.2 and 12.3 indicate that the most of compounds in the oil extracted by HD are more than the compounds present in the oil extracted by steam explosion method. Similarly, limonene present in oil extracted by steam explosion was greater than in oil extracted by HD method.

TABLE 12.2 Composition of EO by Hydro-Distillation

SL. No.	Compound	HD (%)	Retention Time (h)
1.	1-Cyclohexene-1-carboxaldehyde	1.030	0.363
2.	1-Octanol	2.180	0.289
3.	alpha-Pinene	1.170	0.241
4.	Beta-Myrcene	6.080	0.260
5.	Decanal	2.920	0.337
6.	Dodecanal	0.410	0.406
7.	E-Citral	0.660	0.359
8.	L-Linalool	5.130	0.301
9.	Limonene	77.390	0.279
10.	p-Menthon-8-thiol	0.440	0.318
11.	Sabinene	1.410	0.255
12.	Sinensal	0.350	0.491
13.	Z-Citral	0.630	0.349
Total		99.8	–

TABLE 12.3 Essential Oil Composition by Steam Explosion Method

SL. No.	Compound	Steam Explosion Method (%)	Retention Time (h)
1.	Sabinene	0.55	0.112
2.	Alpha-pinene	0.72	0.095
3.	n-Octanal	0.35	0.127
4.	n-Nonane	2.0	0.080
5.	n-Decane	0.40	0.124

TABLE 12.3 *(Continued)*

SL. No.	Compound	Steam Explosion Method (%)	Retention Time (h)
6.	Beta-myrcene	3.41	0.121
7.	Sinensal	0.20	0.220
8.	Limonene	89.13	0.140
9.	Dodecanal	0.82	0.175
10.	p-Menthon-8-thiol	0.22	0.172
Total (%)		97.8	–

12.5.1.4 SOLVENT EXTRACTION (SE)

The SE is also called LLE (liquid–liquid extraction) method and it separates the EO based on solubility of components in the mixture. This method extracts the oil from heat sensitive plant materials using different solvents including hexane, acetone, and petroleum ether, etc. [1]. The citrus peel is added to the solvent and mixture is heated to recover the oil from peels and condenser is installed to collect the oil along with water and it is transferred to Florentine flask, where oil gets separated from water due to the density difference and immiscible properties of oil in water. The most common solvent in use is hexane due to its lower boiling point and easy recovery. The different solvents give different yields of the extract, and these depend on different compositions of the extract [79]. The solvent blend of hexane and isopropanol has given high recovery of EO [46].

12.5.1.5 COLD PRESS METHOD

It is considered as the best method to extract the EO from citrus peels. The extracted oil is considered as pure with retention of all the chemical components. In this method, the plant materials are firstly grounded or chopped and are immersed in water for few hours before pressing. The soaked peels are pressed at 120°F and oil is extracted after pressing, which is further separated by decantation process. It is usually operated at low temperature and pressure. This process is used for heat sensitive plant materials. It is mostly used for oil extraction from flowers, seeds, lemons, and tangerines [59].

12.5.2 MODERN METHODS

12.5.2.1 SUPERCRITICAL FLUID (SCF) EXTRACTION

Supercritical fluid (SCF) extraction method separates the EO from the matrix using SCF as a solvent. Liquid CO_2 is commonly used, which has defined critical pressure (72 bar) and temperature (32°C). This fluid is nontoxic, non-flammable, non-reactive, safe, and can be easily separated from extract [39]. It is also mentioned that this method is best for EO extraction from citrus peel as this process operates at low temperature and no solvent residue is left in this process [64]. The best yield with this method was observed in mandarin fruit while compared to other extraction processes [4]. The oil obtained from this method has shown to have higher antioxidant activity than other extraction methods [24]. The SCF extraction is a better option for oil extraction than other conventional methods, such as, cold press method [62]. The oil extracted from this method has higher percentage of volatile compound, which was not present in oil extracted by HD method [38]. The details on the use of SCF for oil extraction are also indicated in an alternate study [19].

The EO was extracted from lime using SCF extraction and HD [4]. The lime oil composition extracted from two extraction methods were compared to the commercial sample [41].

Similarly, the experiment was performed on extraction of non-polar flavonoids using water as a solvent in orange fruits [43]. The SCWE (supercritical water extraction) was correlated with three conventional methods and the results indicated that SCWE was highly efficient to retain the bioactive components and antioxidant properties of flavonoid compounds. The outcome of combining citrus seed and citrus peel was observed for bioactive properties (i.e., total flavonoid and phenolic contents) of oil extracted by SCF extraction using CO_2 as a solvent [57]. The oil from this mixture (peel + seed) had superior bioactive properties than the oil from citrus peel alone.

The extraction of EO from kabosu (*Citrus sphaerocarpatanaka*) was performed from citrus peel with SCF using CO_2 as a solvent and the efficiency of extraction was compared with CP method [72]. The yield obtained by SCF was 13-times higher than with the CP method. The qualitative analysis of oil was done using GC-MS and it was concluded that oil extracted by SCF had low amounts of monoterpenes and greater retention of oxygenated compound and sesquiterpenes, which are important for aroma in oil. For the extraction of EO with HD and SCF-CO_2 (SC-CO_2) methods, the yield of extraction was compared [4]. It was concluded that best yield was 7.93% with SC-CO_2 followed by 5.45% with HD.

12.5.2.2 HIGH PRESSURE EXTRACTION (HPE)

High pressure extraction (HPE) extracted the phenolic compounds from orange peel [13]. The low-pressure treatment with 300 MPa/10 min in orange and 500 MPa/3 min in lemon was successful in retaining phenolic compounds, and antioxidant properties in both citrus fruit oil; and high severe pressure treatment with 500 MPa/10 min resulted in the reduction of phenolic and antioxidant properties of oil. The lemon peel treated with pressure of 500 MPa/3 min showed higher retention (149.41 mg/100 g) of total antioxidant activity compared with 300 MPa/3 min (131.31 mg/100 g) [44]. The retention of phenolic compound was higher with increase of pressure with enhancing antioxidant effects.

12.5.2.3 MICROWAVE ASSISTED HYDRO-DISTILLATION (MAHD)

Microwave generates the electromagnetic wave at the frequency ranging from 300 MHz to 300 GHz, which usually penetrates the plant materials and interaction was occurred for polar group of substance including water to generate the heat for the extraction of the EO [51]. HD with pre-treatment by microwave for EO extraction was studied and the efficiency of extraction was controlled with water and dielectric constant value of materials [11]. The effect of microwave on HD for oil extraction from *thymus vulgaris* leaves was carried out and results were analyzed with the data from conventional HD process for extraction time and chemical composition of oil [35]. MAHD was efficient with extraction time of 75 min compared with 4 h time by HD process. The GC-MS was used to analyze the bioactive components of extracted EO and it was found that irradiation did not cause the adverse effects on oil composition.

12.5.2.4 HYDRO-DISTILLATION (HD) WITH PRE-TREATMENT OF ENZYMES

The treatment with enzymes to food waste has been considered as a potential alternative giving higher efficiency of extraction compared with conventional methods [8, 70]. The pre-treatment with xylanases enzyme for the oil separation of mandarin by HD process was performed [55]. The enzyme concentrations of 0.1%, 0.2% and 0.3% were used and HD extraction was observed. The yield of oil extraction by HD with pre-treatment was higher and it was concluded that the yield was higher due to action of enzymes on oil extraction process. The pre-treatment with cellulase enzyme was performed to different citrus peels and the recovery of extraction for peel of orange and grape fruit were 2 to 6 times higher while comparing with conventional methods [14]. The pre-treatment with enzyme not only facilitates the release of oils but it is a comprehensive approach for utilization of food waste in industry.

12.5.2.5 MICROWAVE ASSISTED STEAM DISTILLATION (MASD)

Microwave of supplying 800 W power is normally used in MASD for EO extraction from orange peel. In microwave heating, non-ionizing electromagnetic radiation is used with dipole moment of molecules without changing the molecular structures. The commonly used microwave frequency for commercial level is 2,450 MHz. The condenser and steam generator are placed outside of the microwave. The heat generated from the saturated steam gets interacted with microwave heat, which results in release of oil trapped in plant tissue. The extraction is continued for 140 min and condenser collects the mixture of oil and water, which is finally transferred to Florentine flask. The Florentine flask separates oil from the mixture because of density difference and immiscible properties. The decanted oil is dried up in moist-free sodium sulfate and is kept at 4°C.

The extracted oil from citrus fruit with microwave pre-treatment from SD process was compared with the results with only SD process [67]. The better results were observed with MASD in comparison with SD in terms of separation time and energy utilization. The time taken for extraction by MASD was 6 min in comparison with 2 h for SD. The experiment was performed to separate citrus oil from orange skin using microwave assisted steam distillation (MASD), conventional SD process [42]. Also, the processes were compared for yield, extraction time and chemical components of EO. The extracted oil yield from MASD was higher than SD and the extraction time of MASD was 140 min in comparison of 7 h by SD. MASD was also effective to recover maximum quantity of EO compound in comparison with SD, based on the analysis by GC-MS.

Similarly, the extraction of EO was performed from orange peel by CSD (conventional steam diffusion) and MSD techniques [42] and these two techniques were compared for time of extraction, yield of extraction, chemical components (Table 12.4), and quality of EO. It was found that the extraction time taken by MSD was 140 min in comparison with 7 h with SD; and the yield value with MSD was higher than CSD. Oil extracted from MASD had higher amount of retention of nutritional components as compared with SD (Table 12.4).

TABLE 12.4 Orange Oil Components Extracted by SD and MASD

SL. No.	Compounds	SD (%)	MASD (%)
1.	α-Pinene	0.67	0.55
2.	Hexasiloxane	–	0.18
3.	δ-Carene	0.74	–
4.	γ-Gurjunene	–	0.32
5.	n-Decanal	–	0.16
6.	Myrcene	1.87	1.95
7.	α-Terpineol	0.84	0.86
8.	Thiocynic acid	–	0.25
9.	Sabinene	–	0.20
10.	Octyl aldehyde	–	0.19
11.	n-Octanol	–	0.10
12.	n-Nonadecanoic	–	0.50
13.	Limonene	93.39	92.42
14.	β-Fenchyl alcohol	–	0.30
15.	Dimethoxybenzylidene	–	0.20
16.	β-Pinene	0.94	0.69
17.	δ-Guaiene	–	0.12
18.	Cyclotrisiloxane	–	0.79
19.	Cyclohexene	1.55	0.12

The EO was separated from orange, lemon, and mandarin skin with SD and MASD and the two methods were analyzed for separation time, yield, and the compositions were analyzed using GC-MS [68]. The yield obtained by MASD had higher value of 1.15% in 35 min extraction time, compared with 1.095% in 45 min of extraction time for orange oil in SD process. Same kinds of results were observed for lemon and mandarin fruits.

12.5.2.6 ULTRASOUND ASSISTED EXTRACTION (UAE)

In the UAE method, the sample is mixed with water or organic solvent (methanol, ethanol, etc.) and the mixture is treated with ultrasound [3]. The use of green technology, i.e., UAE in oil extraction helps in the retention of nutritional properties of EO [75]. The separation of EO was carried out from orange skin by using conventional Clevenger extraction, which was compared with UACE (ultrasound assisted Clevenger extraction) for separation time and chemical components of the extracted oil [61]. It was concluded that there was reduction in the extraction time by UACE extraction without effecting the chemical composition of the oil.

12.5.2.7 SOLVENT-FREE MICROWAVE EXTRACTION (SFME)

The disadvantages of HD, SD, and SE methods are low amount of volatile compound, time consuming and presence of solvent residue [63]. SFME (solvent free microwave extraction) is beneficial than distillation and SE methods [23]. The SFME is a process of grouping of dry distillation and microwave heating. This method gives higher oil yield as compared with HD [7]. Microwave treatment had reduced time of extraction of 30 min in comparison with 3 h for HD and 1 h for cold press methods [26]. SFME separates the EO from matrix without using any organic solvent [49]. In this method, the extracting materials is dipped in water for 1–2 h and the surplus water is drained out and moist sample is placed in the microwave oven.

The performance of EO extracted from Algerian Myrtle leaf with SFME was compared with HD method; and GC-MS and GC were used for this purpose [49]. The oil extracted from SFME was richer in oxygenated compounds with reduction of time of extraction from 180 min (HD) to 6 min (SFME). The condenser was used to accumulate the condensed liquid and EO was obtained by drying in anhydrous sodium sulfate, which was placed at 4C in the dark. The SFME extracted the D-limonene from lemon and orange peels [5]. The microwave was operated at power and the time of heating was 10 min. The yield of EO was 3.7% for orange and 2% for lemon peel, which was better than all other extraction methods shown in Table 12.4.

12.5.2.8 MICROWAVE HYDRO-DIFFUSION AND GRAVITY (MHDG) METHOD

Microwave hydro-diffusion and gravity (MHDG) separates the EO in the absence of water and organic solvent; and it consumes less amount of energy [20]. Heating is done by microwave because of which the internal water molecules of plant materials undergo agitation that helps to rupture the gland; and phenomena is normally called hydro-diffusion. As a result, extract diffuses outside of the plant materials and because of gravity it is collected in a perforated Pyrex disc. The condenser is setup to collect the mixture that gets separated due to the density difference and immiscible properties. The advantage of these techniques is that it reduced the extraction time (20 min) in comparison with SD (90 min) [74]. The MHDG apparatus is represented in Figure 12.1.

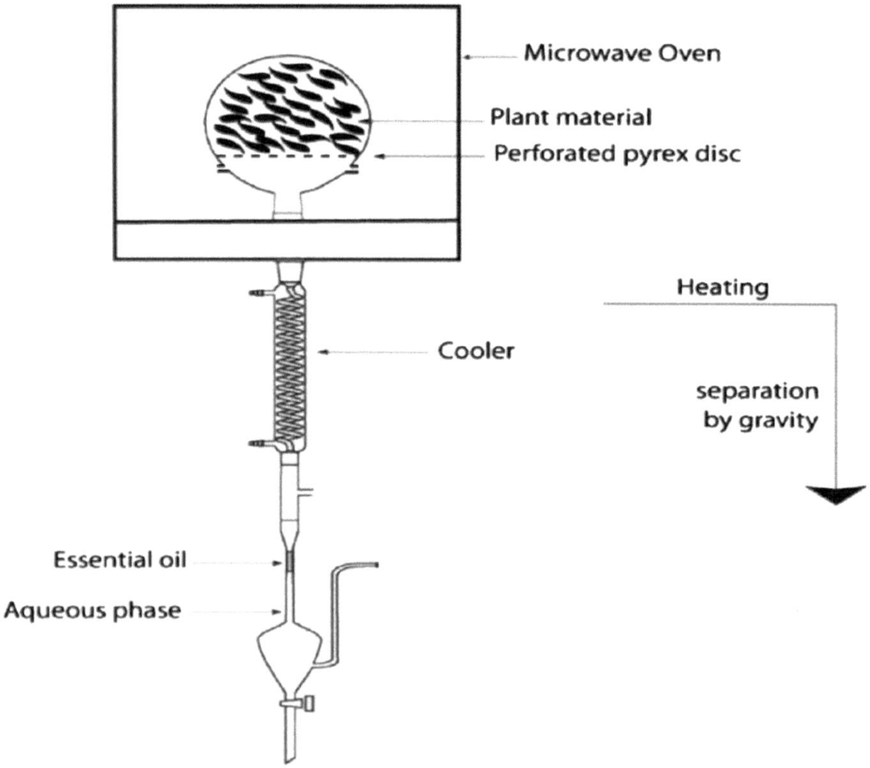

FIGURE 12.1 Microwave hydro-diffusion and gravity apparatus.

Source: Reprinted with permission from Vian et al, 2008.

12.5.2.9 OIL EXTRACTION BY SOLAR ENERGY

The experiment was carried out for oil extraction from *Cymbopogon citrus* by using solar energy and HD method [77]. Also, extraction yield, power, and time consumption were evaluated. The yield obtained was 1.28% for solar energy compared with 1.3% for HD extraction process; but the antioxidant properties of oil extracted from solar energy were superior to the HD method. The power consumption for HD process was higher in comparison with solar energy operating with free power source.

The solar energy apparatus consisted of battery, solar cell, solar energy tube, oil collection unit and temperature controller. It was installed on a right angular wooden frame. The solar energy tube was vacuum packed, and sample was placed in the tube; and no water or solvent were added in the tube. The experiment was done on a sunny day, where the temperature was 27–29°C with light intensity of 20–980 w/m^2. The oil was extracted due to thermal decomposition of leaves and was collected in cooling tank.

12.5.2.10 PULSE ELECTRIC FIELD (PEF)

The oil was extracted from the orange, lemon, and pomelo after the pre-treatment with pulse electric field (PEF) on citrus peel [22]. The electric field capacity of 3 kv/cm and 10 kv/cm was applied to citrus peel and oil extraction yield of only SE method was compared with SE (pre-treated with PEF). The oil yield extracted by SE (pre-treated with PEF) was 59% for lemon, 37% for pomelo and 25% for orange, respectively. The treatment with PEF with field strength of 1–7 kV/cm for 3 μs, after using various pulses (5–50) in orange peel for extraction by press method was used to estimate its effect on total polyphenols and flavonoids contents [50]. The results indicated increment in antioxidant activity and polyphenol quantity.

12.6 PURIFICATION AND STORAGE OF EO

Purification of EO is normally done by winterization in which the extracted oil is kept at 15°C to –5°C for removing the waxes. The purified oil has to be stored in stainless steel containers under inert condition at temperature ranging from 15 to 20°C. The water molecules from the oil have to be completely removed from the oil to prevent oxidation of limonene and other terpene compounds. The purified citrus oil must be stored under dark conditions in opaque bottles, where light should not transfer directly from the bottle. The most commonly used bottle is amber or cobalt blue.

12.7 SUMMARY

The EO of citrus fruit has 85–99% of volatile compounds, i.e., monoterpenes, sesquiterpenes, hydrocarbons, terpenoids, aromatic, and aliphatic components. The EO also contains functional components and bioactive compounds (such as: vitamin C, flavanones (hesperidin and neo hesperidin) and naringin. These compounds are heat and light sensitive. In this chapter, author has reviewed emerging oil extraction techniques like MAE, UAE, SFE extraction, high pressure processing, PEF to extract EO from citrus peel. It was concluded that oil extracted from the emerging techniques gave superior output in terms of extraction time, efficiency, and chemical components of oil as compared with conventional methods, such as: distillation process, SE and cold press. The oil extracted from emerging techniques has higher retention of functional and bioactive compounds as compared with conventional methods.

ACKNOWLEDGMENT

Author is thankful for the guidance and suggestions from professors Biswanath Bhunia and Tarun Kanti Bandyopadhyay.

KEYWORDS

- **citrus peel**
- **distillation**
- **hydro-distillation**
- **microwave hydro-diffusion and gravity**
- **oil extraction**
- **purification**
- **storage**
- **supercritical fluid**

REFERENCES

1. Areias, F., Valentão, P., Andrade, P., Moreira, M., Amaral, J., & Seabra, R., (2000). HPLC/DAD analysis of phenolic compounds from lavender and its application to quality control. *Journal of Liquid Chromatography & Related Technologies, 23*(16), 2563–2572.
2. Asnaashari, S., Delazar, A., Habibi, B., Vasfi, R., Nahar, L., Hamedeyazdan, S., & Sarker, S. D., (2010). Essential oil from *Citrus aurantifolia* prevents ketotifen-induced weight-gain in mice. *Phytotherapy Research, 24*(12), 1893–1897.
3. Assami, K., Pingret, D., Chemat, S., Meklati, B. Y., & Chemat, F., (2012). Ultrasound induced intensification and selective extraction of essential oil from *Carum carvi* L. seeds. *Chemical Engineering and Processing: Process Intensification, 62*, 99–105.
4. Atti-Santos, A. C., Rossato, M., Serafini, L. A., Cassel, E., & Moyna, P., (2005). Extraction of essential oils from lime (*Citrus latifolia*) by hydrodistillation and supercritical carbon dioxide. *Brazilian Archives of Biology and Technology, 48*(1), 155–160.
5. Auta, M., Musa, U., Tsado, D., Faruq, A., Isah, A., Raji, S., & Nwanisobi, C., (2018). Optimization of citrus peels D-limonene extraction using solvent-free microwave green technology. *Chemical Engineering Communications, 205*(6), 789–796.
6. Azmir, J., Zaidul, I., Rahman, M., Sharif, K., Mohamed, A., Sahena, F., Jahurul, M., et al., (2013). Techniques for extraction of bioactive compounds from plant materials: A review. *Journal of Food Engineering, 117*(4), 426–436.
7. Bayramoglu, B., Sahin, S., & Sumnu, G., (2008). Solvent-free microwave extraction of essential oil from oregano. *Journal of Food Engineering, 88*(4), 535–540.
8. Bhat, M., (2000). Cellulases and related enzymes in biotechnology. *Biotechnology Advances, 18*(5), 355–383.
9. Bhatnagar, A., Sillanpää, M., & Witek-Krowiak, A., (2015). Agricultural waste peels as versatile biomass for water purification: A review. *Chemical Engineering Journal, 270*, 244–271.
10. Bozkurt, T., Gülnaz, O., & Kaçar, Y. A., (2017). Chemical composition of the essential oils from some citrus species and evaluation of the antimicrobial activity. *Journal of Environmental Science, Toxicology and Food Technology, 11*(10), 29–33.
11. Brachet, A., Christen, P., & Veuthey, J. L., (2002). Microwave-assisted extraction of cocaine and benzoylecgonine from coca leaves. *Phytochemical Analysis: International Journal of Plant Chemical and Biochemical Techniques, 13*(3), 162–169.
12. Burt, S., (2004). Essential oils: Their antibacterial properties and potential applications in foods: A review. *International Journal of Food Microbiology, 94*(3), 223–253.
13. Casquete, R., Castro, S., Villalobos, M., Serradilla, M., Queirós, R., Saraiva, J., Córdoba, M., & Teixeira, P., (2014). High pressure extraction of phenolic compounds from citrus peels. *High Pressure Research, 34*(4), 447–451.
14. Chávez-González, M. L., López-López, L. I., Rodríguez-Herrera, R., Contreras-Esquivel, J. C., & Aguilar, C. N., (2016). Enzyme-assisted extraction of citrus essential oil. *Chemical Papers, 70*(4), 412–417.
15. Cholke, P., Bhor, A., Shete, A., & Sonawane, R., (2017). Extraction and GC-MS analysis of orange (*Citrus sinensis*) peel oil. *Research Journal of Life Sciences, Bioinformatics, Pharmaceutical and Chemical Sciences, 2*(5), 41–51.

16. Cirico, T. L., & Omaye, S. T., (2006). Additive or synergetic effects of phenolic compounds on human low density lipoprotein oxidation. *Food and Chemical Toxicology, 44*(4), 510–516.
17. Conde-Hernández, L. A., Espinosa-Victoria, J. R., Trejo, A., & Guerrero-Beltrán, J. Á., (2017). CO_2 supercritical extraction, hydrodistillation and steam distillation of essential oil of rosemary (*Rosmarinus officinalis*). *Journal of Food Engineering, 200*, 81–86.
18. Dabbah, R., Edwards, V., & Moats, W., (1970). Antimicrobial action of citrus fruit oils on selected food-borne bacteria. *Applied Microbiology, 19*(1), 27–31.
19. Danielski, L., Brunner, G., Schwänke, C., Zetzl, C., Hense, H., & Donoso, J. P., (2008). Deterpenation of mandarin (*Citrus reticulata*) peel oils by means of countercurrent multistage extraction and adsorption/desorption with supercritical CO_2. *The Journal of Supercritical Fluids, 44*(3), 315–324.
20. Destandau, E., Michel, T., & Elfakir, C., (2013). Microwave-assisted extraction. In: Rostagno, M. A., & Prado, J. M., (eds.), *RCS Green Chemistry – Natural Product Extraction: Principles and Applications* (Vol. 21, pp. 113–156). Cambridge – UK: The Royal Society of Chemistry.
21. Dugo, G., & Di Giacomo, A., (2002). *Citrus: The Genus Citrus* (p. 656). Boca Raton – FL: CRC Press.
22. El-Kantar, S., Boussetta, N., Lebovka, N., Foucart, F., Rajha, H. N., Maroun, R. G., Louka, N., & Vorobiev, E., (2018). Pulsed electric field treatment of citrus fruits: Improvement of juice and polyphenols extraction. *Innovative Food Science & Emerging Technologies, 46*, 153–161.
23. Eskilsson, C. S., & Björklund, E., (2000). Analytical-scale microwave-assisted extraction. *Journal of Chromatography A, 902*(1), 227–250.
24. Fadel, H., Marx, F., El-Sawy, A., & El-Ghorab, A., (1999). Effect of extraction techniques on the chemical composition and antioxidant activity of *Eucalyptus camaldulensis* var. *brevirostris* leaf oils. *Zeitschrift für Lebensmitteluntersuchung und-Forschung A (European Food Research and Technology), 208*(3), 212–216.
25. Farhat, A., Fabiano-Tixier, A. S., El-Maataoui, M., Maingonnat, J. F., Romdhane, M., & Chemat, F., (2011). Microwave steam diffusion for extraction of essential oil from orange peel: Kinetic data, extract's global yield and mechanism. *Food Chemistry, 125*(1), 255–261.
26. Ferhat, M. A., Meklati, B. Y., & Chemat, F., (2007). Comparison of different isolation methods of essential oil from citrus fruits: Cold pressing, hydrodistillation and microwave 'dry' distillation. *Flavor and Fragrance Journal, 22*(6), 494–504.
27. Ferhat, M. A., Meklati, B. Y., Smadja, J., & Chemat, F., (2006). Improved microwave Clevenger apparatus for distillation of essential oils from orange peel. *Journal of Chromatography A, 1112*(1, 2), 121–126.
28. Filly, A., Fernandez, X., Minuti, M., Visinoni, F., Cravotto, G., & Chemat, F., (2014). Solvent-free microwave extraction of essential oil from aromatic herbs: From laboratory to pilot to industrial scale. *Food Chemistry, 150*, 193–198.
29. Fisher, K., & Phillips, C., (2008). Potential antimicrobial uses of essential oils in food: Is citrus the answer? *Trends in Food Science & Technology, 19*(3), 156–164.
30. Francis, A., Shetty, T., & Bhattacharya, R., (1989). Modulating effect of plant flavonoids on the mutagenicity of N-methyl-N′-nitro-N-nitrosoguanidine. *Carcinogenesis, 10*(10), 1953–1955.

31. Friedman, M., Henika, P. R., & Mandrell, R. E., (2002). Bactericidal activities of plant essential oils and some of their isolated constituents against *Campylobacter jejuni, Escherichia coli, Listeria monocytogenes*, and *Salmonella enterica. Journal of Food Protection, 65*(10), 1545–1560.
32. Gamarra, F., Sakanaka, L., Tambourgi, E., & Cabral, F., (2006). Influence on the quality of essential lemon (*Citrus aurantifolia*) oil by distillation process. *Brazilian Journal of Chemical Engineering, 23*(1), 147–151.
33. Gebre, T. Y., (2012). Estimation of solutes in orange peel extract for pectin production. *International Journal of Scientific and Engineering Research, 3*(10), 1–7.
34. Giwa, S. O., Muhammad, M., & Giwa, A., (2018). Utilizing orange peels for essential oil production. *ARPN Journal of Engineering and Applied Sciences, 13*(1), 17–27.
35. Golmakani, M. T., & Rezaei, K., (2008). Comparison of microwave-assisted hydrodistillation with the traditional hydrodistillation method in the extraction of essential oils from *Thymus vulgaris* L. *Food Chemistry, 109*(4), 925–930.
36. Golmohammadi, M., Borghei, A., Zenouzi, A., Ashrafi, N., & Taherzadeh, M. J., (2018). Optimization of essential oil extraction from orange peels using steam explosion. *Heliyon, 4*(11), E-article ID 00893.
37. Guenther, E., (1948). *The Essential Oils: History and Origin in Plants Production Analysis* (Vol. 1, p. 456). New York – USA: Van Nostrand Company, Inc.
38. Hawthorne, S. B., Rickkola, M. L., Screnius, K., Holm, Y., Hiltunen, R., & Hartonen, K., (1993). Comparison of hydrodistillation and supercritical fluid extraction for the determination of essential oils in aromatic plants. *Journal of Chromatography A, 634*(2), 297–308.
39. Iwai, Y., Hosotani, N., Morotomi, T., Koga, Y., & Arai, Y., (1994). High-pressure vapor-liquid equilibria for carbon dioxide[+] linalool. *Journal of Chemical and Engineering Data, 39*(4), 900–908.
40. Kamal, G., Anwar, F., Hussain, A., Sarri, N., & Ashraf, M., (2011). Yield and chemical composition of citrus essential oils as affected by drying pretreatment of peels. *International Food Research Journal, 18*(4), 1275–1282.
41. Koketsu, M., Magalhaes, M., Wilberg, V., & Donalisio, M., (1983). *Óleos Essenciais de Frutos Cítricos Cultivados en Brazil (Essential Oils from Citric Fruits Cultivated in Brazil)* (p. 21). https://ainfo.cnptia.embrapa.br/digital/bitstream/item/65375/1/CTAA-DOCUMENTOS-007-OLEOS-ESSENCIAIS-DE-FRUTAS-CITRICAS-CULTIVADAS-NO-BRASIL-FL-06784A.pdf (accessed on 2 November 2023).
42. Kusuma, H. S., Putra, A. F. P., & Mahfud, M., (2016). Comparison of two isolation methods for essential oils from orange peel (*Citrus auranticum* L.) as a growth promoter for fish: Microwave steam distillation and conventional steam distillation. *Journal of Aquaculture Research and Development, 7*(2), E-article ID 1000409.
43. Lachos-Perez, D., Baseggio, A. M., Mayanga-Torres, P., Junior, M. R. M., Rostagno, M., Martínez, J., & Forster-Carneiro, T., (2018). Subcritical water extraction of flavanones from defatted orange peel. *The Journal of Supercritical Fluids, 138*, 7–16.
44. Lagha-Benamrouche, S., & Madani, K., (2013). Phenolic contents and antioxidant activity of orange varieties (*Citrus sinensis* L. and *Citrus aurantium* L.) cultivated in Algeria: Peels and leaves. *Industrial Crops and Products, 50*, 723–730.
45. Lanças, F. M., & Cavicchioli, M., (1990). Analysis of the essential oils of Brazilian citrus fruits by capillary gas chromatography. *Journal of High-Resolution Chromatography, 13*(3), 207–209.

46. Li, H., Pordesimo, L., & Weiss, J., (2004). High intensity ultrasound-assisted extraction of oil from soybeans. *Food Research International, 37*(7), 731–738.
47. Li, X. J., Wang, W., Luo, M., Li, C. Y., Zu, Y. G., Mu, P. S., & Fu, Y. J., (2012). Solvent-free microwave extraction of essential oil from *Dryopteris fragrans* and evaluation of antioxidant activity. *Food Chemistry, 133*(2), 437–444.
48. Lopresto, C. G., Petrillo, F., Casazza, A. A., Aliakbarian, B., Perego, P., & Calabrò, V., (2014). A non-conventional method to extract D-limonene from waste lemon peels and comparison with traditional Soxhlet extraction. *Separation and Purification Technology, 137*, 13–20.
49. Lucchesi, M. E., Chemat, F., & Smadja, J., (2004). Solvent-free microwave extraction of essential oil from aromatic herbs: Comparison with conventional hydrodistillation. *Journal of Chromatography A, 1043*(2), 323–327.
50. Luengo, E., Álvarez, I., & Raso, J., (2013). Improving the pressing extraction of polyphenols of orange peel by pulsed electric fields. *Innovative Food Science & Emerging Technologies, 17*, 79–84.
51. Mandal, V., Mohan, Y., & Hemalatha, S., (2007). Microwave assisted extraction: An innovative and promising extraction tool in medicinal plant research. *Pharmacognosy Reviews, 1*(1), 7–18.
52. Masango, P., (2005). Cleaner production of essential oils by steam distillation. *Journal of Cleaner Production, 13*(8), 833–839.
53. Mastelić, J., Politeo, O., & Jerković, I., (2008). Contribution to the analysis of the essential oil of *Helichrysum italicum*: Determination of ester bonded acids and phenols. *Molecules, 13*(4), 795–803.
54. Meng, X., & Ragauskas, A. J., (2014). Recent advances in understanding the role of cellulose accessibility in enzymatic hydrolysis of lignocellulosic substrates. *Current Opinion in Biotechnology, 27*, 150–158.
55. Mishra, D., Shukla, A., Dixit, A., & Singh, K., (2005). Aqueous enzymatic extraction of oil from mandarin peels. *Journal of Oleo Science, 54*(6), 355–359.
56. Moncada, J., Tamayo, J. A., & Cardona, C. A., (2016). Techno-economic and environmental assessment of essential oil extraction from oregano (*Origanum vulgare*) and rosemary (*Rosmarinus officinalis*) in Colombia. *Journal of Cleaner Production, 112*(Part 1), 172–181.
57. Ndayishimiye, J., Lim, D. J., & Chun, B. S., (2018). Antioxidant and antimicrobial activity of oils obtained from a mixture of citrus by-products using a modified supercritical carbon dioxide. *Journal of Industrial and Engineering Chemistry, 57*, 339–348.
58. Okunowo, W. O., Oyedeji, O., Afolabi, L. O., & Matanmi, E., (2013). Essential oil of grape (*Citrus paradisi*) fruit peels and its antimicrobial activities. *American Journal of Plant Sciences, 4*(7), 1–12.
59. Pannu, S. K. V., Kaur, H., Kaur, K., & Singh, V., (2018). A comparative study of various oil extraction techniques in plants: A review. *Agriways, 6*(2), 59–65.
60. Périno-Issartier, S., Ginies, C., Cravotto, G., & Chemat, F., (2013). A comparison of essential oils obtained from lavandin with different extraction processes: Ultrasound, microwave, turbohydrodistillation, steam and hydrodistillation. *Journal of Chromatography A, 1305*, 41–47.
61. Pingret, D., Fabiano-Tixier, A. S., & Chemat, F., (2014). An improved ultrasound Clevenger for extraction of essential oils. *Food Analytical Methods, 7*(1), 9–12.

62. Poiana, M., Fresa, R., & Mincione, B., (1999). Supercritical carbon dioxide extraction of bergamot peels: Extraction kinetics of oil and its components. *Flavor and Fragrance Journal, 14*(6), 358–366.
63. Pollien, P., Ott, A., Fay, L., Maignial, L., & Chaintreau, A., (1998). Simultaneous distillation–extraction: Preparative recovery of volatiles under mild conditions in batch or continuous operations. *Flavor and Fragrance Journal, 13*(6), 413–423.
64. Prasanna, M., & Rahul, B., (2018). Techniques available for the extraction of essential oils from plants: A review. *International Journal for Research in Applied Science and Engineering Technology, 6*(3), 2931–2935.
65. Ramanathan, L., & Das, N., (1992). Studies on the control of lipid oxidation in ground fish by some polyphenolic natural products. *Journal of Agricultural and Food Chemistry, 40*(1), 17–21.
66. Routray, W., & Orsat, V., (2012). Microwave-assisted extraction of flavonoids: A review. *Food and Bioprocess Technology, 5*(2), 409–424.
67. Sahraoui, N., Vian, M. A., El Maataoui, M., Boutekedjiret, C., & Chemat, F., (2011). Valorization of citrus by-products using microwave steam distillation (MSD). *Innovative Food Science & Emerging Technologies, 12*(2), 163–170.
68. Shakir, I. K., & Salih, S. J., (2015). Extraction of essential oils from citrus byproducts using microwave steam distillation. *Iraqi Journal of Chemical and Petroleum Engineering, 16*(3), 11–22.
69. Shamsol, Z., Iskandar, M. I., Ariffin, Z. Z., & Safian, M. F., (2021). Chemical constituents, antioxidant, and antimicrobial activities of *kaempferia galanga* rhizome essential oils. *International Journal of Social Science Research, 3*(4), 405–419.
70. Sowbhagya, H. B., Sushma, S., Rastogi, N. K., & Naidu, M. M., (2013). Effect of pretreatments on extraction of pigment from marigold flower. *Journal of Food Science and Technology, 50*(1), 122–128.
71. Stuart, G., Dariva, C., & Oliveira, J. V., (2000). High-pressure vapor-liquid equilibrium data for CO_2 – orange peel oil. *Brazilian Journal of Chemical Engineering, 17*(2), 181–189.
72. Suetsugu, T., Tanaka, M., Iwai, H., Matsubara, T., Kawamoto, Y., Saito, C., Sasaki, Y., et al., (2013). Supercritical CO_2 extraction of essential oil from *Citrus sphaerocarpa Tanaka* peel. *Flavor, 2*(1), 18–25.
73. Valensi, P., Behar, A., De Champvallins, M., Attalah, M., Boulakia, F., & Attali, J. R., (1996). Effects of a purified micronized flavonoid fraction on capillary filtration in diabetic patients. *Diabetic Medicine, 13*(10), 882–888.
74. Vian, M. A., Fernandez, X., Visinoni, F., & Chemat, F., (2008). Microwave hydrodiffusion and gravity, a new technique for extraction of essential oils. *Journal of Chromatography A, 1190*(1, 2), 14–17.
75. Vilkhu, K., Mawson, R., Simons, L., & Bates, D., (2008). Applications and opportunities for ultrasound assisted extraction in the food industry – A review. *Innovative Food Science & Emerging Technologies, 9*(2), 161–169.
76. Yeh, C. C., Kao, S. J., Lin, C. C., Wang, S. D., Liu, C. J., & Kao, S. T., (2007). The immunomodulation of endotoxin-induced acute lung injury by hesperidin *in vivo* and *in vitro*. *Life Sciences, 80*(20), 1821–1831.
77. Yen, H. Y., & Lin, Y. C., (2017). Green extraction of *Cymbopogon citrus* essential oil by solar energy. *Industrial Crops and Products, 108*, 716–721.

78. Zagklis, D. P., & Paraskeva, C. A., (2015). Purification of grape marc phenolic compounds through solvent extraction, membrane filtration and resin adsorption/desorption. *Separation and Purification Technology, 156*(Part 2), 328–335.
79. Zarnowski, R., & Suzuki, Y., (2004). Expedient Soxhlet extraction of resorcinolic lipids from wheat grains. *Journal of Food Composition and Analysis, 17*(5), 649–663.

CHAPTER 13

Extraction and Role of Vitamin B12 in Food Products: Health Effects

Ramesh Sharma, Biswanath Bhunia, and Tarun Kanti Bandyopadhyay

ABSTRACT

With increasing demand for strict vegetarian persons, the research studies for estimation of vitamin B12 (vit B12) from plant sources are increasing day-by-day. Several methods were tested to estimate the vit B12 from plant or food fortified-based samples. The emergence of ultra-performance liquid chromatography (UPLC) after detecting with UV has higher sensitivity and specificity for quantifying vit B12. The immunoaffinity column is being used for purification of vitamin B12 and the method is more selective to estimate vit B12 without measuring the inactive corrinoids as is measured by microbiological assay (MBA). The method with good sensitivity and selectivity is required for the analysis of in-situ produced vit B12 after using the plant-based substrate for consideration as a dietary source. The emergences of optical spectroscopy (such as: chemiluminescence (CL), absorption and fluorescence spectroscopy (AFS), surface plasmon resonance (SPR) and Raman spectroscopy (RS)) have been considered as selective approaches for estimation of vit B12. The study on the stability of vit B12 under light conditions, varying temperature, and pH is equally important to meet the dietary intake of vit B12. The vit B12 has been produced under in-situ conditions while considering the food matrix after using the Propioni bacteria for food-grade conditions. This chapter explores extraction, estimation, and stability analysis of vit B12 with the application of food-grade microorganisms for in-situ production of vit B12.

The Functional Foods: Nutrient and Health Benefits.
Megh R. Goyal, Junaid Ahmad Malik, and Anu Kumari (Eds.)

13.1 INTRODUCTION

Vitamin B12 belongs to the family called cobalamins, which are further classified as: cyano cobalamin (CNCbl), hydroxo cobalamin (OHCbl), adenosyl cobalamin (ADLCbl), and methyl cobalamin (MTLCbl) [55]; and among which the methyl cobalamin and adenosyl cobalamin are bioactive forms. Each form of cobalamin has its biological activity based on its absorption potential and it plays vital role in the growth and proliferation of the human cells. Additionally, vit B12 is an enzyme's major component and performs the function of amino acid metabolism [8]. The lack of Vit B12 has led to anemia and neuropathy problems in our body.

Naturally, Vit B12 is available in dairy and meat products; and the form of cyanocobalamin (CNCbl) is only used to fortify food products. The quantity present in cheese and fish ranged of a few ng/g to 100 ng/g for fortified cereals and liver cells [143]. Traditionally, vit B12 is estimated in food products by microbiological assay (MBA), using Lactobacillus leishmanii [55]. Although the assay was performed with high sensitivity, yet it lacks specificity due to the presence of inactive cobalamin in some foods, which can hinder growth of microorganisms [8]. The other drawback is that MBA method takes more time with less precision for the determination of vitamins.

Radio isotopic dilution was performed to estimate the vit B12 in the processed food [21, 104]; however, it lacks selectivity because the assay also binds to another cobalamin. Assay associated with protein-binding with applications of surface plasma resonance (SPR) spectroscopy can estimate the vit B12 in the milk samples [60]. The method was sensitive for estimation of 2 ng/g of vit B12 in the sample, but the restriction occurred due to the limitation of specificity as the protein used for binding in the assay was reactive with analogs of cobalamin.

Several results have been reported for the quantification of vit B12 using HPLC with elution under isocratic/gradient conditions and detection mode of UV visible conditions [5, 69, 74, 85, 101, 162, 177]. The limitations of UV detection are that it is not efficient to estimate vit B12 in food products, which are non-fortified in nature. Similarly, solid phase extraction (SPE) approach was used for the estimation of vit B12 using LC–UV in fluid [25] and food samples [62] but the method was limited to selective food sample due to the selectivity issues. Based on the doses of cobalt, the other spectrometry detection method are atomic absorption [161] and argon plasma emission [102], which were suggested for the quantification of vit B12. Unfortunately, the limit of detection (LOD) for these methods was 150 mg/g; and the method is only applied to quantify higher vit B12 content in the food sample.

Recently, capillary electrophoresis (CE) [6] and micro-HPLC [24] in combination with ICP-MS have been developed to quantify cobalamins. Although this complex method performed estimation quickly of cobalamin in the food sample, yet the estimated value was higher than the actual vit B12 present in food sample. The detection by fluorometric method is adapted to quantify low content of vit B12 in the range of 3–10 ng/g in food sample, but this method requires to extract vit B12 by enzyme treatment and reaction needs to occur for emitting fluorescent properties, as the vit B12 does not show the fluorescent properties under natural conditions [116].

This chapter emphasizes on brief discussions on characterization, extraction, estimation, stability analysis of vit B12 with applications of food grade microorganisms for producing vit B12 under in-situ conditions.

13.2 OCCURRENCE AND CHEMICAL FEATURES OF VITAMIN B12

Vitamins are considered most essential nutrients for the proliferation and growth of cells and lack of vitamins may cause physiological issues in the human body.

Vit B12 is grouped as a corrinoid compound, a complex ring with a cobalt atom at the ring [49], and this vitamin is stored in the liver for years [20]. The synthetically obtained CNCbl is the most stable form and is mainly used as supplement in pharmaceutical industries, and for fortification purpose in food products.

CNCbl is metabolized in the human body to obtain methyl cobalamin by ilea enterocytes, which are biologically active in our body [92]. Vit B12 functions as a co-enzyme and plays a crucial role in fat and carbohydrate metabolisms in the formation of red blood cells. Besides this, it promotes the nervous system's functioning and helps to translocate the methyl group during the synthesis of DNA [7, 149]. The deficiency of vit B12 is uncommon but is mostly seen among strict vegetarians and deficiency leads to megaloblasts, which can result in abnormal growth of cells; and symptoms include breathlessness, infection in the body, excessive tiredness, and listlessness. The severe deficiency may lead to the degeneration of nerves and unrepairable damage of the nervous system. The lack of vit B12 may be due to an imbalance of intake of the nutrients, malabsorption syndrome, and gastrointestinal (GI) problems [64].

The Institute of Medicine considered the RDA (recommended daily allowance) of 2.4 μg/d for vit B12 [56]. Cobalamin is obtained by de novo synthesis and is produced by some bacteria and archaea species. Hence, these organisms are the vit B12 producing biological sources available in nature. Therefore, cobalamin creates the demand for nutritional requirements for animals and it is not synthesized by plants (as it is not required for their growth) [13]. Even though vit B12 is not present in plant source food materials [119, 135], yet vit B12 is present in the range of 1.09–2.65 μg/100 g dry biomass weight in the edible mushroom species (such as: Craterellus cornucopioides and Cantharellus cibarius) in comparison with other wild species having no vit B12. Although the vit B12 is not present in majority of plants, yet it is available in some plant sources and animal origin sources (Table 13.1).

The corrinoids in the mentioned mushrooms are reported to have vit B12 [175]. However, some actinobacteria (which fix the atmospheric nitrogen) and Frankia alni can synthesize the vit B12 and these bacteria can perform the function for the formation of nodules endophytes that are present in woody trees. Frankia alni functions symbiotically with actinorhizal plants [22]. Hence, the symbiosis may result in the presence of vit B12 in selected plant sources [79]. Therefore, the source of vit B12 may be the fermented foods obtained by using such bacteria, and the plant that functions in the symbiotic

TABLE 13.1 Examples of Plant and Animal Sources for Vitamin B12

Plant Sources [106]	Vitamin B12 in ng/100 g
Black mustard	1.52 ng/100 g
Hippophae rhamnoides (HR) sample 1	37.01 ng/100 g
HR liquid sample	0.10 ng/100 g
HR sample 2	0.11 ng/100 g
HR sample 3	0.08 ng/100 g
HR sample 4	3.10 ng/100 g
HR sample 5	4.58 ng/100 g
Plant extract sample	26.28 ng/100 g
Plant extract sample 1	26.28 ng/100 g
Vit B12 tablets [106]	9583.79 (49.03 μg per tablet)
Vit B12 in Animal Sources	
Lamb, loin cooked (90 g) [129]	2.70 mg
Tinned tuna (90 g)	2.50 mg
Beef mince (90 g)	2.30 mg
Roast turkey (90 g)	1.30 mg
Pork loin (90 g)	0.60 mg
Chicken (90 g)	0.50 mg
Liver (veal) [106]	65.59 ng/100 g)

relation with Frankia alni, and the tissues of animals that had ingested the food containing vit B12. Additionally, the ruminants of animals can be the source of cobalamin because certain groups of bacteria in the microflora may synthesize vitamins and the liver may be a good source of such vitamins.

The mainly vit B12 is found in liver, egg, meat, fish, milk, and oysters (Table 13.1). The quantity present in animal-based materials (such as: liver, cheese, meat, and milk) were reported to be 83, 2.4, 1–2, and 0.9 μg per 100 g, respectively [35]. Certain species of edible cyanobacteria (such as: Spirulina sp., Nostoc, and Aphanizomenon) have been reported with sufficient quantity of vitamin B12 especially pseudo-vitamins, which are biologically inactive, and these vitamins are present at a concentration of 127–244 μg per 100 g of spirulina tablets [173]. Similarly, other widely produced edible algae (such as: Enteromorpha sp. and Porphyra sp.) retained sufficient quantity of vit B12 at a concentration of 133 μg per 100 g of dry weight; but the drawback is that the biological function of algae-derived corrinoids has remained uncleared [96, 174].

Additionally, actinorhizal plants (such as: Hippophae rhamnoides and Myrica) performed symbiotic functions with actinobacteria including Frankia alni; and are considered as important host for the production of corrinoids compound of vit B12 [78].

The analogs of vit B12 are inactive with biological function and may even block the system of vit B12 metabolism in cells of mammal organisms [73]. The low quantity of natural sources for the extraction of vit B12 and uncertainty of biological function of vit B12 have resulted to test the content of vit B12 in the fermented food samples. Normally, the fermented food including tempeh contains vit B12 in the range of 0.7–8 μg per 100 g; and similarly, it is found in sauerkraut at 7.2 μg per 100 g of sample, and the LAB fermentation of fenugreek juice yields the vit B12 of 12.5 μg per 100 ml of sample. Therefore, strict vegetarian people are at greater risk of deficiency with cobalamin than non-vegetarian people; and fortified vit B12 food products need to be consumed to avoid deficiency.

13.3 CHALLENGES FOR EXTRACTION OF VITAMIN B12

The complexity of the extraction of vit B12 in processed food items may be attributed to: (i) as the vit B12 concentration is very low in the range of 3–50 ng per g of product compared with other vitamins; (ii) mainly the fortification of vit B12 is performed in the form of CNCbl but its form is also present in the form of hydroxocobalamin in food products; and (iii) the other restriction is that vit B12 is available in free or bound forms in nature.

To release the bound form, there is a protocol [116], where the sample is digested by the pepsin enzyme. This protocol is applied efficiently to identify the peak of vit B12 in the food sample by representing it on a chromatogram. The results of efficacy obtained after testing the pepsin enzyme for releasing the bounded vit B12 are then mentioned in the food sample [55].

Similarly, the other challenge is that the organism L. leishmanii can consume deoxy-ribosides and deoxy-nucleotides including vit B12 [133]; and it has been reported with a high value in the content of vitamins by MBA due to the availability of analogs vit B12 in the hypoallergenic product.

13.4 EXTRACTION OF VITAMIN B12

The sample preparation steps for vit B12 extraction have been reported in the past [106]. The sample weight of 0.5–10 g based on the percentage of vit B12 in the sample is grounded in the mill and is finally homogenized and is filled into the jar with the addition of liquid nitrogen of 55 to 60 ml. The aliquot is weighed accurately and is collected in a flask with the addition of 60 ml of acetate buffer (50 mM, pH 4), KCN (potassium cyanide; 1% w/v), one g of pepsin; and 300 μl of α-amylase is weighed and is collected in a flask and incubation is performed at 37°C for 1.5 h by providing constant agitation of the sample. The estimation of vit B12 is performed by conversion of cobalamin to CNCbl [28, 55] and this form is declared on the label.

The conversion to the required form is performed by boiling the sample for 30 min at 98°C with the addition of cyanide [41]. Vit B12 is normally present in natural forms, which are sensitive to light; therefore, it is treated with cyanide to change it into more stable form of vit B12 in the analytical methods.

The reaction of the enzyme is restricted by boiling the sample at 100°C for 30 to 35 min, and it is allowed to reach room temperature; and the sample is centrifuged at 4,000 RPM for 25 min at 4°C. After the temperature reaches 100°C, enzymes and proteins including pepsin and amylase (which are added to improve the extraction of vit B12 from plants including polysaccharides and proteins) are finally denatured by immunoaffinity steps [55]. The purpose of treatment with amylase is to increase the filtration efficiency of the sample by achieving clear supernatant, especially for the starch-containing food. For the sample containing high fat (e.g., Hippophae rhamnoides), the filtration is repeated twice to collect less viscous sample. Once the sample is cooled, it is finally centrifuged at 4,000 RPM for 25 min. The collected supernatant is filtered twice with a filter paper and pH of 7 is maintained; and it is made up to 100 ml by adding deionized water.

13.5 APPROACHES FOR ESTIMATION OF VITAMIN B12

The previously reported analytical methods for estimation of vit B12 included: atomic absorption spectrometry [3, 61], radio isotope assay [21, 104, 170], spectrophotometer method [67, 100], chemiluminescence (CL) [123, 184], MBA method [29, 87], analysis by electrochemical approach [68], PDMS (plasma desorption mass spectrometry) [11], NMR analysis [95], and HPLC [1, 5, 54, 100, 115, 177]. Among these, MBA method has been officially considered as the test method for estimation of vit B12, which uses the test organism Lactobacillus leishmanii [29], but the method has drawbacks, such as: time-consuming, tedious, low-specificity, and low-precision.

The estimation by radio isotopic dilution method [21, 170] is simpler and time-saving, but is costlier due to the need for radio-labeled CNCbl. The spectrophotometer method [66, 99] is not recommended for complex matrix samples as it performs with low sensitivity. Atomic absorption spectrometry [3, 61] and CL [123] can measure indirectly the cobalt concentration but the application of this method for estimation of vit B12 is limited because this method has failed to distinguish between the free- and bonded cobalt form of vit B12.

Chromatographic methods and CL methods have been proposed to quantify vit B12 including atomic absorption [35], UV and visible photometry [54, 80, 100, 115, 177], and ICP–MS [24]. However, these methods are not suitable to quantify vit B12 because of low concentration of vit B12 (i.e., 3 to 20 n/g) in the food sample.

The HPLC with detection by fluorescence has been mentioned to determine the vit B12 [89]. This method was inefficient for complex samples including the processed-foods, where estimation was a tedious task. The vit B12 needed to be derivatized into a complex of fluorescent, and detection by fluorescence is performed in HPLC [116]. It was concluded that enrichment and purification steps are necessary while dealing with a complex sample. Therefore, considering the number of estimation approaches for quantification of vit B12, the general methods of estimation are described in this section.

13.5.1 MBA FOR ESTIMATION OF VITAMIN B12 IN FOOD SAMPLE

Traditionally, the vit B12 is estimated with a microbial culture in which certain microorganisms are allowed to grow with exposure to various samples and their growth is compared against the standard growth curve obtained for a specific compound [18]. The traditional approach to determine vit B12 is mainly by MBA with the use of Lactobacillus delbrueckii as a test microorganism and it is referred to as the analytical method by AOAC [132].

The Lactobacillus leichmannii is dependent on the corrinoids compound for their growth and it has been used for estimation of vit B12 as the growth improves after the external supply of vit B12 [18, 39, 57]. An experiment was performed for the microbiological determination of vit B12 using culture of Lactobacillus leichmannii in medium based on liquid skim milk that was incubated for 24 h with the regulation of pH before the addition to refine serum assay and were autoclaved for 15 minutes at 120°C [138]. The results after the microbial growth were compared with obtained growth rate curve of microorganisms with added cobalamin.

Similarly, the vit B12 estimation by MBA has also been described by the BAMC (British Analytical Methods Committee) [28]. In this method, the test organism L. leishmanii was used to quantify the vit B12, and the experiment was carried out in a boiled water with the addition of cyanide. The drawback of the method is that the test organism consumed the incomplete corrinoids, deoxy-ribosides, and deoxy-ribonucleotides for their growth; and hence the yield of vit B12 is higher in MBA than the UPLC [168]. Even though the method has high sensitivity, yet it has poor selectivity; and there has been over estimation by 5–30%, because the test organism mainly got reacted with other forms of vit B12 including analogs and nucleic acid [23].

One of the reported inactive corrinoids was found as pseudo vit B12 in the human body, which is released by Lactobacillus reuteri CRL1098 [158]. The inactive corrinoid is present with adenine in place of DMBI at the site of the lower ligand. Similarly, vit B12 content in meat products was found to be 2.2-times higher in the MBA method compared to the HPLC method [49], indicating poor specificity by MBA. Additionally, the MBA method is time consuming and laborious. The MBA method with poor selectivity has been reported while analyzing tablets of spirulina and shellfish showing a 6–8-folds high concentration than the concentration estimated by the CL method [172].

Mostly, the measured vit B12 are inactive corrinoids. The inactive corrinoids mostly occurred due to the incomplete synthesis of vit B12 as a result of insufficient synthesis of DMBI [98]. In addition to this, the production of lactic acid by LAB facilitates the synthesis process, as lactate is the major source of carbon for PAB production [120]. While baking sourdough, the LAB may promote the production of a compound similar to vit B12-like compounds, which may affect with positive growth of L. delbrueckii during testing with MBA [53].

13.5.2 IMMUNOASSAY METHODS

Immunoassays techniques can be used for the detection of biological molecules in a sample using a special antibody for binding the target molecules of substrate for its estimation [36]. Firstly, the immunoassay was performed to estimate vit B12 in human serum [40]; while further research added the competitive immunoassays to quantify vitamin B12 [48, 50]. The ELISAs were implemented successfully for the quantification of water-soluble vitamins while considering different matrices with little or no interference of matrix [10, 66].

The analysis by UPLC–MS/MS was performed to estimate the vit B12 and the estimated concentration was 0.0024 ng/μl for infant milk and 0.0039 ng/μl for supplementary tablets of vit B12. The obtained value was lower than the value obtained by ELISAs with LOD of 10 ng/ml. Hence, the ELISAs also can be considered equally sensitive to HPLC for the detection of lower limits of vit B12. Additionally, with the application of ELISA, the food sample showed no or little matrix effect than HPLC under the estimation conditions.

13.5.3 CAPILLARY ELECTROPHORESIS (CE) METHOD

In this method, the capillary is added with liquid electrolyte, which is added to the sample before the supply of electric field across it for the separation of ions [37, 65]. According to Lambert et al. [80], the 15 kV was supplied to the glass capillary; and the ions from the solutions could travel across the capillary length as a result of electro-osmosis and cobalamin derivatives capillary. The derivatives of cobalamin are separated through the capillary with detection from UV through absorption at 266 nm. The vit B12 with

LOD for CE was 20 μg/mL; and CE can be used as a good substitute to HPLC for different concentrations and complex matrices [80].

13.5.4 RADIOISOTOPE AND MASS SPECTROMETRY (MS)

The available form of radioactive isotope for vit B12 has been used to detect the measurement of radioactivity. The vit B12 with a radioactive isotope in the form of (57Co) was used [44, 45]. The blood plasma radioactivity [34, 46] of the whole body [17, 126, 139, 150] was measured after correlation with vit B12 for diagnosis purpose [77, 109].

In a study, the form of vit B12 having 14C was attached with MS to estimate the absorption of vit B12 in the human body [19]. MS functions with the ionization of sample molecules with the creation of charge fragments of the chemicals were finally separated with the acceleration of charge particles in the magnetic or electric field. The deflection was occurred by field as per the charge to the mass ratio, which resulted in the separation after arriving at different points for detection [164].

The chemical species have been identified based on the detected mass fragments, which are finally correlated with a known mass profile of chemicals; and 14C isotope was reported to detect the lower detection limit of a biological sample [163]. This method has been implemented for the estimation of vit B12 with a LOD of 100 fg/mL [19] and it was considered for detection of modified form of cobalamin that is detected with complicated set of experiments while comparing to natural form of vit B12.

13.5.5 CHROMATOGRAPHY METHOD FOR ESTIMATION OF VITAMIN B12

The research study for the presence of vit B12 was conducted to promote the supply of vit B12 among strict vegetarian persons through the ingestion of fortified food samples. The extraction and estimation need to be optimized, as it is present in low amounts in the plant source; therefore, there is a need to improve the sensitivity of methods for estimation of vit B12. The liquid chromatography (LC) was used to estimate the vit B12 in different processed food samples (Table 13.2).

The HPLC methods have been reported to quantify vit B12 in the fortified samples, supplementary, and infants' diets [27, 55, 92]. The HPLC was performed for analyzing Vit B12 in food samples [23]. The CNCbl got separated using a column (reversed-phase C18) at 361 nm with separation by UPLC. The CNCbl resolution was obtained by using a water linear-gradient (milli-Q) and acetonitrile, which is present with 0.025% of TFA at a fixed flow rate.

Several studies have been reported on the estimation of vit B12 in food including meat [49] and fermented milk-based products [156]. These HPLC were performed using UV detection method. Similarly, HPLC was used using fluorescence detection and vit B12 was derivatized into α-ribazole (a fluorescent compound) [116]. Campos [186] reported that quantification of vit B12 was 20–30% higher in the MBA method, in comparison with HPLC. The HPLC-UV method is not efficient to quantify vit B12 in the complex samples due to interference from other compounds.

TABLE 13.2 Determination of Vitamin B12 in the Processed Food Products by Liquid Chromatography (LC)

Food-Based Products	Estimation by HPLC (ng/g)		MBA (ng/g)
	Free Vitamin B12	Total Vitamin B12	
Infant cereal (absence of milk sample 1)	–	9.50	9.70
Infant cereal (absence of milk sample 2)	–	12.0	19.70
Infant milk powder 1	13.0	17.30	19.3
Infant milk powder 2	–	13.0	18.0
Infant milk powder 3	–	13.0	18.0
Petfood 1	264.20	301.0	269.10
Petfood 2	34.70	41.30	73.90
Wheat-based breakfast	–	27.70	35.90

Similarly, the SPE with the C18 column is inefficient to purify the vit B12 in food products [116] due to the poor selectivity of SPE cartridges especially C18. Therefore, there is a need for a combined approach of purification and concentration with the application of immunoaffinity columns to estimate the accurate prediction of vit B12. It has been mentioned as a good substitute to the MBA method to quantify vit B12 in food samples including infant formula powder based on milk [55]. The purification of vit B12 with the implication of the immunoaffinity approach has raised the sensitivity of HPLC for the estimation, especially in the supplementary diet and fortified food [15, 55, 103]. Under the supplied column, the protocol was followed by using phosphate buffer (pH 7) as a solution for washing and loading the sample into the column; and a higher recovery was achieved [55]. The treatment with amylase is done to improve the filtration in a sample containing the starch molecules [106]. Usually, centrifugation is not performed on samples containing high fat because a clear sample is not obtained in such a sample. To obtain a clear sample, membrane has been developed to obtain a clear supernatant for the oily samples [55].

Similarly, other research study implemented the immunoaffinity method to quantify vit B12 in the enriched food [92], salami, and meat-based products [49]. The immunoaffinity method was adopted after following the extraction with solid phase and a concentration step was applied to quantify vit B12 in edible mushrooms [171]. However, the emergence of the UPLC technique (which is present with a separation column of less than 2 μm particle size) could improve the sensitivity and resolution of chromatography [148].

In recent times, UPLC-MS method was used to detect vit B12 in dairy-based products [185]. The cells of P. freudenreichii after analyzing by UPLC obtained after fermentation of supplemented whey permeates could yield 620 ng per ml of vit B12 [23]. The P. freudenreichii strain could give a vit B12 concentration of 100–2,500 ng per ml in SWP after adding cobalt and DMBI, and it represented diversity of inter-strain for synthesis of vit B12 [58]. The improved sensitivity of UHPLC and advanced instrument features including sample cleaning and concentration potential of vit B12 have been achieved, but still it has not been fully utilized to analyze vit B12 in natural or processed food samples.

The MBA studies to estimation of vit B12 using P. freudenreichii cells were comparable with UPLC results, but results were higher by 17% for pig liver. Similarly, the content of vit B12 estimated by UPLC and MBA were 67% and 63%, respectively for the rye and barley samples. The malt extract sample fermented with propionic and lactic acid producing bacteria found no concentration of vit B12 while testing with UPLC; but the concentration of 1,200 ng per g was obtained by testing the sample with MBA method. These differences in concentration of vit B12 measured by MBA and UPLC method may be due to the drawback of MBA method.

13.6 OPTICAL DETECTION TECHNIQUES

The optical detection for biological samples has been considered the reliable and sensitive approach to estimate DNA molecules [154], vitamins [14, 51], and proteins [154]. The detection is performed as a result of light interaction with the target molecules or with intermediate molecules by changing the behavior of molecules by the change of light color for the estimation of specified molecules [81, 113, 131, 159]. The estimation of vitamins is performed by Raman Scattering [39, 105, 109] and detection was by fluorescence [12, 26], where incidence light changes the color of the sample due to the presence of a chemical bond on the sample, and these techniques gave promising results [90].

13.6.1 CHEMILUMINESCENCE (CL) METHOD

Vitamins are poor emitters of light with the exception of vitamin A [157]. Similarly, vitamin B12 does not emit light under the conditions of optical excitation. The detection with fluorescence is performed with the interaction of light-emitting molecules and vitamins with the CL method. Cobalt enhances the CL reaction between dissolved oxygen and luminol [140] or between hydrogen peroxide and luminol at LOD of 890 pg/mL [4]. The detection limits with similar results were obtained using the CL technique with the luminol deployed in the laboratory [91].

The conventional approaches for vit B12 estimation are time-consuming, slow, risky, and are performed with low sensitivity; whereas CL analysis can detect enhanced liberation of cobalt during acidification with ultra-sensitive detection and a wider range of vit B12 [123, 130]. The CL was performed with simplicity, and sensitivity for the determination of organic and inorganic molecules [141, 142]. The estimation of vit B12 could be performed using CL with the application of DBS–LDHs (dodecylbenzene sulfonate–layered double hydroxides) for measuring the released ions of cobalt (II) [182]. The CL signal is improved with increasing the concentration of vit B12 with a detection limit of 570 pg/ml.

The research study was constructed to improve specificity of DBS-LDHs for vit B12 by reducing the cross-contamination with other ions of metal that significantly reduced cross-sensitivity with CL detection method with luminol-based concept. The LOD with 5 pg/ml for vit B12 for tablets/multivitamin capsules was observed by CL and HPLC (0.1 ng/ml) [78, 86], AAS (atomic absorption spectroscopy) (42 μg/ml) [160], RDA (5 ng/ml) [128]; by ICPMS (inductively coupled plasma mass spectroscopy) (0.22 μg/ml) [99], fluorometric detection (0.1 ng/ml) [84], and CL (20 μg/ml) [80] and 50 pg/ml [140], respectively.

13.6.2 ABSORPTION AND FLUORESCENCE

The vit B12 emits light poorly and the intrinsic absorption at the concentration present in biological materials is at a low level for detection and its availability can reduce the efficiency of other compounds, which are light emitting in nature. The Rhodamine 6G is used to estimate vit B12 indirectly in the solution and vit B12 had shown the transfer of energy between Rhodamine 6G and Aridine [88]. Based on this concept, the vit B12 was detected with a LOD of 2 μg/ml [180]. The fluorescent probe (called HINBD) was used for detecting vit B12 with a limit of 0.1 μg/ml [136].

The quenching effects of vit B12 had been seen with the fluorescence of CdTe quantum dots; and the quantum dot absorbed energy was transferred resonantly to molecules of vit B12 [155]. Similarly, the compound having carbon also performed a quenching effect with vit B12, and the results were seen clearly by using a layer of graphene oxide [42] or carbon dots [166]. Wang et al. mentioned the detection with LOD of 0.1 μg/mL in the solution after using the carbon dots. The use of the fluorescent method estimated vit B12 with LOD of 0.2–10 μg/ml [169], even though the measurement sensitivity was lower than the RDA (0.05–2 μg/L) or MBA method (0.01–0.2 μg/L) [134]; and this method is mostly used to estimate vit B12 in the food sample, which retains a higher percentage of vitamins because the application of MBA and RDA method need the dilution of the fruit extract for performing the assay.

13.6.3 SURFACE PLASMON RESONANCE (SPR)

The SPR senses the changes in color scattering on the surface of metal with the presence of the compound to be detected [167]. Surface plasmons are oscillatory electrons present on the metal interface and dielectric surface. The oscillations with their frequency work with high sensitivity to the environmental refractive index; therefore, the SPR frequency depends on the sample refractive index under study [146], in the configuration of the sensor. This technique is performed with the requirement of an antibody, which binds the target molecules selectively on the sensor surface due to stronger resonance on the surface. SPR can determine kinetics and affinity between binding of protein-ligand with the formation of complex in biology [105].

The SPR technique was used for the quantification of vit B12 in dietary supplements [165]. This technique has been implemented to study the binding characteristics between Cbl and their transport proteins. As the SPR is performed with higher sensitivity with the mass of interacting molecule [144], therefore radiolabeled cobalamin or fluorescent is not necessary for the analysis. Additionally, binding characteristics can be observed, and parameters of binding can be estimated automatically by using a single surface for the evaluation of Cbl interacting proteins. Besides this, the quantity of protein-ligand complex is measurable at equilibrium conditions under the availability of unbound reactants while avoiding the disturbance of equilibrium conditions [127].

The SPR method has been implemented to quantify the vit B12 after observing vit B12 interaction with their binding proteins with a detection limit of one μg/ml [16]. The binding ability of the SPR method has been considered the promising approach to study the interaction of Cbl with transcobalamin (TCII). This technique is to be also useful for studying the interactions of IF (intrinsic factor) and HC with Cbl and cobalamin analogs once suitable regeneration conditions could be developed [16].

13.6.4 RAMAN SPECTROSCOPY (RS)

The optical detection methods in this chapter can mostly detect vit B12 in the solution and there are various issues for real samples. Selected detection techniques based on luminol include CL, which is highly susceptible to cross the limit of sensitivity due to interference by metal ions from sample. Besides this, the technique requires sample preparation extensively with complexity in a chemical reaction, which has limited its use for development in the real world.

The potential technique to overcome these demerits is the application of Raman Scattering Spectroscopy, which detects the identification of chemical bonds available in the sample [71]. The Raman scattering is the spectroscopic approach, in which the emitted light from the laser source is delivered to the sample by changing the wavelength, i.e., color with a fraction that responded to the change in vibration energy of the chemical bond of a biological sample [76]. The biological sample was retained with chemical bonds named 'Raman fingerprint,' which represented the available chemical bonds of the sample [38].

The optical spectrometer breaks down light sources into components of wavelengths, which are responded by the components of biological molecules [153]. As most the biological samples are available with similar components including nitrogen, carbon, oxygen, etc., and similarity with chemical bonds discriminates the detection by RS [70]. Vit B12 is present with cobalt ions, which are linked to the corrin ring making it a unique compound and is not available in every part of the human body showing a clear peak of vit B12 in the Raman spectrum [30, 94]. Therefore, RS is considered a promising technique to estimate vit B12 [6].

The vit B12 was estimated firstly by RS in 1973 [6, 178] and the spectrum of vitamin B12 was identified with its derivatives in solutions [110, 124, 152]. The experiment was performed repeatedly until 1989 and Raman studies were performed using an improved version [111, 112] of detectors and laser source. The RS studies were mostly performed to comprehend the structure of vit B12 including modes of vibration of cobalamin [112] and their co-enzymes [32]. The scope of this research is focused on understanding the structure of vit B12 mainly modeling and characterization of vibration modes of cobalamin [6, 118, 145] and understanding the changes (chemical and conformation) that occurred in the body mainly during binding with coenzymes for the process of metabolism [33, 59, 75].

The limited studies are available on the detection limit of vit B12 by RS, which can be a promising direction for research in the future. RS was performed with low intensity of signal compared to the fluorescence method with a detection limit of 250 ng/ml [152] but with up-gradation of techniques, such as, SERS [82, 108, 181] increased the intensity of signal with the interaction of signal of metallic substrates with cobalamin [183]. Therefore, RS for vit B12 measurement can be used for obtaining an improved yield of estimation. The application of SERS technique for the detection of vitamin C has shown similar results to that of HPLC after using the structured silver substrate for enhancing the Raman signal [137]; therefore, the SERS is considered a reliable method to detect vitamins [72]. The recent studies added the use of SERS-active-biotags to characterize the sample and it was concluded that the intensity of the signal is higher in SERS compared with fluorescence with additional features of unique identification in the sample [117], which is considered as innovative step to estimate the vit B12 concentration.

13.7 STABILITY OF VITAMIN B12

The vit B12 is unstable under ambient conditions of light and thermal treatment, respectively [125]. Firstly, the vit B12 was quantified at ambient light conditions. The freshly prepared standard solution was measured directly, at

24 h and 48 h time intervals. The results concluded that there was significant loss at LOQ level with 10.80% at 24 h and 21.33% at 48 h time intervals; and with increasing the vitamin concentration, the value was decreased inversely. At a higher concentration of 0.5 μg per ml, the vitamin remains under stable conditions for >2–3 days.

It is mentioned that a thermal treatment for 30 min was optimum at 100°C, to avoid excessive loss of vit B12 [106]. It is also concluded that the presence of carnosine, which is a natural antioxidant, could avoid the degradation of vit B12 with the presence of copper (Cu2+), and vitamin C [107].

Considering various foods, bread is considered a good substrate for fortification or for in situ production of vit B12. For the fortification potential of bread, it is required to study the stability of vit B12 during the baking process. Vit B12 is present with many functional groups, which improve the possibilities for occurrence of chemical reactions. The stability over bound vit B12 had been performed in various products including meat [43] milk-based and fish products [168]; and loss of 30–50% has been reported while considering cooking methods and time. The presence of vit B12 in solutions shows higher sensitivity to light, temperature, oxidizing, and reducing substances.

The sample containing other vitamins also affected the stability of vit B12 [83]. The CNCbl group of cobalamins has shown the highest stability. The AdoCbl and MeCbl showed the characteristics of photosensitivity and hence get converted to OHCbl easily in presence of UV radiation [63]. A pilot program in France was conducted regarding the enrichment of vit B12 in wheat flour [35] and reported an overall loss of 45% CNCbl, which was added to wheat flour during the baking process. The fortification of folic acid and CNCbl in bread samples using baking powder as the leavening agent had resulted in the loss of CNCbl by 19% [176]. A study reported that the prolonged heat treatment accelerated the destruction of OHCbl after adding in buffered solution (pH of 7) [114]. The other research studies had also reported on the degradation of bound vit B12 in food samples especially in milk, fish, and meat due to prolonged heating or higher cooking temperature [43, 114, 168]. The degradation of vitamin B12 was minimum at lower than 100°C and could not destroy it in milk samples [43] and prevented degradation in cooked fish in a vacuum-packed pouch [114].

In addition to heat, cobalamin is showing higher sensitivity to light and fluctuations in pH. The lowering of pH could be the reason for minimum detainment of CNCbl and OHCbl in the sourdough sample while comparing the straight dough and sponge sample [35]. The pH value was stable during the process of making the straight- and sponge-dough and was closed to a pH of 5.7. Additionally, vitamin B6 and thiamine had shown the effect of degradation of CNCbl in an aqueous solution [97]. The presence of a reactive compounds other than cobalamin were resulted, which may degrade the OHCbl with the formation of an oxidized product. Ascorbic acid acts as a reducing agent with the degradation of CNCbl firstly to OHCbl, which is further degraded to form oxidized products mainly at 5 pH [97]. It had been reported that OHCbl was destroyed faster in presence of ascorbic acid than the CNCbl [2]. Besides this, also mentioned on oxidation phenomena for the declining the presence of CNCbl and OHCbl mainly occurred by reducing $Co3+$ to $Co2+$ that resulted in the removal of cobalt from the molecules thereby breaking of corrin ring with forming oxidized products.

13.8 FORTIFICATION OF VITAMIN B12 UNDER IN-SITU CONDITIONS BY FERMENTATION APPROACH

Few microorganisms are involved in synthesis of vit B12 including Bacillus megaterium, Pseudomonas denitrificans, and Propioni bacterium freudenreichii [93]. The animal-based food is the major source of vit B12; and without the supplementary diet or fortification with fermentation, it is not available as a source in plants [168]. The deficiency of vit B12 mainly occurs due to inadequate daily intake or due to malabsorption problems in the body [47]. The consumption of vit B12 fortified plant-based products is the best approach to ensure sufficient intake for the people who do not consume animal-origin products.

The strains of Lactobacillus species are reported for the production of vit B12 [9, 151]. Few studies reported that the lactobacilli strain produced the pseudo vit B12 form [52, 122]. Pseudo vit B12 is different from active vit B12 with the availability of adenine at the place of 5,6-DMBI (5,6-dimethylbenzimidazole) considering a lower ligand that performed with complete inactivity in our body [147, 168]. The potential of Lactobacillus strains over the synthesis of 5,6-DMBI has not been reported to date. P. freudenreichii is recognized as safe for the synthesis of 5,6-DMBI [31] for producing an active form of vit B12 with sufficient amount and it is added in food-based products [121]. The production capacity of 37 ng per g of the active form of vit B12 is produced by P. freudenreichii barley-based matrix at a fermentation period of 7-days [23]. With the fermentation period of seven days using

Propionibacterium freudenreichii, the content of vit B12 had achieved at 33.0 ± 4.0 (durum flour), 87.0 ± 10.0 (wheat flour), and 155.0 ± 17.0 ng/g (wheat bran) [179]. The durum flour accelerates the cell growth rate of P. freudenreichii at higher densities with the efficient production of propionic acid compared to other two matrices. The wheat bran and whole-wheat flour are promising matrices for the production of vit B12 under in situ conditions.

13.9 SUMMARY

Among various methods, the application of immunoaffinity and UPLC methods are most efficient to determine vit B12 in a food sample. Whereas the MBA method was mostly inefficient with over prediction of vit B12. The emergence of optical spectroscopy methods (such as: CL, AFS, SPR, and RS methods) are improved selective approaches for estimation of vit B12. The synbiotic relations with Frankia alni concluded that there are chances of occurrence of vit B12 in some plant species. European grass varieties (called Elymus, Inula helenium, and Brassica nigra (black mustard)) were featured with the presence of sufficient amount of vit B12, which can be the potential source of plant-based vit B12 for vegetarians [106].

KEYWORDS

- **capillary electrophoresis**
- **chemiluminescence**
- **corrinoids**
- **extraction**
- **Frankia alni**
- **microbiological assay**
- **vitamin B12**

REFERENCES

1. Adjalla, C., Lambert, D., Benhayoun, S., Berthelsen, J., Nicolas, J., Guéant, J., & Nexo, E., (1994). Forms of cobalamin and vitamin B12 analogs in maternal plasma, milk, and cord plasma. The Journal of Nutritional Biochemistry, 5(8), 406–410.

2. Ahmad, I., Qadeer, K., Zahid, S., Sheraz, M. A., Ismail, T., Hussain, W., & Ansari, I. A., (2014). Effect of ascorbic acid on the degradation of cyanocobalamin and hydroxocobalamin in aqueous solution: A kinetic study. AAPS Pharm. Sci. Tech., 15(5), 1324–1333.
3. Akatsuka, K., & Atsuya, I., (1989). Determination of vitamin B12 as cobalt by electrothermal atomic absorption spectrometry using the solid sampling technique. Fresenius' Zeitschrift für analytische Chemie (Analytical and Bioanalytical Chemistry), 335(2), 200–204.
4. Akbay, N., & Gök, E., (2008). Determination of vitamin B12 using a chemiluminescence flow system. Journal of Analytical Chemistry, 63(11), 1073–1077.
5. Albala, S., Veciana, M. T., & Marine, A., (1997). Determination of water-soluble vitamins in infant milk by high-performance liquid chromatography (HPLC). Journal of Chromatography A, 778(1, 2), 247–253.
6. Andruniow, T., Zgierski, M. Z., & Kozlowski, P. M., (2002). Vibrational analysis of methyl cobalamin. The Journal of Physical Chemistry A, 106(7), 1365–1373.
7. Baker, S., & Miller, N., (2000). Determination of cobalamins using capillary electrophoresis inductively coupled plasma mass spectrometry. Spectrochimica Acta Part B: Atomic Spectroscopy, 55(12), 1823–1832.
8. Ball, G. F., (2013). Bioavailability and Analysis of Vitamins in Foods (p. 569). Cham: Springer International.
9. Bhushan, B., Tomar, S., & Chauhan, A., (2017). Techno-functional differentiation of two vitamin B12 producing Lactobacillus plantarum strains: An elucidation for diverse future use. Applied Microbiology and Biotechnology, 101(2), 697–709.
10. Bigio, I. J., & Mourant, J. R., (1997). Ultraviolet and visible spectroscopies for tissue diagnostics: Fluorescence spectroscopy and elastic-scattering spectroscopy. Physics in Medicine & Biology, 42(5), 803–810.
11. Blankenship, J., VanStipdonk, M., & Schweikert, E., (1997). Matrix effects on the fragmentation of vitamin B12 in plasma desorption mass spectrometry. Rapid Communications in Mass Spectrometry, 11(1), 143–147.
12. Brochon, J. C., & Valeur, B., (2001). New Trends in Fluorescence Spectroscopy: Applications to Chemical and Life Sciences (Vol. 1, p. 515). Cham: Springer International.
13. Burgess, C. M., Smid, E. J., & vanSinderen, D., (2009). Bacterial vitamin B2, B11 and B12 overproduction: An overview. International Journal of Food Microbiology, 133(1, 2), 1–7.
14. Cai, T., Gu, H., Yuan, X., & Liu, F., (2011). Normal Raman and SERS spectroscopy of the vitamin E. Journal of Physics: Conference Series 15, 277(1), E-article ID 012010.
15. Campos-Gimnez, E., Fontannaz, P., Trisconi, M. J., Kilinc, T., Gimenez, C., & Andrieux, P., (2008). Determination of vitamin B12 in food products by liquid chromatography/UV detection with immunoaffinity extraction: Single-laboratory validation. Journal of AOAC International, 91(4), 786–793.
16. Cannon, M. J., Myszka, D. G., Bagnato, J. D., Alpers, D. H., West, F. G., & Grissom, C. B., (2002). Equilibrium and kinetic analyses of the interactions between vitamin B12 binding proteins and cobalamins by surface plasmon resonance. Analytical Biochemistry, 305(1), 1–9.
17. Capps, B. F., Hobbs, N. L., & Fox, S. H., (1949). A method for the microbiological assay of vitamin B12. The Journal of Biological Chemistry, 178(1), 517–518.

18. Cardarelli, J. A., Slingerland, D. W., Burrows, B. A., & Miller, A., (1985). Measurement of total-body cobalt57 vitamin B12 absorption with gamma camera. Journal of Nuclear Medicine, 26(8), 941–943.
19. Carkeet, C., Dueker, S. R., Lango, J., Buchholz, B. A., Miller, J. W., Green, R., Hammock, B. D., et al., (2006). Human vitamin B12 absorption measurement by accelerator mass spectrometry using specifically labeled 14Ccobalamin. Proceedings of the National Academy of Sciences, 103(15), 5694–5699.
20. Carmel, R., (1996). Prevalence of undiagnosed pernicious anemia in the elderly. Archives of Internal Medicine, 156(10), 1097–1100.
21. Casey, P. J., Speckman, K. R., Ebert, F. J., & Hobbs, W. E., (1982). Radioisotope dilution technique for determination of vitamin B12 in foods. Journal of the Association of Official Analytical Chemists (AOAC), 65(1), 85–88.
22. Chaia, E. E., Wall, L. G., & Huss, K., (2010). Life in soil by the actinorhizal root nodule endophyte Frankia: A review. Symbiosis, 51(3), 201–226.
23. Chamlagain, B., Edelmann, M., Kariluoto, S., Ollilainen, V., & Piironen, V., (2015). Ultra-high-performance liquid chromatographic and mass spectrometric analysis of active vitamin B12 in cells of Propioni bacterium and fermented cereal matrices. Food Chemistry, 166, 630–638.
24. Chassaigne, H., & Lobiński, R., (1998). Determination of cobalamins and cobamamides by microbore reversed-phase HPLC with spectrophotometric, ion-spray ionization MS and inductively coupled plasma MS detection. Analytica Chimica Acta, 359(3), 227–235.
25. Chatzimichalakis, P. F., Samanidou, V. F., Verpoorte, R., & Papadoyannis, I. N., (2004). Development of a validated HPLC method for the determination of B-complex vitamins in pharmaceuticals and biological fluids after solid phase extraction. Journal of Separation Science, 27(14), 1181–1188.
26. Chen, J., Li, B. Q., Cui, Y. Q., Yu, E., & Zhai, H. L., (2015). Fast and effective method of quantitative analysis of B1, B2 and B6 in B-vitamins tablets based on three-dimensional fluorescence spectra. Journal of Food Composition and Analysis, 41, 122–128.
27. Chen, P., Wolf, W. R., Castanheira, I., & Sanchez, A., (2010). LC/UV/ Vis method for determination of cyanocobalamin (B12) in multivitamin dietary supplements with on-line sample clean-up. Analytical Methods, 2(8), 1171–1175.
28. Committee on Analytical Methods, (1956). The estimation of vitamin B12. Analyst, 81(960), 132–136.
29. Cooper, B. A., (1959). Studies with a more rapid method of vitamin B12 assay utilizing Euglena gracilis. Journal of Clinical Pathology, 12(2), 153–157.
30. Day, P., (1967). Theory of the optical properties of vitamin B12 and its derivatives. Theoretica Chimica Acta, 7(4), 328–341.
31. Deptula, P., Kylli, P., Chamlagain, B., Holm, L., Kostiainen, R., Piironen, V., Savijoki, K., & Varmanen, P., (2015). BluB/ CobT2 fusion enzyme activity reveals mechanisms responsible for production of active form of vitamin B12 by Propioni bacterium freudenreichii. Microbial. Cell Factories, 14(1), 1–12.
32. Dong, S., Padmakumar, R., Banerjee, R., & Spiro, T. G., (1999). Co-C bond activation in B12-dependent enzymes: Cryogenic resonance Raman studies of methylmalonyl-coenzyme A mutase. Journal of the American Chemical Society, 121(30), 7063–7070.
33. Dong, S., Padamkumar, R., Maiti, N., Banerjee, R., & Spiro, T. G., (1998). Resonance spectra show that coenzyme B12 binding to methylmalonyl-coenzyme A mutase

changes the corrin ring conformation but leaves the Co-C bond essentially unaffected. Journal of the American Chemical Society, 120(38), 9947–9948.
34. Doscherholmen, A., Hagen, P. S., & Liu, M., (1957). Radioactive vitamin B12 absorption studies: Results of direct measurement of radioactivity in the blood. Blood, 12(4), 336–346.
35. Edelmann, M., Chamlagain, B., Santin, M., Kariluoto, S., & Piironen, V., (2016). Stability of added and in situ-produced vitamin B12 in breadmaking. Food Chemistry, 204, 21–28.
36. Engvall, E., & Perlmann, P., (1971). Enzyme-linked immunosorbent assay (ELISA) quantitative assay of immunoglobulin-G. Immunochemistry, 8(9), 871–874.
37. Ewing, A. G., Wallingford, R. A., & Olefirowicz, T. M., (1989). Capillary electrophoresis. Analytical Chemistry, 61(4), 292A–303A.
38. Ferraro, J. R., Nakamoto, K., & Brown, C., (2003). Introductory Raman Spectroscopy (p. 434). San Diego – CA: Academic Press.
39. Ford, J., (1952). The microbiological assay of vitamin B12. British Journal of Nutrition, 6(1), 324–330.
40. Frater-Schroder, M., Kierat, L., Andres, R. Y., & Römer, J., (1982). Solid-phase immunoassay for the vitamin B12-binding protein transcobalamin II in human serum. Analytical Biochemistry, 124(1), 92–101.
41. Frenkel, E. P., Prough, R., & Kitchens, R. L., (1980). Measurement of tissue vitamin B12 by radio isotopic competitive inhibition assay and quantitation of tissue cobalamin fractions. Methods in Enzymology, 67, 31–40.
42. Gholami, J., Manteghian, M., Badiei, A., Javanbakht, M., & Ueda, H., (2015). Label-free detection of vitamin B12 based on fluorescence quenching of graphene oxide nanolayer. Fullerenes, Nanotubes and Carbon Nanostructures, 23(10), 878–884.
43. Gille, D., & Schmid, A., (2015). Vitamin B12 in meat and dairy products. Nutrition Reviews, 73(2), 106–115.
44. Glass, G. B. J., Boyd, L. J., & Gellin, G. A., (1955). Surface scintillation measurements in humans of the uptake of parenterally administered radioactive vitamin B12. Blood, 10(2), 95–114.
45. Glass, G. B. J., Boyd, L. J., Gellin, G. A., & Stephanson, L., (1954). Uptake of radioactive vitamin B12 by the liver in humans: Test for measurement of intestinal absorption of vitamin B12 and intrinsic factor activity. Archives of Biochemistry and Biophysics, 51(1), 251–257.
46. Goldberg, S. R., Trivedi, B. K., & Oliner, L., (1957). Radioactive vitamin B12 studies: Experience with the urinary excretion test and the measurement of absorbed plasma radioactivity. The Journal of Laboratory and Clinical Medicine, 49(4), 583–589.
47. Green, R., Allen, L. H., Bjørke-Monsen, A. L., Brito, A., Guéant, J. L., Miller, J. W., Molloy, A. M., et al., (2017). Vitamin B12 deficiency. Nature Reviews Disease Primers, 3(1), 1–20.
48. Greibe, E., & Nexo, E., (2011). Vitamin B12 absorption judged by measurement of holotrans cobalamin, active vitamin B12: Evaluation of a commercially available EIA kit. Clinical Chemistry and Laboratory Medicine, 49(11), 1883–1885.
49. Guggisberg, D., Risse, M., & Hadorn, R., (2012). Determination of vitamin B12 in meat products by RP-HPLC after enrichment and purification on an immunoaffinity column. Meat Science, 90(2), 279–283.

50. Hampel, D., Shahab-Ferdows, S., Domek, J. M., Siddiqua, T., Raqib, R., & Allen, L. H., (2014). Competitive chemiluminescent enzyme immunoassay for vitamin B12 analysis in human milk. Food Chemistry, 153, 60–65.
51. Hancewicz, T. M., & Petty, C., (1995). Quantitative analysis of vitamin A using Fourier transform Raman spectroscopy. Spectrochimica Acta Part A: Molecular and Biomolecular Spectroscopy, 51(12), 2193–2198.
52. Hazra, A. B., Tran, J. L., Crofts, T. S., & Taga, M. E., (2013). Analysis of substrate specificity in Cob-T homologs reveals widespread preference for DMB, the lower axial ligand of vitamin B12. Chemistry & Biology, 20(10), 1275–1285.
53. Herbert, V., (1988). Do not ignore low serum cobalamin (vitamin B12) levels. Archives of Internal Medicine, 148(8), 1705–1707.
54. Heudi, O., Kilinç, T., & Fontannaz, P., (2005). Separation of water-soluble vitamins by reversed-phase high performance liquid chromatography with ultra-violet detection: Application to poly-vitamin premixes. Journal of Chromatography A, 1070(1, 2), 49–56.
55. Heudi, O., Kilinc, T., Fontannaz, P., & Marley, E., (2006). Determination of vitamin B12 in food products and in premixes by reversed-phase high performance liquid chromatography and immunoaffinity extraction. Journal of Chromatography A, 1101(1, 2), 63–68.
56. Ho, C., Kauwell, G. P., & Bailey, L. B., (1999). Practitioners' guide to meeting the vitamin B12 recommended dietary allowance for people aged 51 years and older. Journal of the American Dietetic Association, 99(6), 725–727.
57. Hoffmann, C., Stokstad, E., Hutchings, B., Dornbush, A., & Jukes, T. H., (1949). The microbiological assay of vitamin B12 with Lactobacillus leichmannii. Journal of Biological Chemistry, 181(2), 635–644.
58. Hugenschmidt, S., Schwenninger, S. M., Gnehm, N., & Lacroix, C., (2010). Screening of a natural biodiversity of lactic and propionic acid bacteria for folate and vitamin B12 production in supplemented whey permeate. International Dairy Journal, 20(12), 852–857.
59. Huhta, M. S., Chen, H. P., Hemann, C., Hille, C. R., & Marsh, E. N. G., (2001). Protein–coenzyme interactions in adenosyl cobalamin dependent glutamate mutase. Biochemical Journal, 355(1), 131–137.
60. Indyk, H. E., Persson, B. S., Caselunghe, M. C., Moberg, A., Filonzi, E. L., & Woollard, D. C., (2002). Determination of vitamin B12 in milk products and selected foods by optical biosensor protein-binding assay method. Journal of AOAC International, 85(1), 72–81.
61. Ivanovska, M., Stafilov, T., & Stojanoski, K., (1996). Determination of vitamin B12 in pharmaceutical preparations by electrothermal atomic absorption spectrometry. Acta Pharmaceutica (Zagreb), 46(1), 75–81.
62. Iwase, H., & Ono, I., (1997). Determination of cyano cobalamin in foods by high-performance liquid chromatography with visible detection after solid-phase extraction and membrane filtration for the precolumn separation of lipophilic species. Journal of Chromatography A, 771(1, 2), 127–134.
63. Juzeniene, A., & Nizauskaite, Z., (2013). Photodegradation of cobalamins in aqueous solutions and in human blood. Journal of Photochemistry and Photobiology B: Biology, 122, 7–14.
64. Kals, J., (2019). Effect of vitamin B12 on the alleviation of nutritional anemia in common sole (Solea solea). Aquaculture Nutrition, 25(1), E-article ID 12871.

65. Karger, B. L., Cohen, A. S., & Guttman, A., (1989). High-performance capillary electrophoresis in the biological sciences. Journal of Chromatography, 492, 585–614.
66. Karmi, O., Zayed, A., Baraghethi, S., Qadi, M., & Ghanem, R., (2011). Measurement of vitamin B12 concentration: A review on available methods. IIOAB J, 2(2), 23–32.
67. Karşilayan, H., (1996). Quantitation of vitamin B12 by first-derivative absorption spectroscopy. Spectrochimica Acta Part A: Molecular and Biomolecular Spectroscopy, 52(9), 1163–1168.
68. Karube, I., Wang, Y., Tamiya, E., & Kawarai, M., (1987). Microbial electrode sensor for vitamin B12. Analytica Chimica Acta, 199, 93–97.
69. Klejdus, B., Petrlová, J., Potěšil, D., Adam, V., Mikelová, R., Vacek, J., Kizek, R., & Kubáň, V., (2004). Simultaneous determination of water-and fat-soluble vitamins in pharmaceutical preparations by high-performance liquid chromatography coupled with diode array detection. Analytica Chimica Acta, 520(1, 2), 57–67.
70. Kneipp, K., Kneipp, H., Itzkan, I., Dasari, R. R., & Feld, M. S., (1999). Ultrasensitive chemical analysis by Raman spectroscopy. Chemical Reviews, 99(10), 2957–2976.
71. Kneipp, K., Kneipp, H., Itzkan, I., Dasari, R. R., & Feld, M. S., (2002). Surface-enhanced Raman scattering and biophysics. Journal of Physics: Condensed Matter, 14(18), R597–R602.
72. Kokaislová, A., & Matějka, P., (2012). Surface-enhanced vibrational spectroscopy of B-vitamins: What is the effect of SERS-active metals used? Analytical and Bioanalytical Chemistry, 403(4), 985–993.
73. Kondo, H., Binder, M. J., Kolhouse, J. F., Smythe, W. R., Podell, E. R., & Allen, R. H., (1982). Presence and formation of cobalamin analogues in multivitamin-mineral pills. The Journal of Clinical Investigation, 70(4), 889–898.
74. Kozhanova, L., Fedorova, G., & Baram, G., (2002). Determination of water-and fat-soluble vitamins in multivitamin preparations by high-performance liquid chromatography. Journal of Analytical Chemistry, 57(1), 40–45.
75. Kozlowski, P. M., Andruniow, T., Jarzecki, A. A., Zgierski, M. Z., & Spiro, T. G., (2006). DFT analysis of Co−alkyl and Co−adenosyl vibrational modes in B12 cofactors. Inorganic Chemistry, 45(14), 5585–5590.
76. Krafft, C., & Sergo, V., (2006). Biomedical applications of Raman and infrared spectroscopy to diagnose tissues. Spectroscopy, 20(5, 6), 195–218.
77. Kristensen, H., & Hald, T., (1962). Measurement of plasma radioactivity following oral administration of 57Co-labelled B12. A simple diagnostic test. Danish Medical Bulletin, 9, 167–170.
78. Kumar, S. S., Chouhan, R. S., & Thakur, M. S., (2009). Enhancement of chemiluminescence for vitamin B12 analysis. Analytical Biochemistry, 388(2), 312–316.
79. Kysil, O. A., (2014). Grundlegende Untersuchungen zur mikrobiellen Synthese von Vitamin B12 in symbiontischen System der Frankia: Entwicklung neuartiger pflanzlicher Extracte mit hohem essentiellen Vitamin B12 – Gehalt (Basic Studies on the Microbial Synthesis of Vitamin B12 in the Symbiotic System of Frankia: Development of Novel Plant Extracts with a High Essential Vitamin B12 Content). (p. 140). Doctoral Thesis, University of Hanover, Hanover, Leibniz.
80. Lambert, D., Adjalla, C., Felden, F., Benhayoun, S., Nicolas, J., & Gueant, J., (1992). Identification of vitamin B12 and analogues by high-performance capillary

electrophoresis and comparison with high-performance liquid chromatography. Journal of Chromatography A, 608(1, 2), 311–315.
81. Lee, B., (2003). Review of present status of optical fiber sensors. Optical Fiber Technology, 9(2), 57–79.
82. LeRu, E., & Etchegoin, P., (2009). Principles of Surface Enhanced Raman Spectroscopy (pp. 663). London: Oxford University Press.
83. Lešková, E., Kubíková, J., Kováčiková, E., Košická, M., Porubská, J., & Holčíková, K., (2006). Vitamin losses: Retention during heat treatment and continual changes expressed by mathematical models. Journal of Food Composition and Analysis, 19(4), 252–276.
84. Li, H. B., & Chen, F., (2000). Determination of vitamin B12 in pharmaceutical preparations by a highly sensitive fluorometric method. Fresenius' Journal of Analytical Chemistry, 368(8), 836–838.
85. Li, H. B., & Chen, F., (2001). Simultaneous determination of twelve water-and fat-soluble vitamins by high-performance liquid chromatography with diode array detection. Chromato Graphia, 54(3), 270–273.
86. Li, H. B., Chen, F., & Jiang, Y., (2000). Determination of vitamin B12 in multivitamin tablets and fermentation medium by high-performance liquid chromatography with fluorescence detection. Journal of Chromatography A, 891(2), 243–247.
87. Lichtenstein, H., Beloian, A., & Reynolds, H., (1959). Vitamin B12 in foodstuffs, comparative vitamin B12 assay of foods of animal origin by Lactobacillus leichmannii and Ochromonas malhamensis. Journal of Agricultural and Food Chemistry, 7(11), 771–774.
88. Liu, B., Gao, J., & Yang, G., (2005). Determination of vitamin B12 concentration by fluorescence quenching with acridine orange-rhodamine 6G energy transfer system. Guang Pu Xue Yu Guang Pu Fen xi Guang Pu (GU Case GPU Snowy GU Case GPU Analysis: GU Case GPU), 25(7), 1080–1082.
89. Luo, X., Chen, B., Ding, L., Tang, F., & Yao, S., (2006). HPLC-ESI-MS analysis of vitamin B12 in food products and in multivitamins-multimineral tablets. Analytica Chimica Acta, 562(2), 185–189.
90. Marazuela, M., & Moreno-Bondi, M., (2002). Fiber-optic biosensors: An overview. Analytical and Bioanalytical Chemistry, 372(5), 664–682.
91. Mark, D., Haeberle, S., Roth, G., Stetten, F. V., & Zengerle, R., (2010). Microfluidic lab-on-a-chip platforms: Requirements, characteristics, and applications. Chemical Society Reviews, 39(3), 1153–1182.
92. Marley, E. C., Mackay, E., & Young, G., (2009). Characterization of vitamin B12 immunoaffinity columns and method development for determination of vitamin B12 in a range of foods, juices and pharmaceutical products using immunoaffinity clean-up and high-performance liquid chromatography (HPLC) with UV detection. Food Additives & Contaminants: Part A, 26(3), 282–288.
93. Martens, J. H., Barg, H., Warren, M. A., & Jahn, D., (2002). Microbial production of vitamin B12. Applied Microbiology and Biotechnology, 58(3), 275–285.
94. Mayer, E., Gardiner, D., & Hester, R., (1973). Resonance Raman spectra of vitamin B12 and di-cyano-cobalamin. Biochimica et Biophysica Acta (BBA): General Subjects, 297(2), 568–570.

95. Medek, A., & Frydman, L., (2000). A multinuclear solid-state NMR analysis of vitamin B12 in its different polymorphic forms. Journal of the American Chemical Society, 122(4), 684–691.
96. Miyamoto, E., Yabuta, Y., Kwak, C. S., Enomoto, T., & Watanabe, F., (2009). Characterization of vitamin B12 compounds from Korean purple laver (Porphyra sp.) products. Journal of Agricultural and Food Chemistry, 57(7), 2793–2796.
97. Monajjemzadeh, F., Ebrahimi, F., Zakeri-Milani, P., & Valizadeh, H., (2014). Effects of formulation variables and storage conditions on light protected vitamin B12 mixed parenteral formulations. Advanced Pharmaceutical Bulletin, 4(4), 329–332.
98. Moore, S. J., & Warren, M. J., (2012). The anaerobic biosynthesis of vitamin B12. Biochemical Society Transactions, 40(3), 581–586.
99. Morelli, B., (1995). Determination of a quaternary mixture of vitamins B6, B1, and B12 and uridine 5'-triphosphate by derivative spectrophotometry. Journal of Pharmaceutical Sciences, 84(1), 34–37.
100. Morelli, B., (1996). High-resolution/higher-order derivative spectrophotometry for determination of ternary mixtures of B-complex vitamins in pharmaceuticals. Fresenius' Journal of Analytical Chemistry, 354(1), 97–102.
101. Moreno, P., & Salvado, V., (2000). Determination of eight water- and fat-soluble vitamins in multi-vitamin pharmaceutical formulations by high-performance liquid chromatography. Journal of Chromatography A, 870(1, 2), 207–215.
102. Morita, M., Uehiro, T., & Fuwa, K., (1980). Speciation and elemental analysis of mixtures by high performance liquid chromatography with inductively coupled argon plasma emission spectrometric detection. Analytical Chemistry, 52(2), 349–351.
103. Movasaghi, Z., Rehman, S., & Rehman, I. U., (2007). Raman spectroscopy of biological tissues. Applied Spectroscopy Reviews, 42(5), 493–541.
104. Muhammad, K., Briggs, D., & Jones, G., (1993). Comparison of a competitive binding assay with Lactobacillus leichmannii ATCC 7830 assay for the determination of vitamin B12 in foods. Food Chemistry, 48(4), 431–434.
105. Myszka, D. G., & Rich, R. L., (2000). Implementing surface plasmon resonance biosensors in drug discovery. Pharmaceutical Science & Technology Today, 3(9), 310–317.
106. Nakos, M., Pepelanova, I., Beutel, S., Krings, U., Berger, R. G., & Scheper, T., (2017). Isolation and analysis of vitamin B12 from plant samples. Food Chemistry, 216, 301–308.
107. Naumann, D., (2001). FT- Infrared and FT- Raman spectroscopy in biomedical research. Applied Spectroscopy Reviews, 36(2, 3), 239–298.
108. Negri, P., & Dluhy, R. A., (2013). Ag nanorod based surface-enhanced Raman spectroscopy applied to bioanalytical sensing. Journal of Biophotonics, 6(1), 20–35.
109. Nelp, W., McAfee, J., & Wagner, Jr. H., (1963). Single measurement of plasma radioactive vitamin B12 as a test for pernicious anemia. Journal of Laboratory and Clinical Medicine, 61, 158–165.
110. Nestor, J., Spiro, T. G., & Klauminzer, G., (1976). Coherent anti-stokes Raman scattering (CARS) spectra, with resonance enhancement, of cytochrome-c and vitamin B12 in dilute aqueous solution. Proceedings of the National Academy of Sciences, 73(10), 3329–3332.
111. Nie, S., Marzilli, L. G., & Yu, N. T., (1989). Near-infrared Fourier transform Raman spectroscopy of photolabile organocobalt B12 and model compounds, Part 1: Detection

of the cobalt-carbon stretching mode in the solid state and in solution. Journal of the American Chemical Society, 111(26), 9256–9258.
112. Nie, S., Marzilli, P. A., Marzilli, L. G., & Yu, N. T., (1990). Near-IR Fourier transform Raman spectroscopy of photolabile organocobalt B12 and model compounds. identification of the Co–C bond stretch in cobalamins. Journal of the Chemical Society, Chemical Communications, (10), 770–771.
113. Nie, S., & Zare, R. N., (1997). Optical detection of single molecules. Annual Review of Biophysics and Biomolecular Structure, 26(1), 567–596.
114. Nishioka, M., Kanosue, F., Yabuta, Y., & Watanabe, F., (2011). Loss of vitamin B12 in fish (round herring) meats during various cooking treatments. Journal of Nutritional Science and Vitaminology, 57(6), 432–436.
115. Okba-Michael, M., & Sañudo, S. A., (2004). New method for the determination of vitamin B12 in seawater. Analytica Chimica Acta, 517(1, 2), 33–38.
116. Pakin, C., Bergaentzlé, M., Aoudé-Werner, D., & Hasselmann, C., (2005). α-Ribazole, a fluorescent marker, for the liquid chromatographic determination of vitamin B12 in foodstuffs. Journal of Chromatography A, 1081(2), 182–189.
117. Pallaoro, A., Braun, G. B., & Moskovits, M., (2015). Biotags based on surface-enhanced Raman can be as bright as fluorescence tags. Nano Letters, 15(10), 6745–6750.
118. Park, K., & Brunold, T. C., (2013). Combined spectroscopic and computational analysis of the vibrational properties of vitamin B12 in its Co3+, Co2+, and Co1+ oxidation states. The Journal of Physical Chemistry B, 117(18), 5397–5410.
119. Pawlak, R., Parrott, S. J., Raj, S., Cullum-Dugan, D., & Lucus, D., (2013). How prevalent is vitamin B12 deficiency among vegetarians? Nutrition Reviews, 71(2), 110–117.
120. Piveteau, P., (1999). Metabolism of lactate and sugars by dairy propionibacteria: A review. Le Lait, 79(1), 23–41.
121. Piwowarek, K., Lipińska, E., Hać-Szymańczuk, E., Bzducha-Wróbel, A., & Synowiec, A., (2018). Research on the ability of propionic acid and vitamin B12 biosynthesis by Propioni bacterium freudenreichii strain T82. Antonie Van Leeuwenhoek (Anthony Van Leeuwenhoek), 111(6), 921–932.
122. Piwowarek, K., Lipińska, E., Hać-Szymańczuk, E., Kieliszek, M., & Ścibisz, I., (2018). Propioni bacterium spp.: Source of propionic acid, vitamin B12, and other metabolites important for the industry. Applied Microbiology and Biotechnology, 102(2), 515–538.
123. Qin, W., Zhang, Z., & Liu, H., (1997). Chemiluminescence flow sensor for the determination of vitamin B12. Analytica Chimica Acta, 357(1, 2), 127–132.
124. Rajoria, D., & Nath, A., (1977). IR spectroscopic studies of some cobalamins. Journal of Inorganic and Nuclear Chemistry, 39(7), 1291–1294.
125. Raju, C. S. K., Lee, L. Y., Schiel, J. E., & Long, S. E., (2013). A simple and sensitive LC-ICP-MS method for the accurate determination of vitamin B12 in fortified breakfast cereals and multivitamin tablets. Journal of Analytical Atomic Spectrometry, 28(6), 901–907.
126. Reizenstein, P., Cronkite, E. P., & Cohn, S., (1961). Measurement of absorption of vitamin B12 by whole-body gamma spectrometry. Blood, 18(1), 95–101.
127. Rich, R. L., & Myszka, D. G., (2000). Survey of the 1999 surface plasmon resonance biosensor literature. Journal of Molecular Recognition, 13(6), 388–407.
128. Sahni, M. K., Spanos, S., Wahrman, M. Z., & Sharma, G. M., (2001). Marine corrinoid-binding proteins for the direct determination of vitamin B12 by radio-assay. Analytical Biochemistry, 289(1), 68–76.

129. Santoyo-Sánchez, A., Aponte-Castillo, J., Parra-Peña, R., & Ramos-Peña, C., (2015). Dietary recommendations in patients with deficiency anemia. Revista Médica Del Hospital General De México (Medical J. of General Hospital of Mexico), 78(3), 144–150.
130. Sato, K., Muramatsu, K., & Amano, S., (2002). Application of vitamin B12 targeting site on Lactobacillus helveticus B-1 to vitamin B12 assay by chemiluminescence method. Analytical Biochemistry, 308(1), 1–4.
131. Schartner, E. P., Tsiminis, G., François, A., Kostecki, R., Warren, S. C., Nguyen, L. V., Heng, S., et al., (2015). Taming the light in micro structured optical fibers for sensing. International Journal of Applied Glass Science, 6(3), 229–239.
132. Schmidt, A., Call, L. M., Macheiner, L., & Mayer, H. K., (2019). Determination of vitamin B12 in four edible insect species by immunoaffinity and ultra-high performance liquid chromatography. Food Chemistry, 281, 124–129.
133. Schneider, Z., (1987). Non-enzymatic vitamin B12 binding proteins in man and animals. In: Schneider, Z., & Stroinski, A., (eds.), Comprehensive Vitamin B12 (pp. 259–266). Berlin – Germany: De Gruyter.
134. Schneider, Z., & Stroinski, A., (2011). Comprehensive Vitamin B12 (p. 421). Berlin – Germany: De Gruyter.
135. Selvakumar, L. S., & Thakur, M. S., (2012). Dipstick – based immunochemiluminescence biosensor for the analysis of vitamin B12 in energy drinks – novel approach. Analytica Chimica Acta, 722, 107–113.
136. Shang, Z. B., Wen, Y. J., Yan, X. Q., Sun, H. H., Wang, Y., & Jin, W. J., (2014). Synthesis of a novel fluorescent probe based on 7-nitrobenzo-2-oxa-1, 3-diazole skeleton for the rapid determination of vitamin B12 in pharmaceuticals. Luminescence, 29(6), 598–602.
137. Shuhai, J., Jun, W., & Yang, J., (2013). Ordered silver nanoparticle arrays as surface-enhanced Raman spectroscopy substrates for label-free detection of vitamin – C in serum. Sensors and Actuators A: Physical, 201, 416–420.
138. Skeggs, H. R., Nepple, H. M., Valentik, K. A., Huff, J. W., & Wright, L. D., (1950). Observations on the use of Lactobacillus leichmannii in the microbiological assay of vitamin B12. Journal of Biological Chemistry, 184(1), 211–221.
139. Smith, T., & Hesp, R., (1979). Measurement of 57Co-labelled vitamin B12 using a liquid-scintillator whole-body counter. The British Journal of Radiology, 52(622), 832–835.
140. Song, Z., & Hou, S., (2003). Sub-picogram determination of vitamin B12 in pharmaceuticals and human serum using flow injection with chemiluminescence detection. Analytica Chimica Acta, 488(1), 71–79.
141. Song, Z., Hou, S., & Zhang, N., (2002). New green analytical procedure for monitoring sub-nanogram amounts of chlorpyrifos on fruits using flow injection chemiluminescence with immobilized reagents. Journal of Agricultural and Food Chemistry, 50(16), 4468–4474.
142. Song, Z., & Wang, L., (2001). Reagent-less chemiluminescence flow sensor for the determination of riboflavin in pharmaceutical preparations and human urine. Analyst, 126(8), 1393–1398.
143. Souci, S. W., Fachmann, W., & Kraut, H., (2000). Food Composition and Nutrition Tables (7th edn., p. 1300). Berlin – Germany: MedPharm GmbH Scientific Publishers.

144. Stenberg, E., Persson, B., Roos, H., & Urbaniczky, C., (1991). Quantitative determination of surface concentration of protein with surface plasmon resonance using radiolabeled proteins. Journal of Colloid and Interface Science, 143(2), 513–526.
145. Stich, T. A., Brooks, A. J., Buan, N. R., & Brunold, T. C., (2003). Spectroscopic and computational studies of Co3+- corrinoids: Spectral and electronic properties of the B12 cofactors and biologically relevant precursors. Journal of the American Chemical Society, 125(19), 5897–5914.
146. Strobbia, P., Languirand, E. R., & Cullum, B. M., (2015). Recent advances in plasmonic nanostructures for sensing: A review. Optical Engineering, 54(10), E-article ID 100902.
147. Stupperich, E., & Nexø, E., (1991). Effect of the cobalt-N coordination on the cobamide recognition by the human vitamin B12 binding proteins intrinsic factor, transcobalamin and haptocorrin. European Journal of Biochemistry, 199(2), 299–303.
148. Swartz, M. E., (2005). UPLC™: An introduction and review. Journal of Liquid Chromatography & Related Technologies, 28(7, 8), 1253–1263.
149. Szterk, A., Roszko, M., Małek, K., Czerwonka, M., & Waszkiewicz, B., (2012). Application of the SPE reversed phase HPLC/MS technique to determine vitamin B12 bio-active forms in beef. Meat Science, 91(4), 408–413.
150. Tait, C. E., & Hesp, R., (1976). Measurement of 57Co-vitamin B12 uptake using a static whole-body counter. The British Journal of Radiology, 49(587), 948–950.
151. Taranto, M. P., Vera, J. L., Hugenholtz, J., De Valdez, G. F., & Sesma, F., (2003). Lactobacillus reuteri CRL1098 produces cobalamin. Journal of Bacteriology, 185(18), 5643–5647.
152. Tsai, C. W., & Morris, M. D., (1975). Application of resonance Raman spectrometry for the determination of vitamin B12. Analytica Chimica Acta, 76(1), 193–198.
153. Tsiminis, G., Chu, F., Warren-Smith, S. C., Spooner, N. A., & Monro, T. M., (2013). Identification and quantification of explosives in nanoliter solution volumes by Raman spectroscopy in suspended core optical fibers. Sensors, 13(10), 13163–13177.
154. Uzunbajakava, N., Lenferink, A., Kraan, Y., Volokhina, E., Vrensen, G., Greve, J., & Otto, C., (2003). Non-resonant confocal Raman imaging of DNA and protein distribution in apoptotic cells. Biophysical Journal, 84(6), 3968–3981.
155. Vaishnavi, E., & Renganathan, R., (2013). Cd-Te quantum dot as a fluorescence probe for vitamin B12 in dosage form. Spectrochimica Acta Part A: Molecular and Biomolecular Spectroscopy, 115(C), 603–609.
156. Van-Wyk, J., & Britz, T. J., (2010). Rapid HPLC method for the extraction and quantification of vitamin B12 in dairy products and cultures of Propionibacterium freudenreichii. Dairy Science & Technology, 90(5), 509–520.
157. Vanexan, R. J., (1979). Localization of vitamin – A by autofluorescence during induced metaplastic changes in cultures of skin. In Vitro, 15(8), 631–640.
158. Varmanen, P. K., Deptula, P., Chamlagain, B. S., & Piironen, V. I., (2016). Letter to the editor on 'Enhancing vitamin B12 content in soy-yogurt by Lactobacillus reuteri. IJFM (International Journal of Food Microbiology), 206, 56–59.
159. Velasco-Garcia, M., (2009). Optical biosensors for probing at the cellular level: A review of recent progress and prospects. Seminars in Cell & Developmental Biology, 20(1), 27–33.
160. Vinas, P., Campillo, N., Garcia, I. L., & Córdoba, M. H., (1996). Speciation of vitamin B12 analogues by liquid chromatography with flame atomic absorption spectrometric detection. Analytica Chimica Acta, 318(3), 319–325.

161. Vinas, P., Campillo, N., Lopez-Garcia, I., & Hernandez-Cordoba, M., (1996). Identification of vitamin B12 analogues by liquid chromatography with electrothermal atomic absorption detection. Chromato Graphia, 42(9), 566–570.
162. Viñas, P., López-Erroz, C., Balsalobre, N., & Hernández-Córdoba, M., (2003). Reversed-phase liquid chromatography on an amide stationary phase for the determination of B-group vitamins in baby foods. Journal of Chromatography A, 1007(1, 2), 77–84.
163. Vogel, J. S., & Turteltaub, K. W., (1998). Accelerator mass spectrometry as a bioanalytical tool for nutritional research. Advances in Experimental Medicine and Biology, 445, 397–410.
164. Vogel, J. S., Turteltaub, K. W., Finkel, R., & Nelson, D. E., (1995). Accelerator mass spectrometry. Analytical Chemistry, 67(11), 353A-359A.
165. Vyas, P., & O'kane, A. A., (2011). Determination of vitamin B12 in fortified bovine milk-based infant formula powder, fortified soya-based infant formula powder, vitamin premix, and dietary supplements by surface plasmon resonance: Collaborative study. Journal of AOAC International, 94(4), 1217–1226.
166. Wang, J., Wei, J., Su, S., & Qiu, J., (2015). Novel fluorescence resonance energy transfer optical sensors for vitamin B12 detection using thermally reduced carbon dots. New Journal of Chemistry, 39(1), 501–507.
167. Wang, J., Zhang, H. Z., Li, R. S., & Huang, C. Z., (2016). Localized surface plasmon resonance of gold nanorods and assemblies in the view of biomedical analysis. Trends in Analytical Chemistry (TrAC), 80, 429–443.
168. Watanabe, F., (2007). Vitamin B12 sources and bioavailability. Experimental Biology and Medicine, 232(10), 1266–1274.
169. Watanabe, F., Abe, K., Takenaka, S., Fujita, T., & Nakano, Y., (1997). Method for quantitation of total vitamin B12 in foods using a highly fluorescent vitamin B12 derivative. Journal of Agricultural and Food Chemistry, 45(12), 4661–4663.
170. Watanabe, F., Nakano, Y., Stupperich, E., Ushikoshi, K., Ushikoshi, S., Ushikoshi, I., & Kitaoka, S., (1993). A radioisotope dilution method for quantitation of total vitamin B12 in biological samples using isolated euglena pellicle fragments as a solid-phase vitamin B12 binding material. Analytical Chemistry, 65(5), 657–659.
171. Watanabe, F., Schwarz, J., Takenaka, S., Miyamoto, E., Ohishi, N., Nelle, E., Hochstrasser, R., & Yabuta, Y., (2012). Characterization of vitamin B12 compounds in the wild edible mushrooms black trumpet (Craterellus cornucopioides) and golden chanterelle (Cantharellus cibarius). Journal of Nutritional Science and Vitaminology, 58(6), 438–441.
172. Watanabe, F., Takenaka, S., Abe, K., Tamura, Y., & Nakano, Y., (1998). Comparison of a microbiological assay and fully automated chemiluminescent system for the determination of vitamin B12 in food. Journal of Agricultural and Food Chemistry, 46(4), 1433–1436.
173. Watanabe, F., Takenaka, S., Katsura, H., Masumder, S. Z. H., Abe, K., Tamura, Y., & Nakano, Y., (1999). Dried green and purple lavers (Nori) contain substantial amounts of biologically active vitamin B12 but less of dietary iodine relative to other edible seaweeds. Journal of Agricultural and Food Chemistry, 47(6), 2341–2343.
174. Watanabe, F., Takenaka, S., Kittaka-Katsura, H., Ebara, S., & Miyamoto, E., (2002). Characterization and bioavailability of vitamin B12 compounds from edible algae. Journal of Nutritional Science and Vitaminology, 48(5), 325–331.

175. Watanabe, F., Yabuta, Y., Tanioka, Y., & Bito, T., (2013). Biologically active vitamin B12 compounds in foods for preventing deficiency among vegetarians and elderly subjects. Journal of Agricultural and Food Chemistry, 61(28), 6769–6775.
176. Winkels, R. M., Brouwer, I. A., Clarke, R., Katan, M. B., & Verhoef, P., (2008). Bread co-fortified with folic acid and vitamin B12 improves the folate and vitamin B12 status of healthy older people: A randomized controlled trial. The American Journal of Clinical Nutrition, 88(2), 348–355.
177. Wongyai, S., (2000). Determination of vitamin B12 in multivitamin tablets by multimode high-performance liquid chromatography. Journal of Chromatography A, 870(1, 2), 217–220.
178. Wozniak, W. T., & Spiro, T. G., (1973). Resonance Raman spectra of vitamin B12 derivatives. Journal of the American Chemical Society, 95(10), 3402–3404.
179. Xie, C., Coda, R., Chamlagain, B., Edelmann, M., Deptula, P., Varmanen, P., Piironen, V., & Katina, K., (2018). In situ fortification of vitamin B12 in wheat flour and wheat bran by fermentation with Propionibacterium freudenreichii. Journal of Cereal Science, 81, 133–139.
180. Xu, H., Li, Y., Liu, C., Wu, Q., Zhao, Y., Lu, L., & Tang, H., (2008). Fluorescence resonance energy transfer between acridine orange and rhodamine 6G and its analytical application for vitamin B12 with flow-injection laser-induced fluorescence detection. Talanta, 77(1), 176–181.
181. Zeng, Z., Liu, Y., & Wei, J., (2016). Recent advances in surface-enhanced Raman spectroscopy (SERS): Finite-difference time-domain (FDTD) method for SERS and sensing applications. Trends in Analytical Chemistry, 75, 162–173.
182. Zhang, L., Rong, W., Lu, C., & Zhao, L., (2014). Organo-modified layered double hydroxide-catalyzed Fenton-like ultra-weak chemiluminescence for specific sensing of vitamin B12 in egg yolks. Talanta, 129, 126–131.
183. Zhang, Z., Wang, B., Yin, Y., & Mo, Y., (2009). Surface-enhanced Raman spectroscopy of vitamin B12 on silver particles in colloid and in atmosphere. Journal of Molecular Structure, 927(1–3), 88–90.
184. Zhou, Y., Li, H., Liu, Y., & Liang, G., (1991). Chemiluminescence determination of vitamin B12 by a flow-injection method. Analytica Chimica Acta, 243, 127–130.
185. Zironi, E., Gazzotti, T., Barbarossa, A., Devicienti, C., Scardilli, M., & Pagliuca, G., (2013). Development and validation of a method using ultra performance liquid chromatography coupled with tandem mass spectrometry for determination of vitamin B12 concentrations in milk and dairy products. Journal of Dairy Science, 96(5), 2832–2836.
186. Zironi, E., Gazzotti, T., Barbarossa, A., Farabegoli, F., Serraino, A., & Pagliuca, G., (2014). Determination of vitamin B12 in dairy products by ultra-performance liquid chromatography-tandem mass spectrometry. Italian Journal of Food Safety, 3(4), E-article ID 4513.

CHAPTER 14

Extraction Methods for Phytochemicals from Annatto (*Bixa orellana* L.): Natural Food Colorant

P. Sudha, Jikky Jayakumar, and V. Manoja

ABSTRACT

Annatto (*Bixa orellana* L.) is a promising natural food colorant that is used widely because of its numerous functional and health properties. Extraction of annatto for its bioactive compounds needs advanced technologies. This chapter focuses on the nature of the annatto, its chemical composition and on the enhanced technologies for extraction of bixin and norbixin from annatto. To meet out the market demand, improved or enhanced techniques should be followed. Extraction methods either as single or combination could improve the productivity and quality with minimum loss in bioactive and functional compounds of annatto.

14.1 INTRODUCTION

Extraction is a process to separate the desired compounds from the solid matrix [97]. Ancient civilization used different extraction techniques to extract natural bioactive compounds applied as nutritional, esthetic, and spiritual purposes. The process as solid–liquid extraction started over 4,000 years ago. The methods included solvent extraction (SE), sublimation, distillation, and compression. Among these, solvent method is most commonly used for its simplicity. Greek and Roman civilization laid the basis of distillation process [24]. The extraction process involves four different phases

The Functional Foods: Nutrient and Health Benefits.
Megh R. Goyal, Junaid Ahmad Malik, and Anu Kumari (Eds.)

[97]: (i) penetration of solvents into matrix; (ii) dissolution of solute into the solvents; (iii) diffusion of solute from the matrix; and (iv) collection of extracted solutes.

Desirability of extraction of the targeted compounds depends on factors, such as, selection of solvents. It is based on availability, cost, safety, and suitability. One should select the solvents near to the polarity value of the solute for better performance of extraction. Other factors include smaller (size) of the solid matrix, greater the penetration of solvent and diffusion of solute. Tiny matrix leads to difficulty in filtration. Ratio of solvent to solid should be taken in accountability. Higher the quantity of solvents, longer is the time to concentrate the compound. Temperature is one of the sensitive factors for considered for extraction. Optimum temperature gives better extraction of pigments or natural compound. Too-high temperature may have the possibility of evaporation of solvents, charring of the compounds, evaporation of volatile compounds these gives out undesirable products. Definite residence time is also a required factor to increase the extraction efficiency. However, longer duration after saturation of extraction is not recommended [97].

Reflux extraction, percolation, and maceration methods are conventional methods of extraction requiring large quantity of organic solvents, which are costly, not eco-friendly, difficult to recycle, and large volume of solvent usage increases the process duration [96]. To overcome these issues, "Green chemistry" concepts were introduced by Anastas and Warner. "Green extraction" term was proposed based on low energy consumptions during extraction, improved product yield, high purity of desired compounds, less usage of solvents, less time for extraction, and cost-effective downstream process [7, 22].

Colors are vital characteristic features of food. A colored food product set a good impression besides its taste. Processing of foods at high temperature may lead to loss of color, its volatile compounds, etc. [57]. To maintain the attractiveness and appealing nature of the food products, food colorants as additives are recommended by Food and Drug Administration (FDA). The colorant may be synthetic or natural. FDA permitted seven types of synthetic colorants as food additives. Inappropriate use of synthetic colorant causes lethal health issues. Consumers are very consciousness on their health, and thus showing interest in natural colorants. This fashioned the extraction of pigments from the natural sources. The natural pigments are chlorophylls, carotenoids, betalains, and anthocyanin [86].

Bixin, apo-carotenoids are only source from annatto (Bixa orellana L.) [9]. Annatto is recognized and designated as natural food colorant E106b by European Commission [71]. Annatto is a red-yellow-orange pigment [41]. Bixin a major coloring compound a fat-soluble substance is extracted using polar solvents like ethanol, acetone, chloroform, methanol, ethyl acetate, or with vegetable oil. Norbixin, water-soluble compounds impart yellow color [39].

Annatto is widely grown in the Central and South America [4, 5]. Owing to interest on natural colorants and its health benefits, their production has extended in Asian and African countries. The largest producer and exporters are Peru followed by Brazil, Kenya, and India [3, 77]. The current annual production of annatto seeds is about 30,000 tons. This contributes 11.8% to the global carotenoid market share [43]. The Latin American countries alone share 61% of the total world production, and the rest is by Africa (27%) and Asia (12%) [53].

Normally, Soxhlet, air condensing, mechanical agitation methods are used for the extraction of bixin and norbixin [5]. These methods need large volume of solvents and are time-consuming processes. The concept of "green extraction" is being followed for extraction of annatto and other bioactive compounds. The enhanced techniques are ultrasound assisted extraction (UAE), supercritical fluid extraction, microwave extraction, accelerated extraction, pressurized extraction, pulsed electric field extraction and spouted bed extraction. These eco-friendly techniques help to yield higher pigments and biocompounds with better quality [22, 57].

This chapter deliberates on the annatto (Bixa orellana L.): origin, cultivation practices, morphological nature, chemical composition, and the emerging extraction techniques for extraction of bixin and norbixin.

14.2 ORIGIN AND HISTORY OF ANNATTO

Annatto or achiote trees or shrubs produce fruits called achiotes. In honor of conqueror Francisco de Orellana, annatto is scientifically named as Bixa orellana L. [19, 89]. Five different species of annatto are recognized by native tribes like B. orellana, B. arborea, B. excelsa, B. platycarpa, and B. urucurana. The differences between these species are their nature of habitats [52]. Annatto is indigenous to Brazil but is also populated in Central and South America. In Brazil, the annatto is popularly known as "urucum" [89]. Native people of Americas used the annatto pigment (Red-orange-yellow) to paint their skin to protect against insects, to heal the wound and also for different ritual contexts. Annatto as condiment is chiefly used for cooking.

The seeds are soaked in oil for cooking to enhance the flavor and color to food items. During 16–19th century, annatto was considered as an important item of production for export to Europe [52].

Artificial colorant has varied applications in foods, cosmetics, pharmaceutical, and textile industries in recent years. The indiscriminate use of artificial colorants (such as: tartrazine (E102), sunset yellow (E110), Allura red (E129)) in food products causes carcinogenic illnesses [53]. Annatto seed is composed of pigments called bixin and norbixin. These pigments are carotenoids one of which is solvent (bixin) and other is water-soluble (Norbixin) [41]. These strong dyes extraction process was simple and reliable. In view of their solid antioxidant properties and high scavenging activity (antitumor), they are mainly used as food colorants [47].

Annatto, a safe and secure pigment, has been valued as a commercial colorant in food industries. Therefore, annatto was recommended as an alternative by Joint FAO/WHO Expert Committee on Food Additives (JECFA: 2006). These pigments were designated as E160b by European Union (EU) and were permitted as safe food colorants [71]. In food industries, it is mainly used to tint dairy products (like butter), margarine, confectionery commodities, soft drinks, meat products, snacks, etc. Annatto is used in pharmaceutical, cosmetics, and textile industries. The extended usage of this pigment is due to its less toxicity, eco-friendly, better biodegradability, compatibility with the environment [73].

Due to the demand of natural colorant in food and other industries, using advanced technologies like plant breeding and biotechnology, different cultivars of annatto are being produced and cultivated in countries like Asia and Africa. In India, annatto production is dominated by Southern Indian states like Andhra Pradesh, Tamil Nadu, some parts of Karnataka, and Kerala followed by some regions of Madhya Pradesh and Gujarat [3, 4].

Annatto, native to Central and South America is cultivated in the tropics of Asia and Africa due their eco-friendly behavior and for their functional properties. Currently, India contributes major share in the production and exporting of annatto to European countries.

14.3 CULTIVATION PRACTICES FOR ANNATTO

Annatto cultivation is best suited to lowlands, well-drained plains, mountain areas, and lands with high elevations. The ideal soil types are red and alluvial soils with a pH of 6–7.5. It can thrive in temperatures ranging from 20 to 30°C having an annual rainfall ranging from 1,250 to 2,000 mm. It is vulnerable to frost. The annatto plant grows best in open, sunny, humid conditions with no shade for a high yield. This plant's economic lifetime is 20 to 25 years [44].

The land is tilled, leveled, and prepared before planting. Before the rainy season starts, pits are dug and stuffed with a mixture containing soil and compost. Healthy seedlings are planted in the field as soon as rain begins to fall. The ideal time to raise the nursery is from March to June. The opted season to raise the seedlings and for germination is the months of November and December. For planting, the ideal season is from June to September. Irrigation should be done depending on the type of soil, its water holding capacity and temperature; 2–3 irrigations will be given soon after planting.

Flowering begins three years after planting. The capsules begin to form after 30 days of flowering. The fully formed fruits are seen between September and February (Figure 14.1(A)). Pruning is recommended to obtain good yield of fruits (Figure 14.1(B)). Crop is ready to harvest after 135–150 days following seeding. Fruit harvesting may commence in October. The fruit is collected in bunches manually. The appropriate harvest stage is identified by the moisture content (dryness) and fractures on the capsules. The yield of seeds varies based on the plant type, soil type, and climate [44, 70].

FIGURE 14.1 Annatto tree. (A) Annatto tree with fruit; and (B) pruned annatto tree.

14.4 MORPHOLOGICAL CHARACTERISTICS OF ANNATTO

The Annatto tree or shrub can grow to a height of 3–5 m or even to 10 m. They have different names in India like "japhara" in Tamil and Telugu, "latkan" in Hindi, "sindhuri" in Sanskrit but locally known as lipstick tree [4, 64, 65]. The trunk of the tree is short. The bark of the tree is tough or smooth of dark nature. The leaves are ovate, wide, sharp (heart shaped) with puffy and lengthy petioles. The flowers are pinkish-white in color. The fruits are mostly heart shaped bi-valved (capsules: Figure 14.2(A)) possessing seeds on each side of the capsule (Figure 14.2(B) and 14.2(C)) [4].

FIGURE 14.2 Annatto fruits and seeds. (A) Fresh annatto fruit; (B) dried annatto fruit; and (C) annatto seeds.

Source: Authors

The capsules are covered with soft bristles. An average of 30–60 numbers of seeds may be found in each capsule [89]. The average length and breadth of the seeds are 0.3–0.5 cm and 0.2–0.3 cm, respectively. The seeds have a wide range of forms, from pyramidal to almost conical [89]. Once, the fruits get matured and dried, the capsules split open for easy removal of the seeds. The dried reddish-orange seeds are the economic part of the tree (Figure 14.2(C)) [49, 89].

The pericarp or arils of the seeds is responsible for yellow-red-orange pigment [4, 9]. The 80% of the seed is composed of main pigment called bixin. Norbixin hydrophilic compound is chiefly used for fat and protein rich food products [41]. Bixin content of seeds depends on the area of cultivation, cultivar, post-harvest handling. The commercial value of seed increases with increase in bixin content [89]. To achieve higher yield of pigment, advanced biotechnology, plant breeding technologies are used to enhance the pigment content in seed. This advancement greatly affects the morphological characteristics of annatto tree, fruit, and seeds.

14.5 CHEMICAL COMPOSITION OF ANNATTO SEEDS

The red colorant annatto extract (bixin) is extracted from the pericarp of annatto seeds. The chemical formula for bixin (9′cisbixin (methyl hydrogen 9′-cis-6,6′-diapocarotene-6,6′-dioate: minor class of apocarotenoids) is

$C25H30O4$ having molecular weight of about 394.51 [89]; and it is the most important coloring component that accounts for about 80% of the carotenoid present in annatto seed [45, 60]. The strong crimson color comes due to the presence of nine-carbon double bond. A carboxylic acid group is bonded with one end of the chain, while a methyl ester group is attached with the other [71].

The seeds must contain more than 2.5% of bixin to be commercially viable. Bixin occurs in nature in the cis form; However due to isomerization, it is changed to the trans-form, which is known as isobixin. The 9′-cis-bixin is soluble in polar solvents and produces an orange hue but is not soluble in the vegetable oil. The most stable isomer, Trans–Bixin, is highly soluble in oil, and gives the red color.

The characteristics of cis and trans-bixins are similar [70]. Cis-norbixin ($C24H28O4$) is a demethylated bixin derivative that occurs naturally during the saponification process and is referred to water-soluble saponification product of bixin [3]. Bixin can be hydrolyzed into norbixin and can easily be converted into potassium and sodium salts when the pH is optimized. Cis-norbixin makes up less than 5% of the overall pigment content [2]. Other important components include [80]: cellulose (40–45%), proteins (13–16%), carbohydrates (3.5–5.5%), essential oils (3%), pigments (1–4.5%), and tannins, saponins, and alpha-carotene and beta-carotene.

More than 100 volatile chemicals have been identified in the annatto extracts. Geranylgeraniol, an oil molecule in dry seeds, accounts for 1% of their weight. The lipid portion of annatto seeds yields tocotrienols, primarily δ-tocotrienol [59]. Beta-carotene, cryptoxanthin, lutein, zeaxanthin, orellin, tomentosic acid, ishwarane, ellagic acid, salicylic acid, bixein, bixol, crocetin, threonine, tryptophan, and phenylalanine are among the other substances contained in the seeds [89]. Some hydrocarbon compounds (such as: α-pinene (9.3%), β-pinene (4.8%), α-elemene (3.3%), valencene (2.7%), and amorphene (0.2%)) have also been identified [70].

14.6 APPLICATIONS

The bark, leaves, seeds, and roots of annatto tree possess various functional properties, such as: antimicrobial activity, anti-inflammatory activity, anti-diabetic activity, anti-venom activity, Anti-oxidant activity and shows less allergic reactions [46, 55, 65, 89]. This increases the usage of annatto in pharmaceutical, textile, and cosmetic sectors [3, 70]. For medicinal purposes, it is used to cure smallpox, rashes; digestive issues like as diarrhea, abdominal pain, indigestion, and dysentery; headache, and sore throat.

It showed antiseptic and anti-bacterial effects against malaria and to heal wounds. The anti-inflammatory effect was found in leaves and seed extracts [93]. The extract of seeds can be used to prevent osteoporosis [25]. Seed extract with solvents showed repellent effect on mosquitoes [35].

Smith [77] has also approved the annatto extract as safe food [9]. Since, it is an antioxidant and bactericidal potential and non-toxic [1], therefore it can prevent cancer, heart diseases, vision problem, etc. [9, 14, 49]. For many years, dyes from the annatto have been used in dairy industries and dairy products and bakery products, vegetable oils, beverages (cool drinks) and dietary supplement, etc. [49]. The dyes are used in textile, painting, and cosmetic industries mainly for the production of sunscreen lotion to prevent against UV-rays [49, 53, 54].

Further research and studies are needed to utilize all medicinal benefits of annatto extract that can used in the application of pharma, cosmetics, textile with future vision. Future technologies will help to unmask the further applications of annatto.

14.7 EXTRACTION OF PIGMENTS FROM ANNATTO SEEDS

Traditionally, oil and water were used to extract the pigment from the annatto seeds. Solvent based extraction is being followed generally to obtain the major biocompound called bixin. Solvents such as ethanol, acetone, chloroform, propylene glycol, methanol are recommended to extract bixin. European Commission has laid a common definition for food colorant that has separate regulations of bixin and norbixin extraction methods and purity criteria: (i) Solvent extracted bixin and norbixin-indirect extraction method; (ii) alkali-extracted annatto-direct extraction method; and (iii) oil extracted annatto-direct extraction method. The purity is based on the product with limited residual solvents and undesirable compounds proposed by regulation board (FAO). Various methods (Figure 14.3) to produce different types of annatto pigments are given in the process flow chart [71, 94].

Traditional techniques involve long processing duration, exposing the bioactive substances to degradation thus resulting in low efficiency because the extraction methodology impacts the functional activities of the extract. Furthermore, emerging extraction technologies, mechanical techniques, spouted bed, sonication, microwaves, and supercritical fluids (CO2) have focused solely on the extraction of bixin and norbixin [6, 62, 67, 79, 92]. The majority of these research studies focus on the coloring capacity of extract, while some can assess the extraction technique's effect for antioxidant and antibacterial activities [62, 67, 87].

14.7.1 SOLVENT EXTRACTION (SE)

Solvents play key role for the extraction of natural biocompounds. Best suited solvents based on the nature of material, solvents to solid ratio, process duration, temperature are main factors for the best end-product. Usually, it is carried out using in reflux and Soxhlet methods. Ethanol is a common solvent used for Reflux and Soxhlet methods of extraction of pigments. Reflux method is not suitable for heat-sensitive compounds. Soxhlet, an automatic repetitive process, uses less amount of solvent and works on the principle of reflux and siphoning. High temperatures with longer duration may char the product [97].

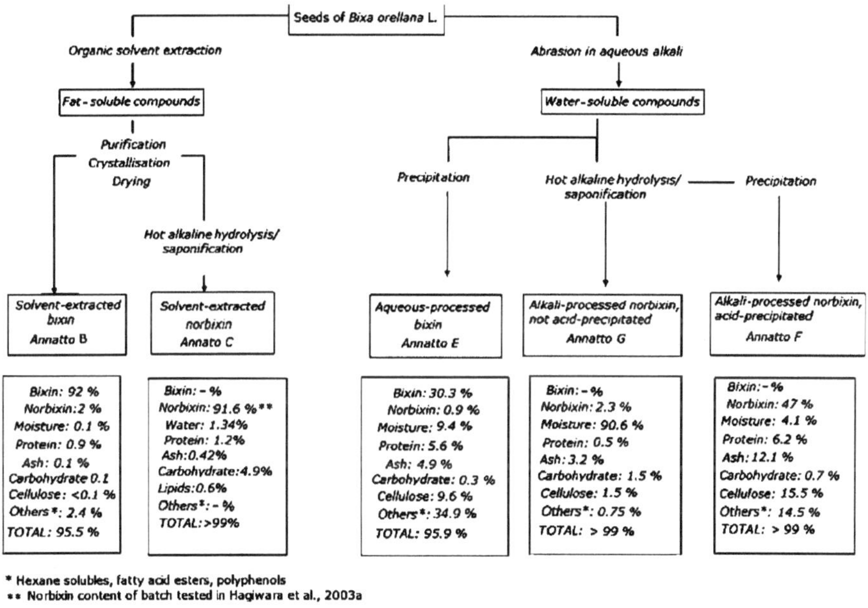

FIGURE 14.3 Process flow for extraction of annatto.

Source: Reprinted with permission from EFSA Panel on Food Additives and Nutrient Sources added toFood, 2016. https://creativecommons.org/licenses/by-nd/4.0/

Van Chuyen reported that combined extraction of sodium hydroxide and soybean oil improved the bixin yield and lowered the concentration of undesirable compounds compared to acetone (used as the single solvent) used for extraction [83]. Reflux method using ethyl acetate yields better bixin content than Soxhlet method of extraction [68]. Extraction of Annatto with acetone by Soxhlet method produces higher bixin content but also yields undesirable compounds (Figure 14.4(B)). To reduce the production of xylene, sodium hydroxide (alkaline extraction) is recommended [27]. Pure-ethanol at ambient temperature can produce high bixin content with high concentration of δ-tocotrienol and antioxidants [71].

14.7.2 AQUEOUS EXTRACTION

Water with addition of potassium or sodium hydroxide [71] produce trans-form of bixin (Figure 14.4). Alkali concentration and temperature are factors for removal of pigments from the seeds. The extracts are acidified to pH below 6 to get precipitates of bixin and norbixin. Annatto seeds with strong alkaline aqueous solution at high temperature (nearly 70°C) produce high percentage of norbixin. Cold aqueous solution can yield 25% bixin (Figure 14.4(A)) [5].

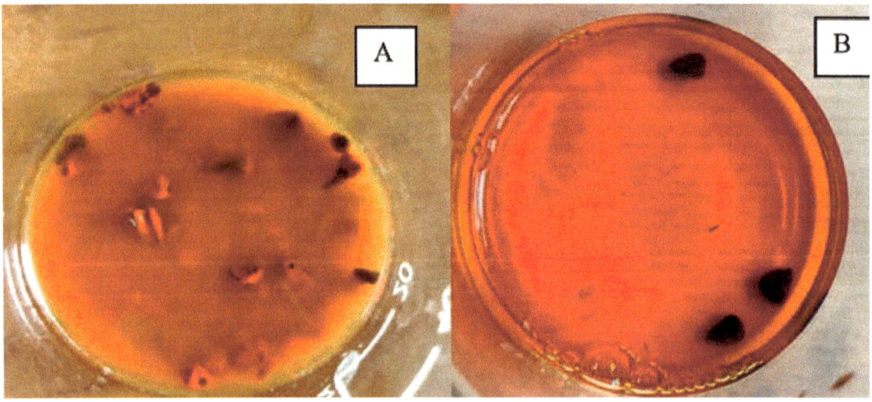

FIGURE 14.4 Annatto extract. (A) Extract using water; and (B) extract using solvent.

Cabrera and Barahona found that 5.58% concentration of norbixin formulation present in aqueous medium is suitable for delivering annatto dye for industrial applications having a shelf-life long enough to go through its self-storage, distribution, and use [17]. Aqueous alkali extraction at pH of 13 recorded maximum of 0.4–1.25% of norbixin [18]. The 0.25% of alkaline solution may give the optimum color of the extract with 41% of bixin recovery [26]. Extraction of annatto by lowering the pH of distilled water to 4 at 80 and 90°C inferred higher pigment content with moderate antibacterial property [39].

14.7.3 SPOUTED BED EXTRACTION

Particle attrition and impact are two alternative ways for extracting the pigment from annatto seeds. Simple equipments, namely ball mill or spouted bed, are used in these approaches [12, 32]. A spouted bed has the advantage of greater control over the process [12, 69]. Spouted beds, in addition to their ability to handle coarse particles, have structural and cyclic flow patterns with effective fluid–solid interaction, making them an excellent alternative to mechanical extraction [12, 75].

Spouted bed extraction has been used in variety of industrial operations, such as: granular material drying, polymer chip blending, tablet coating, and fertilizer granulation, etc. [12, 28]. Spouted beds are classified into three distinct zones: a spout in the center (where gas and particles ascend at a high rate); and the particle as a fountain zone, in which particles reach their highest elevation and then drop down onto the annulus' surface; and an annulus zone (As a thick phase, particles migrate slowly downward [12, 13]. As a result, a cyclic trend of solid movement emerges. The gas passes the column through an orifice, then flows fast upward through the spout and move gradually upward through the annulus zone's interstices [12]).

Barrozo et al. studied the extraction of bixin from Bixa orellana seeds using a spouted bed and found that a rise of 10% of flow rate resulted in 30% increase in the extracted mass, while the 25% increase in seed charge resulted in 33% increase in extracted mass [12].

14.7.4 ULTRASOUND ASSISTED EXTRACTION (UAE)

Ultrasound assisted extraction (UAE) technique is the most popular technique for extraction of different natural products. It is also as sonication or ultra-sonication technique that uses high frequency sound waves for extraction. The efficiency of the extraction process is improved due to the cavitation caused by the wave rupture of the membrane, and it helps to diffuse the solvent into the materials. This accelerates the dissolution and causes diffusion at faster rate. The heat that produces during process enhances the extraction activity. This technique needs less solvent, temperature, and time. The high frequency makes the solvents to be more efficient when compared to the conventional extraction method. It can be recommended for heat-sensitive and unstable compounds [56, 97].

Ultrasound at 40 kHz with ethanol and annatto seed can yield high bixin content [58]. Combined effect of turbulence and cavitation with 800 watts, 19 kHz at 60C for 40-minute, ethanol as a solvent contributed to higher extraction efficiency of bixin and phenol when compared with an individual method of extraction [82]. Annatto extraction using ultrasound with high power-short time/low power-long-time experiment showed that high power-short time was best treatment to yield high bixin content [30].

Annatto seed powder extracted with double-distilled water in combination with ultrasound at 40 kHz and 300 W for one-hour at 30C yielded 40.33 ± 0.32 mg/g of bixin. Application of extract on pork patties showed antimicrobial effects. The 0.25% annatto seed was recommended as an additive for meat products [84]. Combined effect of solvents (ethyl acetate, isopropyl acetate, ethanol) and sonication succeeded in average bixin yield and higher δ-tocotrienol content [47].

The theory of ultrasonic assisted extraction is based on the cavitation phenomena that increase mass transfer and strong interaction between the plant tissues and solvent [8, 23, 90, 91]. Microjets are created when cavitation bubbles collide near the tissue surfaces, causing tissue rupture; and extensive penetration of solvent into the tissue matrix takes place [8, 15]. As cavitation effects primarily influence the microbiological, nutritional, and physicochemical properties of fluid foods, ultrasonic-aided extraction technology is widely used in the processing of liquid food [8, 10, 31, 98].

Quiroz et al. reported that a pH of 7.0, a seed-to-solvent ratio of 1:7, and a treatment time of 20 min were optimum, resulting in an annatto extract with 0.62% bixin, 3.81 mg gallic acid/mg equivalent to polyphenol compounds (ABTS 1035.7, FRAP 424.7, and DPPH 1161.5 μM trolox/L); and minimum inhibitory concentration against Bacillus cereus and Staphylococcus aureus of 32 and 16 mg/L, respectively [61]. Using ethyl acetate, a solvent to feed ratio of 0.05 L/g, and minimum 30 min extraction time as optimum extraction parameters for annatto seeds, the yield of bixin was about 85% with ultrasound and 95% with dielectric heating (microwave) [8].

14.7.5 MICROWAVE ASSISTED EXTRACTION (MAE)

The oil extraction can be carried out with microwave technique using the principle of ionic conduction and dipole rotation. The interaction of microwaves with polar solvents produces heat resulting in mass transfer. Usage of solvents is very low, and it is regarded as a green technology. Two possible methods in microwave assisted extraction (MAE) are [97]: (i) solvent-free extraction for volatile compounds; and (ii) SE for non-volatile compounds.

MAE is based on the theory of dielectric heating of material with a time-varying electric field [8, 20, 88] that will align induced and intrinsic dipole moments within the element [8, 81]. The heat produced by this alignment is subsequently dispersed into the surrounding solution. Because of the moisture content, microwave power is absorbed quickly by the inside of plant cells [8, 16]. Plant cells inflate as moisture evaporates, forcing the cell-wall from inside, straining, and eventually rupturing of cell-wall occurs. This action makes it easier for the active component in the ruptured cells to leach into the surrounding solution [8, 96]. When compared to leaching extraction, MAE improved the amount of polyphenols and bixin and their antioxidant activity [61]. A significant improvement in bixin content was observed in MAE using ethanol as a solvent by Grillitsch [38].

14.7.6 ACCELERATED SOLVENT EXTRACTION (ASE)

The accelerated solvent extraction (ASE) method employs organic solvents at high pressures and temperatures over the boiling point. A solid sample is encased in a sample cartridge that is filled with an extraction fluid that is used to effectively extract the sample in a short period of time (5–10 min) at elevated temperature (50–200°C) and pressure (500–3,000 psi). The sample extract is expelled from the cell into a collection receptacle using compressed gas. ASE can currently be used for solid or semisolid samples that can be kept in the cell throughout extraction. Rahmalia et al. [63] reported that the combinations of cyclohexane and acetone were useful for obtaining a polarity range and determining the best solvent ratio for bixin extraction; and authors concluded that the cyclohexane: acetone solution (60%: 40%) gave the best bixin extraction yield [63].

14.7.7 PRESSURIZED EXTRACTION TECHNIQUE

This extraction technique is also called by different names by different researchers as ASE, high pressure SE, and pressurized fluid extraction, etc. The pressure produced during the extraction keeps the solvents in liquid even above their boiling point causing high diffusion and soluble rate. According to the research studies, this technique is not suitable for heat-sensitive materials at high temperatures. Therefore, high temperature and short processing time have been suggested [56, 97].

Based on the potential increase in solubility of the substances to be extracted and the acceleration of the desorption kinetics of these substances from the matrix, a certain liquid can be pressurized and heated at pressures and temperatures less than its critical point; and then can be used in the extraction of several compounds. The liquid extraction process at pressures greater than ambient pressure and medium to high temperatures is known as "pressurized liquid extraction (PLE)" [67].

14.7.8 SUPERCRITICAL FLUID EXTRACTION

Supercritical fluid extraction uses carbon dioxide (CO2) as solvents. This lowers the usage of organic solvents. Under critical temperature and pressure, the CO2 behaves like liquid and gas. SCFCO2 (supercritical fluid carbon dioxide) has similar solubility to liquid and diffusivity to gas. Since the SCFCO2 is less polar, it can be used for extraction of non-polar materials. The extraction efficiency can be increased by providing a modifier solvent. The critical conditions of CO2 can be reached even at the temperature of 30°C and 7 MPa [56, 97].

To recover higher bixin contents, the oil from the seeds has to be removed. SCF technique is widely used as a pre-treatment to remove the oil from the seeds [66]. The supercritical CO2 extraction can be suggested for extraction of cis- and trans-bixin. The advantage of using SFE is no pre-treatment is needed, and it can be used under median temperatures, limit the usage of solvents, etc. [5]. Annatto extract can be obtained at the temperature range of 30 to 50°C and the pressure between 10 and 35 MPa. Supercritical extraction is only suitable for extraction of bixin [76]. Oil extracted from annatto using SCF showed higher δ-tocotrienol and γ-tocotrienol [78, 95].

Sequential SCF extraction of annatto seeds resulted in the production of geranyl geraniol and tocotrienol [85]. Due to minimal thermal effects on products, this method has several advantages than the conventional methods for extracting compounds from complex food systems [21, 78], such as: excellent quality of recovered products, low-energy required for solvent recovery, and good selectivity in the separation process, use of supercritical CO2 with various pressures and temperatures.

14.8 FOOD APPLICATIONS OF ANNATTO

14.8.1 OIL SOLUBLE EXTRACT

Bixin extracted using soybean, grapeseed, and sunflower oils has a yellow-orange-red tint. The stable trans-bixin is produced by hot vegetable oil extraction and has an orange to yellow tone. Further heating generates a thermally deteriorated "Yellow C-17" with a yellow hue. Because of its limited solubility in oil, only 0.05–1% of bixin can be recovered from annatto seeds by oil extraction. This annatto extract is used in the manufacturing of Cheese, salads, and dressings [40, 74]. The extracts are also used in cosmetic [36] and paint production industries [48].

14.8.2 OIL SUSPENSION

The annatto powder is mixed with oil to make an oil suspension. Oil temperatures below 100C produce an orange tint, whereas temperatures over this limit provide a yellow color. The oil suspension contains at least 4% cis- and trans-bixin that can be directly used for cooking [40].

14.8.3 WATER-SOLUBLE EXTRACT

The pigment in water soluble cis-norbixin ranges from 1 to 15%. They are primarily used in the dairy industry (particularly for cheese), as well as in beverages. Because of the interaction and connection of proteins that impart persistent color, this extract is favored in cheese coloring. According to Giuliano et al., 1% norbixin in one-L can color 16 tons of cheese [36].

14.8.4 EMULSIFIED PIGMENT

Emulsifiers (such as: propylene glycol and polysorbates) are used to make emulsions based on bixin and norbixin [40]. It aids in the miscibility of the extract in the food oil and water phase. Annatto extracts of this sort are found in ice cream, margarine, mayonnaise, creamy sauces, various chocolates, and a variety of baked products. Bixin is present in the emulsified pigment in the range of 1 to 2.5% [77].

14.8.5 MICRO- AND NANO-ENCAPSULATED ANNATTO EXTRACT

Temperature, light, and oxygen have significant impact on the pigments. The color changes from red to yellow after thermal degradation [50, 51] that may result in the release of m-xylene and toluene [72]. Micro- and nano-encapsulations are preferred to combat the instability under unfavorable circumstances. Barbosa et al. investigated the light stability of microencapsulated bixin using spray drying procedure with different matrix, such as, gum Arabic or maltodextrin and bixin as core materials [11]. The encapsulation efficiency of 16.92 ± 0.16 lg/mL for bixin concentrations with a mean diameter of 195 ± 27 nm was 100% [11]. The nano-encapsulated bixin was stable at room temperature for 119 days [29].

14.9 REGULATIONS

The acceptable daily intake (ADI) is 6–12 mg of bixin/kg body weight and ≤ 0.6 mg of norbixin/kg body weight was suggested by the Joint FAO/WHO Expert Committee on Food Additives (JECFA) [77] review [71, 94]. The safety of annatto extracts (E 160b) as an additive in foods was evaluated by Aguilar et al. [2, 71, 94].

Younes et al. [94] indicated, "The Scientific Panel on Food Additives and Flavoring (FAF) evaluated the safety of annatto E and the exposure to the annatto colorants bixin and norbixin (E 160b) when used as food additives in response to a request from the European Commission to the European Food Safety Authority (EFSA). According to Annex II of Regulation (EC) No. 1333/2008 on food additives, Annatto, Bixin, and Norbixin (E 160b) are permitted as food additives in the European Union (EU), and particular purity standards have been established by Commission Regulation (EU) No 231/2012. A scientific opinion on the safety of annatto extracts (E 160b) as a food additive was adopted by the EFSA Panel on Food Additives and Nutrient Sources Added to Food (ANS) in 2016. On the safety of annatto E, the Panel was unable to draw a conclusion. The maximum permitted levels (MPLs) specified in Annex II of Regulation (EC) No. 1333/2008, the use levels provided by the food industry, and the proposed use levels from an applicant for an extension of use were taken into consideration when estimating dietary exposure for annatto (E 160b), bixin- and norbixin-based annatto extracts. All demographic categories and improved exposure scenarios, including the suggested expansion of usage, had bixin exposure estimations below the ADI. For various demographic groups, norbixin exceeded the 95th percentile for the duration of usage."

Bixin should only be prepared in two forms, according to FSSAI 2011 [33, 34, 94]: (i) oil soluble for use in butter and other food products; and (ii) water soluble for use in cheese and other food products. The suggested form varies depending on the type of food to be applied.

14.10 MARKET FOR ANNATTO

The United States, Western Europe, and Japan are currently most important markets for annatto, while there is a large seed trade between suppliers in South and Central America, where annatto is used as a food colorant and condiment [5, 37]. Although annatto was grown in many developing nations, yet only about a third (about 27%) was processed locally as the majority of processing plants were in Peru, Kenya, and Brazil [5, 42]. The global annatto market was worth US$ 196.55 million in 2019 and is expected to grow to US$ 281.97 million by 2027, with a CAGR (compound annual growth rate) of 5% throughout that time.

14.11 SUMMARY

Annatto is a safe and eco-friendly natural colorant. The widespread applications for different uses have increased its demand. Limitations witnessed in the extraction of annatto by conventional methods are rectified or lowered by adopting improved extraction techniques. Annatto a heat and light sensitive compound need a standard protocol using improve techniques. In this chapter, extraction of annatto using different enhanced techniques are briefed. Several future works in the field of extraction of annatto using enhanced technologies would make the availability of pure and quality worldwide.

KEYWORDS

- **accelerated solvent extraction**
- **annatto**
- **Bixa orellana L.**
- **bixin**
- **enhanced extraction**
- **natural food colorant**
- **norbixin**
- **oil suspension**

REFERENCES

1. Abayomi, M., Adebayo, A. S., Bennett, D., Porter, R., Campbell, J. S., & Dawkin, G., (2014). Phytochemical testing and in vitro antibacterial activity of Bixa orellana (annatto) seed extract. British Journal of Pharmaceutical Research, 4(11), 1387–1399.
2. Aguilar, F., Crebelli, R., Di Domenico, A., Dusemund, B., Frutos, M. J., Galtier, P., & Younes, M., (2016). The safety of annatto extracts (E160b) as a food additive: EFSA panel on food additives and nutrient sources added to food (ANS). EFSA Journal, 14(8), 1–87.
3. Akshatha, V., Giridhar, P., & Ravishankar, G. A., (2011). Food, ethnobotanical and diversified applications of Bixa orellana L.: A Scope for its improvement through biotechnological mediation. Indian Journal of Fundamental and Applied Life Sciences, 1(4), 9–31.
4. Akshatha, V., Giridhar, P., & Ravishankar, G. A., (2011). Morphological diversity in Bixaorellana L. and variations in annatto pigment yield. The Journal of Horticultural Science and Biotechnology, 86(4), 319–324.
5. Albuquerque, C. L. C., & Meireles, M. A. A., (2011). Trends in annatto agroindustry: Bixin processing technologies and market. Recent Patents on Engineering, 5(2), 94–102.
6. Alcázar-Alay, S. C., Osorio-Tobón, J. F., Forster-Carneiro, T., & Meireles, M. A. A., (2017). Obtaining bixin from semi-defatted annatto seeds by a mechanical method and solvent extraction: Process integration and economic evaluation. Food Research International, 99(1), 393–402.
7. Anastas, P. T., & Warner, J. C., (1998). Principles of Green Chemistry (p. 30). New York – USA: Oxford University Press. http://masrurichemistry.lecture.ub.ac.id/files/2013/09/3-Prinsip-Green-Chemistry.pdf (accessed on 2 November 2023).
8. Bachtler, S., & Bart, H. J., (2021). Increase the yield of bioactive compounds from elder bark and annatto seeds using ultrasound and microwave assisted extraction technologies. Food and Bioproducts Processing, 125, 1–13.
9. Balakrishnan, M., Gayathiri, S., Preetha, P., Pandiselvam, R., Jeevarathinam, G., Delfiya, D. A., & Kothakota, A., (2021). Microencapsulation of bixin pigment by spray drying: Evaluation of characteristics. LWT- Food Science and Technology, 145, E-article ID 111343.
10. Barba, F. J., Galanakis, C. M., Esteve, M. J., Frigola, A., & Vorobiev, E., (2015). Potential use of pulsed electric technologies and ultrasounds to improve the recovery of high-added value compounds from blackberries. Journal of Food Engineering, 167(4), 38–44.
11. Barbosa, M. I. M. J., Borsarelli, C. D., & Mercadante, A. Z., (2005). Light stability of spray-dried bixin encapsulated with different edible polysaccharide preparations. Food Research International, 38(8, 9), 989–994.
12. Barrozo, M. A. S., Santos, K. G., & Cunha, F. G., (2013). Mechanical extraction of natural dye extract from Bixa orellana seeds in spouted bed. Industrial Crops and Products, 45(2013), 279–282.
13. Barrozo, M. A., Duarte, C. R., Epstein, N., Grace, J. R., & Lim, C. J., (2010). Experimental and computational fluid dynamics (CFD) study of dense-phase, transition region, and dilute-phase spouting. Industrial & Engineering Chemistry Research, 49(11), 5102–5109.

14. Beni, A. A., Rodrigues, R. F., Conte, L., Costa, I. F., Delalibera, É. A., & Somacal, S., (2020). Dietary supplementation with annatto food-coloring extracts increases the resistance of human erythrocytes to hemolysis. Nutrition Research, 76(3), 71–81.
15. Both, S., Chemat, F., & Strube, J., (2014). Extraction of polyphenols from black tea–conventional and ultrasound assisted extraction. Ultrasonics Sonochemistry, 21(3), 1030–1034.
16. Budarin, V. L., Shuttleworth, P. S., Farmer, T. J., Gronnow, M. J., Pfaltzgraff, L., Macquarrie, D. J., & Clark, J. H., (2015). The potential of microwave technology for the recovery, synthesis, and manufacturing of chemicals from bio-wastes. Catalysis Today, 239, 80–89.
17. Cabrera, G. C., & Barahona, R. A., (2015). Stability study of an aqueous formulation of the annatto dye. International Food Research Journal, 22(5), 2149–2154.
18. Castello, M. A., Chandra, N. A., Phatak, A. N., & Sharon, M. A., (2004). Estimation of bixin in seeds of Bixa orellana L. from different locations in Western Maharashtra. Indian Journal of Plant Physiology, 9, 185–188.
19. Cevallos, C. E. S., Rodríguez, I. D. J., & Santana, L. D., (2021). Financial evaluation of annatto paste production from ancestral knowledge in manta Blanca. International Journal of Economic Perspectives, 15(1), 38–48.
20. Chan, C. H., Yusoff, R., Ngoh, G. C., & Kung, F. W. L., (2011). Microwave-assisted extractions of active ingredients from plants. Journal of Chromatography A, 1218(37), 6213–6225.
21. Chao, R. R., Mulvaney, S. J., Sanson, D. R., Hsieh, F. H., & Tempesta, M. S., (1991). Supercritical CO2 extraction of annatto (Bixa orellana) pigments and some characteristics of the color extracts. Journal of Food Science, 56(1), 80–83.
22. Chemat, F., Abert-Vian, M., Fabiano-Tixier, A. S., Strube, J., Uhlenbrock, L., Gunjevic, V., & Cravotto, G., (2019). Green extraction of natural products. Origins, current status, and future challenges. Trends in Analytical Chemistry, 118(3), 248–263.
23. Chemat, F., Rombaut, N., Sicaire, A. G., Meullemiestre, A., Fabiano-Tixier, A. S., & Abert-Vian, M., (2017). Ultrasound assisted extraction of food and natural products. Mechanisms, techniques, combinations, protocols, and applications: A review. Ultrasonics Sonochemistry, 34, 540–560.
24. Chemat, F., Vian, M. A., Fabiano-Tixier, A. S., Nutrizio, M., Jambrak, A. R., Munekata, P. E., Lorenzo, J. M., et al., (2020). A review of sustainable and intensified techniques for extraction of food and natural products. Green Chemistry, 22(8), 2325–2353.
25. Chin, K. Y., & Ima-Nirwana, S., (2019). The role of tocotrienol in preventing male osteoporosis – A review of current evidence. International Journal of Molecular Sciences, 20(6), 1–20.
26. Chowdhury, A. I., Molla, A. I., Sarker, M., Rana, A. A., Ray, S. K., Nur, H. P., & Karim, M. M., (2006). Preparation of edible grade dye and pigments from natural source Bixa orellenae L. International Journal Basic Applied Science, 10(4), 7–22.
27. Chuyen, H. V., & Eun, J. B., (2021). Effects of different extraction methods on the extraction yield, degradation of bixin and formation of harmful volatile compounds in the extracts from annatto seeds. Food Research, 5(6), 42–48.
28. Conceição, F. R. S., Barrozo, M. A. S., Limaverde, J. R., & Ataíde, C. H., (1998). The use of a spouted bed in the fertilizer coating of soybean seeds. Drying Technology, 16(9), 2049–2064.

29. De Sousa, L. K. B., Paese, K., Forgearini, J. C., Guterres, S. S., Jablonski, A., & De Oliveira, R. A., (2013). Characterization and stability evaluation of bixin nanocapsules. Food Chemistry, 141(4), 3906–3912.
30. DeFreitas, G., Silva, E., Strieder, M., Vardanega, R., & Petenate, M., (2019). Effect of high-intensity ultrasound specific energy on the bixin and geranylgeraniol recovery from defatted annatto seeds. Revista dos Trabalhos de Iniciação Científica da UNICAMP (Journal of Scientific Initiation Works at UNICAMP), 27, 1–9.
31. Deng, Q., Zinoviadou, K. G., Galanakis, C. M., Orlien, V., Grimi, N., & Barba, F. J., (2015). The effects of conventional and non-conventional processing on glucosinolates and its derived forms, isothiocyanates: Extraction, degradation, and applications. Food Engineering Reviews, 7(3), 357–381.
32. Duarte, C. R., Neto, J. V., Lisboa, M. H., Santana, R. C., Barrozo, M. A. S., & Murata, V. V., (2004). Experimental study and simulation of mass distribution of the covering layer of soybean seeds coated in a spouted bed. Brazilian Journal of Chemical Engineering, 21(1), 59–67.
33. EC (1995). Commission directive 95/45/EC, of 26 July 1995 laying down specific criteria of purity concerning colors for use in foodstuffs. Official Journal of the European Communities, 226, 1–44.
34. European Parliament and Council Directive, (1994). European Parliament and Council Directive 94/36/EC of 30 June 1994 on colors for use in foodstuffs. Official Journal of the European Communities, 237(10), 13–29.
35. Giorgi, A., De Marinis, P., Granelli, G., Chiesa, L. M., & Panseri, S., (2013). Secondary metabolite profile, antioxidant capacity, and mosquito repellent activity of Bixa orellana from Brazilian Amazon region. Journal of Chemistry, 2013(4–6), E-article ID 409826.
36. Giuliano, G., Rosati, C., & Bramley, P. M., (2003). To dye or not to dye: Biochemistry of annatto unveiled. Trends in Biotechnology, 21(12), 513–516.
37. Green, C. L., (1995). Natural colorants and dyestuffs: A review of production, markets, and development potential. Journal of Non-Wood Forest Products (FAO), 1997(4) 1–116.
38. Grillitsch, T., (2019). A Fast, Simple, And Green Method for the Extraction and Purification of Bixin from the Seeds of Bixa Orellana L. by Microwave-Assisted Extraction Coupled with Box-Behnken Design (p. 83). Diploma Thesis, Ubon Ratchathani University, Thailand.
39. Handayani, I., Haryanti, P., & Sulistyo, S. B., (2021). Color and antibacterial activity of annatto extracts at various pH of distilled water solvent and extraction temperature. Food Research, 5(6), 247–253.
40. Hendry, G. A. F., (1996). Natural pigments in biology. In: Hendry, G. A. F., & Houghton, J. D., (eds.), Natural Food Colorants (2nd edn., pp. 1–39). New York – USA: Springer Science & Business Media.
41. Husa, N. N., Hamzah, F., & Said, H. M., (2018). Characterization and storage stability study of bixin extracted from Bixa orellana using organic solvent. IOP Conference Series: Materials Science and Engineering, 358(1), E-article ID: 012035.
42. Iqbal, M., (1993). International Trade in non-Wood Forest Products: An Overview (p. 111). https://www.fao.org/3/x5326e/x5326e00.htm (accessed on 2 November 2023).
43. Kapoor, L., & Ramamoorthy, S., (2021). Strategies to meet the global demand for natural food colorant bixin: A multidisciplinary approach. Journal of Biotechnology, 338, 40–51.

44. Kumaran, K., (2014). Production potential of annatto (Bixa orellana L.) as a source of natural edible dye. In: The Proceedings of the 2014 International Workshop on Natural Dyes (pp. 1–13), Hyderabad, India.
45. Lauro, G. J., (1991). A primer on natural colors. Cereal Foods World, 36(11), 949–953.
46. Lim, T. K., (2012). Edible Medicinal and Non-Medicinal Plants (1st edn., Vol. 1, pp. 835). New York – US: Springer.
47. Linton, G. D. R., Ocaña, N. A., Guerrero, R. A., Alavez, S., López, P. L., Espinoza, M. J. A., & Flores, P. L. J., (2021). Environmentally friendly achiote seed extracts with higher δ-tocotrienol content have higher in vitro and in vivo antioxidant activity than the conventional extract. Journal of Food Science and Technology, 58(7), 2579–2588.
48. Lourido, P. H., & Martınez, S. G., (2010). The Bixa orellana L. in treatment of stomatology affections: A subject that has not been studied yet. Revista Cubana de Farmacia (Cuban Journal on Pharmacy), 44(2), 231–244.
49. Madrid, R. R., Espinosa, A. M., Conejo, C. Y., & Caligaris, G. L. E., (2016). Carotenoid derivates in achiote (Bixa orellana) seeds: Synthesis and health promoting properties. Frontiers in Plant Science, 7, E-article ID 1406.
50. McKeown, G. G., (1965). Composition of oil-soluble annatto food colors. III. Structure of the yellow pigment formed by the thermal degradation of bixin. Journal of the Association of Official Agricultural Chemists (AOAC), 48(4), 835–837.
51. McKeown, G. G., & Mark, E., (1962). The composition of oil-soluble annatto food colors. Journal of Association of Official Agricultural Chemists, 45(3), 761–766.
52. Moreira, P. A., Lins, J., Dequigiovanni, G., Veasey, E. A., & Clement, C. R., (2015). The domestication of annatto from Bixa urucurana in Amazonia. Economic Botany, 69(2), 127–135.
53. Mota, R. D., Flores, P. L. J., Carrari, F., Espinoza, M. J. A., Sánchez, L. F. D., López, P. L., & Cabrera, R. F., (2017). Achiote (Bixa orellana L.): A natural source of pigment and vitamin E. Journal of Food Science and Technology, 54(6), 1729–1741.
54. Mutmainah, F. Y., Puspitaningrum, I., & Kusmita, L., (2020). Sunscreen activity on fruit skin extract of annatto (Bixa orellana L.) in vitro. Indian Journal of Science and Technology, 13(45), 4506–4512.
55. Nathan, V. K., Rani, M. E., Rathinasamy, G., & Narayanan, K., (2019). Antioxidant and antimicrobial potential of natural coloring pigment derived from Bixa orellana L. seed Aril. Proceedings of the National Academy of Sciences – India Section B: Biological Sciences, 89(1), 137–143.
56. Neves, M. I. L., Silva, E. K., & Meireles, M. A. A., (2019). Trends and challenges in the industrialization of natural colorants. Food Public Health, 9(2), 33–44.
57. Ngamwonglumlert, L., Devahastin, S., & Chiewchan, N., (2017). Natural colorants: Pigment stability and extraction yield enhancement via utilization of appropriate pretreatment and extraction methods. Critical Reviews in Food Science and Nutrition, 57(15), 3243–3259.
58. Nguyen, C. K., Tran, N. Q., Nguyen, T. P., & Nguyen, D. H., (2017). Biocompatible nanomaterials based on dendrimers, hydrogels, and hydrogel nanocomposites for use in biomedicine. Advances in Natural Sciences: Nanoscience and Nanotechnology, 8(1), E-article ID 015001.
59. Parvin, K., Aziz, M. G., Yusof, Y. A., Sarker, M. S. H., & Sill, H. P., (2011). Degradation kinetics of water-soluble annatto extract and sensory evaluation of annatto colored yoghurt. Journal of Food, Agriculture & Environment, 9(3), 139–142.

60. Preston, H. D., & Rickard, M. D., (1980). Extraction and chemistry of annatto. Food Chemistry, 5(1), 47–56.
61. Quiroz, J. Q., Duran, A. M. N., Garcia, M. S., Gomez, G. L. C., & Camargo, J. J. R., (2019). Ultrasound-assisted extraction of bioactive compounds from annatto seeds, evaluation of their antimicrobial and antioxidant activity, and identification of main compounds by LC/ESI-MS analysis. International Journal of Food Science, 2019, Article ID 3721828.
62. Quiroz, J. Q., Torres, A. C., Ramirez, L. M., Garcia, M. S., Ciro Gomez, G., & Gomez, G. L., & Camargo, J. J., (2019). Optimization of the microwave-assisted extraction process of bioactive compounds from annatto seeds (Bixa orellana L.). Antioxidants, 8(2), 1–37.
63. Rahmalia, W., Fabre, J. F., & Mouloungui, Z., (2015). Effects of cyclohexane/acetone ratio on bixin extraction yield by accelerated solvent extraction method. Procedia Chemistry, 14, 455–464.
64. Ravindran, P. N., (2017). The Encyclopedia of Herbs and Spices (1st edn., Vol. 1, pp. 1218). Boston-MA: CAB International.
65. Ravindran, P. N., Divakaran, M., & Pillai, G. S., (2012). Other herbs and spices: Achiote to Szechuan pepper. In: Peter, K. V., (ed.), Handbook of Herbs and Spices (2nd edn., Vol. 2, pp. 534–556). London – UK: Woodhead Publishing.
66. Rodrigues, L. M., Alcázar-Alay, S. C., Gomes, M. T. M. S., & Meireles, M. A. A., (2013). Pressurized liquid extraction (PLE) of Bixin from defatted annatto seeds. In: III Iberoamerican Conference on Supercritical Fluids (pp. 1–6). Cartagena de Indias – Colombia.
67. Rodrigues, L. M., Alcázar-Alay, S. C., Petenate, A. J., & Meireles, M. A. A., (2014). Bixin extraction from defatted annatto seeds. Comptes Rendus Chimie (Chemistry Reports), 17(3), 268–283.
68. Sabuz, A. A., Khan, M. H. H., Rahman, M. T., Rana, R., & Brahma, S., (2020). Stability of organic food colorant extracted from annatto seeds on food matrix. International Journal of Food Science and Nutrition, 5(6), 10–16.
69. Santos, K. G., Murata, V. V., & Barrozo, M. A., (2009). Three-dimensional computational fluid dynamics modelling of spouted bed. The Canadian Journal of Chemical Engineering, 87(2), 211–219.
70. Satyanarayana, A., Prabhakara-Rao, P. G., & Rao, D. G., (2003). Chemistry, processing, and toxicology of annatto (Bixa orellana L.). Journal of Food Science and Technology, 40(2), 131–141.
71. Scotter, M., (2009). The chemistry and analysis of annatto food coloring: A review. Food Additives and Contaminants, 26(8), 1123–1145.
72. Scotter, M. J., Wilson, L. A., Appleton, G. P., & Castle, L., (2000). Analysis of annatto (Bixa orellana) food coloring formulations, Part II: Determination of aromatic hydrocarbon thermal degradation products by gas chromatography. Journal of Agricultural and Food Chemistry, 48(2), 484–488.
73. Shabbir, M., Islam, S. U., Bukhari, M. N., Rather, L. J., Khan, M. A., & Mohammad, F., (2017). Application of Terminalia chebula natural dye on wool fiber: Evaluation of color and fastness properties. Textiles and Clothing Sustainability, 2(1), 1–9.
74. Sharma, P., Segat, A., Kelly, A. L., & Sheehan, J. J., (2020). Colorants in cheese manufacture: Production, chemistry, interactions, and regulation. Comprehensive Reviews in Food Science and Food Safety, 19(4), 1220–1242.

75. Silva, D. O., Tamiozzo, L. M., Duarte, C. R., Murata, V. V., & Barrozo, M. A. S., (2011). Modeling of seed coating in a spouted bed. Drying Technology, 29(3), 286–294.
76. Silva, G. F., Gamarra, F., Oliveira, A. D., & Cabral, F. A., (2008). Extraction of bixin from annatto seeds using supercritical carbon dioxide. Brazilian Journal of Chemical Engineering, 25(2), 419–426.
77. Smith, J., (2006). Annatto extracts. In: 67th JECFA – Chemical and Technical Assessment (CTA) (pp. 1–21). Rome – Italy: FAO.
78. Taham, T., Cabral, F. A., & Barrozo, M. A., (2015). Extraction of bixin from annatto seeds using combined technologies. The Journal of Supercritical Fluids, 100, 175–183.
79. Taham, T., Silva, D. O., & Barrozo, M. A., (2016). Improvement of bixin extraction from annatto seeds using a screen-topped spouted bed. Separation and Purification Technology, 158, 313–321.
80. Taylor, L., (2002). Technical data report for bitter melon (Momordica charantia). In: Herbal Secrets of the Rainforest, (2nd edn. pp. 1–10). Austin – US: Sage Press Inc.
81. Thostenson, E. T., & Chou, T. W., (1999). Microwave processing: Fundamentals and applications. Composites Part A: Applied Science and Manufacturing, 30(9), 1055–1071.
82. Torres, R. A. C., Santos, D. T., & Meireles, M. A. A., (2015). Novel extraction method to produce active solutions from plant materials. Food and Public Health, 5(2), 38–46.
83. Van, C. H., Hoi, N. T. N., & Eun, J. B., (2012). Improvement of bixin extraction yield and extraction quality from annatto seed by modification and combination of different extraction methods. International Journal of Food Science & Technology, 47(7), 1333–1338.
84. Van, C. T., & Chin, K. B., (2016). Effects of annatto (Bixa orellana L.) seeds powder on physicochemical properties, antioxidant, and antimicrobial activities of pork patties during refrigerated storage. Korean Journal for Food Science of Animal Resources, 36(4), 476–486.
85. Vardanega, R., Nogueira, G. C., Nascimento, C. D., Faria-Machado, A. F., & Meireles, M. A., (2019). Selective extraction of bioactive compounds from annatto seeds by sequential supercritical CO2 process. The Journal of Supercritical Fluids, 150, 122–127.
86. Vargas, F. D., & López, O. P., (2003). Natural Colorants for Food and Nutraceutical Uses (1st edn., pp. 343). Boca Raton – Florida: CRC press.
87. Vasu, S., Palaniyappan, V., Kothandam, H. P., & Badami, S., (2010). Microwave facilitated extraction of bixin from Bixa orellana and its in-vitro antioxidant activity. Der Pharmacia Lettre, 2(2), 479–485.
88. Veggi, P. C., Martinez, J., & Meireles, M. A. A., (2013). Fundamentals of microwave extraction. In: Chemat, F., & Cravotto, G., (eds.), Microwave- Assisted Extraction for Bioactive Compounds: Theory and Practice (Vol. 4, pp. 15–52). New York – USA: Springer Science & Business Media.
89. Vilar, D. D. A., Vilar, M. S. D. A., Raffin, F. N., Oliveira, M. R. D., Franco, C. F. D. O., De Athayde-Filho, P. F., & Barbosa-Filho, J. M., (2014). Traditional uses, chemical constituents, and biological activities of Bixa orellana L.: A review. The Scientific World Journal, 2014, E-article ID 857292.
90. Vinatoru, M., (2001). An overview of the ultrasonically assisted extraction of bioactive principles from herbs. Ultrasonics Sonochemistry, 8(3), 303–313.
91. Vinatoru, M., (2015). Ultrasonic assisted extraction (UAE) of natural products: Some guidelines for good practice and reporting. Ultrasonics Sonochemistry, 25(1), 94–95.

92. Yolmeh, M., Najafi, M. B. H., & Farhoosh, R., (2014). Optimization of ultrasound-assisted extraction of natural pigment from annatto seeds by response surface methodology (RSM). Food Chemistry, 155, 319–324.
93. Yong, Y. K., Zakaria, Z. A., Kadir, A. A., Somchit, M. N., Cheng-Lian, G., & Ahmad, Z., (2013). Chemical constituents and antihistamine activity of Bixa orellana leaf extract. BMC Complementary and Alternative Medicine, 13(1), 1–7.
94. Younes, M., Castle, L., Engel, K. H., Fowler, P., Frutos, F. M. J., & Aquilina, G., (2019). Safety of annatto E and the exposure to the annatto coloring principles bixin and norbixin (E 160b) when used as a food additive. EFSA Journal (EFSA Panel on Food Additives and Flavorings (EFSA – FAF Panel)), 17(3), E-article ID 05626.
95. Zabot, G. L., Moraes, M. N., & Meireles, M. A. A., (2018). Process integration for producing tocotrienols-rich oil and bixin-rich extract from annatto seeds: A techno-economic approach. Food and Bioproducts Processing, 109, 122–138.
96. Zhang, H. F., Yang, X. H., & Wang, Y., (2011). Microwave assisted extraction of secondary metabolites from plants: Current status, and future directions. Trends in Food Science & Technology, 22(12), 672–688.
97. Zhang, Q. W., Lin, L. G., & Ye, W. C., (2018). Techniques for extraction and isolation of natural products: A comprehensive review. Chinese Medicine, 13(1), 1–26.
98. Zinoviadou, K. G., Galanakis, C. M., Brnčić, M., Grimi, N., Boussetta, N., Mota, M. J., & Barba, F. J., (2015). Fruit juice sonication: Implications on food safety and physicochemical and nutritional properties. Food Research International, 77(Part 4), 743–752.

CHAPTER 15

Foodborne Pathogens and Food-Related Microorganisms: An Overview

Probir Kumar Ghosh, Pijush Kanti Ghosh, and Nupur Ghosh

ABSTRACT

Food has been one of the best thriving media for microorganisms. The infection by microorganisms makes the food biologically unsafe and could invite a host of diseases. There are many Foodborne pathogens that are critical for consumers as well as experts. Many of these micrograms can also result in death of the consumer. Even crops, post-harvest products, products in transit, at warehouses and final packaged products need to be scrutinized. The intent has been to sensitize the scholarly community about the developments that have been made in the understanding of means of detection, aversion techniques and also educate about the role of probiotics, the benevolent microorganisms. Role of modern techniques both in detection and aversion has been discussed in this chapter to enlighten on the developments made and act as a guide for future work.

15.1 INTRODUCTION

Microorganisms have been cohabiting with humans, ever since our inception on the Earth. In fact, they predate the human advent. Microorganisms have both spoiled our food, and also been a part of it. Astrobiologists crave to find any of microbial colonies on any meteorite, as sign of life beyond the Earth, attesting their capacity to survive extremities of nature, and their resilience towards hostilities. It also highlights the concern that we as a species should have from them (microorganisms), since they have the capacity to eliminate

The Functional Foods: Nutrient and Health Benefits.
Megh R. Goyal, Junaid Ahmad Malik, and Anu Kumari (Eds.)

us from the Earth. On the contrary, it's also true that many of our savories would be incomplete without non-harmful microbial incorporation. Particularly the fermentation industry of food science would be crippled without it. In the past, when canning was relatively new and thus was amateurishly conducted without adequate sterilization measures, the disease of botulism was frequently witnessed, which is cause by *Clostridium botulinum.*

Now-a-days thankfully to aseptic practices and knowledge, such cases are almost nil. Lately, due to upsurge in contaminations there has been an increasing awareness among consumers and food processors about microbial load in food products and its effects. The said load regulates the immediate and long-term safety of raw materials, semi-finished and finished products.

United States Food Safety Authority (USFDA), EFSA, and Food Safety and Standards Authority of India (FSSAI) have been commissioned to ensure safety of public health through vigilance of food products and its ingredients, production, and storage practices. Their usefulness in fact is felt more in today's world than ever before. These agencies have released several guidelines and mandates for manufacturers, traders, importers/exporters, and they also conduct timely consumer awareness initiatives. Many producers manufacturing food products, both existing and new, make such compliance a priority in their quality assessment processes. As a vigilant customer, the onus also falls upon us to make even small vendors aware of the risks of foodborne pathogenicity.

When contemplating on microorganism being a friend or foe, it is important to understand their categorization by survival mode. Broadly, as is well known, microorganisms have been categories into psychrophiles, mesophiles, and thermophiles; and those affecting food quality are mainly mesophiles. Since mesophiles are active at near ambient temperatures (20–45°C), it is obvious that most spoilage occurs in tropical and temperature environments. No wonder refrigeration has been a successful technique for households to preserve food products. Of course, exceptions of resilient strains affecting food at cold/hot conditions also constitute a share, but broadly speaking mesophylls are in majority. Hence, owing to the diversity in usefulness on the upside and the risks on the downside, it is important to appraise ourselves on the risks/gains they offer.

The wealth of knowledge created by researchers, and the newer observations they make are very intriguing. It is true that from the earliest times of human civilization, food processing has benefited from controlled microbial fermentation. Till date, numerous indigenous savories are a product of such a practice. However, several bacteria and fungi do produce toxins that affect the quality of food and health of consumer. Therefore, microbes can create a serious havoc in our lives or, be an important salutatory ingredient in our diet, depending upon who between us materializes the opportunity first. For example, even a portion of neglected curd or probiotic milk at ambient temperature does turn sour. Bacterial endotoxin and exotoxin are common notations today. Exotoxins are mostly released from gram positive bacteria when it is alive; whereas endotoxins are released by gram negative bacteria after its death. This could be better understood with examples of *Staphylococcal* enterotoxin released by *Staphylococcus aureus* [34], and listeriolysin O produced by *Listeria monocytogenes* that has RBC destroying capability [9]. Fungal infections are also significant in outbreak intensity and spread, with fungal mycotoxins being a nightmare [119].

Toxins are equally concerning as the original producer, since the microorganism can be targeted by processing steps and antibiotics, but toxins remain and carry forward the risk [71]. This further affirms the need for strong analytical procedures to be in place for early monitoring of spoilage microorganisms before the spread of the disease. Apart from bacteria, one of the common foodborne viruses has been Hepatitis E virus and has been a major contributor to liver diseases, with meat being commonest source. The virus thrives best where sanitation is poor [96]. Parasites that have caused concern for food safety include *Cryptosporidium* spp., *Trichinella* spp., and *Toxoplasma* spp., and are found in poorly sanitized food products [1, 13].

This chapter collates the recent findings made by researchers and discusses their impact on our present understanding of microbial utilization and risk assessment. This exposition has been segregated into several subsections based on subtopics, under the broad section of foodborne diseases and related microorganisms. The subsections explain the research conducted recently to delineate the current scenario of our knowledge on pathogenicity by vectors in our food chain. Besides, microorganism that have been employed to aid the health of consumers, and in food processing have also been described. The aim is to enrich our perspective on modern day food science with regards to utilization of these microorganisms, and design adequate aversion steps wherever necessary to avoid contamination.

15.2 FOODBORNE ILLNESS: CRITICAL POINTS TO CONSIDER

Foodborne diseases have been troublesome across the globe. The risk assessment of foodborne pathogens is essential to ascertain the impact in case of an outbreak. Chrisidis et al. [19] in Canada found *Salmonella* spp. to be one of the premier contagions in the country in common food products, followed by *Campylobacter* spp., when analyzing disease prevalence in an adult cohort, between 2005 and 2015. The diseased state also has a cost to be borne by the patient. Researchers have highlighted the impact of cost of disease in economic terms since health of a nation and its food surplus influence its productivity, consumption, and economy directly. An integration of epidemiological data and economic pattern through various models has revealed the relevance of these two parameters. Combining epidemiological and economic models to have an integrated understanding on the dynamics of burden of disease financially has aided perspective development on the subject from even a socio-economic angel [5].

Not always does the cost materialize as cost for cure and rehabilitation, but it also manifests as ill-effect on work efficiency and quality of life. The burden of illness and cost of disease are some of the relevant parameters that have been considered while assessing costs accrued. The European Union (EU) and Russia have considered the case seriously and have rolled out few regulatory and legislative frameworks. The authorities are concerned about toxic chemicals, microbial load in food products and pre-processing and post processing hygiene [106]. Many incidences have been witnessed wherein the microbial adulteration is deliberate; while, in certain cases it is accidental. It is important to be vigilant about work place hygiene, and cleanliness in food industry to avert episodes of food adulteration, owing to microbial contamination. Such an unfortunate incident was indeed witnessed in South Africa owing to *Listeria monocytogenes* in processed meat, but how such an event occurred remains a mystery. The outbreak affected 1,060 patients among whom 216 died [11]. These cases attest the severity of pathogenicity of microorganisms and the toll it takes on consumers.

The food processing experts opine that it would be appropriate to have certain provisions as preventive steps to decontaminate raw unprocessed food reserves, hygiene maintenance during transit (to and from the factory), in factory cleanliness and safe packaging practices [78]. A rapid alert system, the rapid alert system for food and feed (RASSF) raises alarms on food safety issues and pathogen outbreaks in food processing and supply chain has been worked upon. Ishwika diagrams are also often used to assess the nature of the outbreak in the food industries in terms of 'what happened, when, where, why, and how,' and aids in predicting the potential hazards at processing steps [48, 97]. Additionally, curtailment of contamination before harvest and thereafter should be best avoided. Inadequate GMP facilities at food manufacturing units also raises concern of microbiological safety. In a reported incident, a truck hauling waste culled cantaloupes to a cattle farm got infested with *Listeria monocytogenes* and the produce entered the food processing unit [72]. Scientists have been particularly alerted by *Listeria monocytogenes* owing to its ability to survive cold temperatures, and its protection strategy wherein it creates biofilms around it ensuring its survival in dairy, meat, fish, poultry, and ready to eat factory settings [97].

Provisions should also be made decontaminate agricultural water from dissolved foodborne pathogens that can ultimately affect the plant and its products. An incident linked to this phenomenon has been documented in St. Paul in USA in 2008, wherein contaminated agricultural water affected jalapeno and Serrano pepper that affected 1,500 people with salmonellosis. A similar spread was reported in 2010 in alfalfa sprouts with a morbidity of 140 consumers [86]. Certain bacteria such as *Escherichia coli* O121 and *Salmonella* spp. have even been found to be heat resistant, that too in heat treated (baked) food products, such as muffins [73].

15.3 POST-HARVEST RISKS IN AGRO-PROCESSING RELATED TO MICROBIAL CONTAMINATION

Considering the prevalence and incidence of newer cases of microbial attack in agro-products, it is important to discuss the findings on the same. Farm to fork concept assumes safe handling of food materials while in transit and in the supply chain. However, many a times this is not the reality. Despite best practices in place today, there still are caveats.

A systematic review has analyzed microbial contamination in fresh farm produce consisting of fruits and vegetables, owing to unsafe post-harvest processing [69]. These reviewers have assessed the prime microbial accused to be *Listeria monocytogenes* that thrives well in fields, packing houses and even in packaged products. The microorganism even changes its survival strategy according to the food product and endures many hostile conditions. It thrives best in mesophilic conditions, and affects apples, asparagus, blue berries, melon, cantaloupe, celery, cucumber, jalapeno, lettuce, mango, mushroom, sprouts, spinach, peach, and strawberries. The carrying capacity, i.e., the maximum colonization by the microorganism in a given environment, also seems to vary. This attribute can be influenced by parameters even as the surface contour of the food product, as reported by said reviewers.

Factors of relative humidity and storage practices also have their influence in ensuring a safe food processing and storage practice (a modified atmosphere packaging (MAP) does seem to avert the microbial growth) [69]. Another find has deliberated on the risk of post-harvest microbial attack and the concerns it poses. The group working on pears c.v. Conference observed fungal attack by *Peniccilium expansum* and *Rhizopus stolonifer* with storage [104]. Similarly, characterization of microbial load in raw vegetables in China has revealed *Salmonella* spp. attack, some by serotypes that are even drug resistant [116]. These studies indicate the threat of microbial attack in common households and eateries, if processed slackly. Since microorganism have been so prevalent in food systems, it is important to detect them effectively to prevent diseases. There are many which are advanced techniques to aid in quantifiable detection effectively. The subsequent sections in this chapter elaborate on the usefulness and results of research studies that have been made.

15.4 DETECTION OF MICROORGANISMS IN FOOD SYSTEMS

15.4.1 MICROBIOLOGY-BASED ASSAYS

Microbiological assays (MBAs) depend on culture medium, and an appropriate choice of the same would largely predict the efficiency of detection. Considering this fact, Andritsos et al. evaluated the performance of four culture media to test their effectiveness in monitoring spoilage that is effected by *Campylobacterium* spp. among infected chicken samples [4]. They employed enumeration and detection methods in their work. The culture media they worked upon were CASA, BCCA, mCCDA, and Karmali agar (KA), and the assay was conducted according to ISO guidelines. Although bacteria were detected at >90% of samples by enumeration method, only CASA media was able to detect *Campylobacterium* spp. at 87% (rest being at < 40%). The authors have strongly suggested employing CASA chromogenic media, next to the ISO recommended mCCDA media to detect the said microorganism in food samples; or better, employing two parallel plating methods (enumeration and detection in tandem) for a better detection. This brings into light the importance of screening of effectiveness of microbial broth in such assays for accurate identification target microbial group, besides having a generic preference. Dairy industry products are also prone to microbial attack. The prime risk stems from *Escherichia coli* that contaminates milk during processing. Total agar plate count by the classical surface spreading technique method has been engaged for enumeration of bacteria in ewe's and goat milk in Switzerland. The target microbes producing *Escherichia coli, Staphylococcus aureus* and *Salmonella* spp. and *Enterobacter* spp. that produce were Shiga toxin [35].

While working on bacteriophages (viruses that infect and kill bacteria) as a counter mechanism against *Escherichia coli,* researchers have gained substantial insight into the pattern of inhibition caused by phages. The team working on the subject made substantial progress in targeting heat-tolerant bacteria (*Mycobacterium* spp., *Paratuberculosis* spp.), and bacterial endospores that endure even pasteurization. Bacteriophages are known to be specific with regards to type of bacteria they kill (genus and species); hence, they can be used as detection aid to identify the kind of bacterial contamination present in milk, quicker than plate count method that requires at least 24 h. This results in phage-mediated biocontrol of infection [80]. These authors also emphasize that milk, being a liquid, aids in an easy spread of phage particle within the volume of the product. Given the fact that few phages are already in GRAS list, phages could be a useful bacterial detection and elimination asset for dairy scientists. Phages can also infect biofilms and packaged product, thus facilitating detection of bacteria concealed in a post-processing packaged food pack or, under its biofilm, and a promise for its possible elimination [42].

Reporter phages is yet another arena that has been developed wherein reporter genes is injected into the bacteria. Upon being translated into a bioluminescent protein, the gene alarms bacterial presence by bioluminescence. Such a phage is *Listeria* phage A511 encoding luminescent luciferase gene. Inclusion of *Listeria* spp. specific phage lysin immobilized onto beads for isolation *Listeria* spp. in Camembert and red smear cheese has achieved 99% efficiency [56]. Similarly, 85% efficiency has been obtained with phages employed to recover *Campylobacter jejuni* in milk [85]; whereas *Salmonella* spp. has been separated from milk and chocolate milk using phage S16 [26]. Moreover, phages conveniently identify *Staphylococcus aureus*, *Coronobacter* spp., *Salmonella* spp., *Enterococcus* spp., *Yerisinia* spp. in various food samples [80]. Phages could also be commissioned as bacterial biosensors, given their specificity in target bacteria.

15.4.2 DETECTION BY MOLECULAR BIOLOGY TECHNIQUES

Contamination of spices is a nuisance in food process industry. Processors and food manufacturers have been concerned about the quality of spices used as raw material in food product development. In this regard, a major development has been the development of identification procedure of *Salmonella* spp. in garlic, onion, cinnamon, red chilies, pepper powder and green tea with molecular biology technique. The method involved automated DNA extraction and real time polymerase chain reaction (RT-PCR). PCR is able to capture foreign presence by amplifying the target pathogenic DNA and making it detectable. This detection otherwise may not be possible due to low quantity of the bacterium. The research group working on the subject also spiked the microbial broth with amino acids, as L-cysteine and DL-serine, to improve detection performance by neutralizing antimicrobial activity of few spices. The group claimed that detection was possible within 24 h, unlike classical techniques requiring at least overnight incubation [7]. These authors have also emphasized on the importance of their results since *Salmonella* sp. contaminate about 6% of the global spices as per USFDA.

In another study, a neoteric method for assessment of *Salmonella* spp. in edible products has been employed. In the said research, pork as model food has been worked upon, and is assessed for contamination by polymerase spiral reaction (PSR). Herein, an amplification technique for target DNA such as *Salmonella* spp. has been conducted [74]. The said technique is claimed to be 10 times more accurate in sensing *Salmonella* spp. contamination in samples, compared with end point-PCR. Therefore, the method is equivalent in efficiency to real time PCR (RT-PCR), with a 100% accuracy. The PSR technique is also much cost-effective, time-efficient, hassle-free, and equipment-less, as declared.

Microbial enrichment has been an essential step in reliable detection of microorganisms, since they are present in sample at very low concentrations. The process is, however, time consuming, cumbersome, and makes the process costly. Even without microbial enrichment, Kim et al. have devised a technique for reliable microbial identification. The group targeted identifying *Listeria monocytogenes*, *Salmonella typhimurium* and *Escherichia coli* O157:H7 in lettuce and beef. The said technique involves a simple pretreatment strategy by combining concentrations of bacterial DNA, by cutting upon the DNA elution volumes with modified PrepMan method for achieving an overall low limit of detection (LOD). As a result, there was a 100-fold decrease in LOD for *Escherichia coli* and *Salmonella typhimurium*; while three was 10 times decrease in LOD in *Listeria monocytogenes*. The entire process time was also greatly reduced to just 2 h for DNA extraction, filtration, and RT-PCR processes [50]. This find can be a great boon for quick food analysis for microbial contamination assessment, without the conventional need for microbial enrichment.

Sequencing is an important step to amplify a given set of genes. It could be made useful to amplify genes of pathogenic nature in food samples. Modern techniques have developed this science to a great extent, and few finds showcase the opportunities thus available. The next generation sequencing is a recently explored aspect in sequencing science that allows targeting microorganisms that cannot be subcultured or, pose difficulties for the same. Lewis et al. document that the said sequencing technique is successful in targeting *Salmonella* spp., and phage MS2 (a norovirus surrogate) with next generation Illumina MiSeq: 16S amplicon sequencing, and with RNA-seq, using ScriptSeq (Illumina) and NEBNext (New England BioLabs) kits [61]. These authors further noted that ScriptSeq was a more sensitive technique for detection of microbes having a LOD at 10^4 colony forming units (CFU)/reaction for bacteria, and 10^5 plaque forming unit (PFU)/reaction for phage. The NEBNext kit had 10^6 CFU/reaction for bacteria and 10^7 PFU/reaction for phage.

Multiple identification of microorganisms by multiplex PCR (phosphocreatine) assay using gold nanoparticles has been made possible. This technique has been successful in tandem detection of *Salmonella typhimurium*, *Listeria monocytogenes* and *Escherichia coli* O157:H7, the common foodborne pathogens. The gold nanoparticles that were flower shaped served as a chromogenic particle for colorimetric sensing. The annealing temperature for three genotypes together in PCR was at 531°C. LOD of three strains was 10 pg/ml for *Salmonella typhimurium* and *Listeria monoctogenes* and 50 pg/μl for *Escherichia coli*. A lower LOD would ensure greater sensitivity of the technique for even traceable amounts of pathogen, avoiding a breakout. The said technique has been described to be simpler, quicker, and allowing easy colorimetric assay. This method could therefore serve as a mobile testing kit for food sample analysis [29].

Foodborne bacteria *Vibrio parahaemolyticus* particularly affects food in Asian countries. It has been detected by a novel technique using sodium deoxycholate-propidium 3 monoazide-qPCR in shrimps (as a model food) using quantitative PCR (qPCR) [64]. These researchers employed sodium deoxycholate (SD)-propidium 42 monoazide (PMA)-qPCR (SD-PMA-qPCR) assay that could well identify viable cells from a mix of alive and dead cells. The said group used the tox R gene for identification. LOD for the assay for virus was 10 CFU/ml. In another attempt, to identify *Salmonella* spp. in food products, few researchers have developed a tandem approach of recombinase polymerase amplification (RPA) and lateral flow immunoassay (LFD-RPA). The targeted gene was a conserved fragment (*fim Y*). The LOD was 1.29×10^2 CFU/ml for *Salmonella* spp. [63].

15.4.3 DETECTION OF PATHOGENS BY ANALYTICAL TOOLS AND TECHNIQUES

Colorimetric methods have been quite popular in analytical chemistry because of the presence of production of a colored complex that could be quantified based on the color intensity. The said method holds merit in prudent identification of microorganisms in food samples as well, by colorimetric aptasensor. This method has been useful in quantifiable estimation of common food pathogens *Salmonella* spp., *Escherichia coli* and *Listeria monocytogenes* simultaneously. The group working on the subject used gold nanoparticles and aptameter aptasensor [59]. They have delineated that colloidal gold nanoparticles labeled the target sites and tagged by Ap6 DNA aptameter (screened amongst 25 DNA aptameters choices). The spectrophotometric identification for tagged microbes was found best at 525 and 625 nm, with LOD < 10^5 CFU/ml. The sensing technique was tried on market samples of ham and chicken, and findings were 96% accurate, 96% specific and 100% sensitive. This technique therefore can be safely relied upon for dynamic analysis of food samples in warehouses, markets or in analytical settings.

Targeted identification of *Listeria monocytogenes* with DNA amplification with quartz crystal microbalance at high temperatures has also been carried out [107]. The role of quartz has been to be a part of quartz crystal microbalance on which thiol-modified-ss-DNA probe is immobilized that specifically targets the target microorganism's DNA (phosphotidylcholine-phospholipase C, *plcB*). The process is operated at 50–55°C for DNA probe immobilization, and for DNA targeted hybridization. This method could identify even small quantity of pathogen in the products. The LOD is 0.3–3 CFU/ml, with high sensitivity and with 100% selectivity. The sensitivity however could be further improved two-fold by co-immobilizing L-cysteine because the amino acid increases the probed to probe distance by acting as a spacer. This technique could be useful for quick and reliable assay of *Listeria monocytogenes* contamination in food products.

Salmonella spp. contamination in milk is also a matter of anxiety for the food industry. But the bacteria is quite elusive to be detected with ease. A novel biosensor-based method has been recently developed using multiunit sensing probe of poly-L-lysine (PLL), diethylene diamine pentacetate (DTPA), streptavidin (SA) and gadolinium (Gd). The functional unit involved in the study was PLL-DTPA-SA-Gd-biotin-antibody. The bound complex of *Salmonella* spp. and the sensing probe were analyzed by nuclear magnetic resonance (NMR) imaging. This sensing method achieved a LOD of 2.40 ×

10^4 CFU/ml, with the process turnaround time at 2 h. The method is direct and eliminates any unwanted interference during sample detection, plausibly owing to identification of magnetic signature emanating from the target. This makes the process a suitable choice for analysis. SA improves specificity of the probe, and biotin improves binding capacity of the probe to target. Further, use of contrasting agents such as Gd made detection more accurate [112].

The advanced technique of matrix associated laser desorption/ionization time of flight (MALDI-TOF) chromatography has been also very useful in detecting foodborne pathogens. This technique is considered to produce soft ionization of particles based on proteomics for bio-mass spectrometry, even with a simple pretreatment.

Yan et al. have elaborated on the simplified use of the said instrument for a facile analysis of contamination with a whole cell method [114]. The group employed principal component analysis (PCA) to depict data points graphically for better visualization and analysis. They worked on six foodborne pathogenic bacteria, viz., *Shigella flexneri, Escherichia coli, Staphylococcus aureus, Salmonella enteritidis, Pseudomonas aeruginosa,* and *Listeria monocytogenes.* Their results asserted that direct estimation of whole cell bacteria, skipping protein extraction step, provided better traceability (peaks in analysis) and reproducibility. However, completion of the current protein databank has been recommended by these researchers as a suggestion for future work since current identification techniques depend on protein extraction from microbial sources.

15.4.4 NANOTECHNOLOGY-BASED DETECTION

Detection of microorganisms at the macro-level is challenging, especially considering their low concentration in food samples. Thus, there is a requirement of shifting from macro to micro, and previous sections have elaborated on various 'micro' based techniques that have been applied for analysis. Since the efficiency of detection increases as scale of analysis decreases, this principle has been applied on pathogen identification as well. In order to attain greater specificity and accuracy in assays, in lesser time and with lower product as well as microbial concentration, further downscaling has been opined. Nanotechnology is a suitable obvious choice for miniaturized analytical procedures.

Staphylococcus aureus is a foodborne pathogen that has high severity in a using food poising, affecting meat, poultry, eggs, marine, and dairy products.

It can also produce heat resistant toxins [105]. In order to facilitate easy identification of the microorganism, Yao et al. have developed a colorimetric immunoassay. The assay highlights the use of magnetic carbon dots and silver nanoclusters such as o-phenylenediamine (OPD). This magnetic molecular complex causes rapid enrichment of target microorganism and their quick separation from sample mix (such as: meat in the said study). Thereafter, silver nanoclusters react with OPD as an artificial enzyme, producing a yellow-colored complex with an absorption maxima at 421 nm. The process of detection is sensitive for concentration ranging from 10^1–10^6 CFU/ml, with LOD at 4.9 CFU/ml, and attaining successful recovery of microorganisms from food sample in > 85% of cases [117].

For detection of *Salmonella enteridis* in milk, He et al. employed variable heavy chain only antibody (VHH) or nanobodies, in a first of its kind study. These nanobodies specific for the bacteria were developed after the panning-process and exhibited appreciable thermal stability and specificity. Working with the nanobodies in a sandwich enzyme linked immunosorbent assay (ELISA) mode, after 10 h of enrichment, a LOD of 6 CFU/ml in milk has been obtained for these microorganisms. This technique has considerably cut short the conventional labor and time (2–3 days) involved in assessing microbiological data. The accuracy of such antibodies is also very high compared with conventional techniques since these are pathogenic specific and are raised from immunochemistry-based antibody library, the camelid library [44].

For identification and separation of *Escherichia coli* in food systems, a biosensor-based approach has been worked upon. Aptameter embedded magnetic nanoparticles (MNPs) and cDNA, and up-conversion nanoparticles (UCNPs). The change in concentration of the conjugated complex of MNP-cDNA-UCNP with increase in *Escherichia coli* could be quantified at 662 nm spectrophotometrically. He et al. worked on this technique using suing pork as sample. They reported that the colored complex serves as a colored fluorescence chromophore. In a sample concentration of bacterium at 58×10^{-6} CFU/ml, the technique allowed achieving a LOD of 10 CFU/ml [62]. Hence, as compared with standard plate count method, the performance of the biosensor has been appreciable. The scheme of reaction has been shown in Figure 15.1.

Gold nanoparticles have also been useful in quick and accurate estimation of foodborne pathogens. These nanoparticles are easy to produce, electron rich, optically measurable, target specific and offer great stability [83]. The nanospheres of gold have an absorption maxima at about 520 nm, while other shapes have absorption red-shifted to near infrared (NIR) wavelength.

They can be synthesized as sphere, rods, cube, and star to name a few. They have around 1,000 times higher extinction coefficient than organic dyes that increases their sensitivity [27]. To improve specificity of gold nanoparticles, attachment of oligonucleotides, antibodies, and aptameters (specific oligo nucleotides or special single stranded DNA sequence) are often practiced on the nanoparticle surface. Such an application has been done by Ma et al. developing a biosensor for monitoring *Salmonella typhimurium* presence in food samples, and it achieved LOD 56 CFU/ml [67], gold nanoparticle with recombinase polymerase was able to sense *Salmonella* spp. in milk with LOD 50 CFU/ml [17], *Campylobacter* spp. has been identified in chicken with gold nanoparticle based aptasensor [53], oligonucleotide functionalized gold nanoparticles for scrutiny of *Listeria monocytogenes* [28].

In food toxin/foreign particle analysis, newer biosensors have been designed lately as reported by Saravanan et al. that could be segregated into various categories based on functional roles such as those dependent on recognition elements (based on enzyme, antibody-protein, microorganism), those depended on measurable signals (piezoelectric, optical, electrochemical, bioluminescence), transducer based (grating, resonator, interferometer,

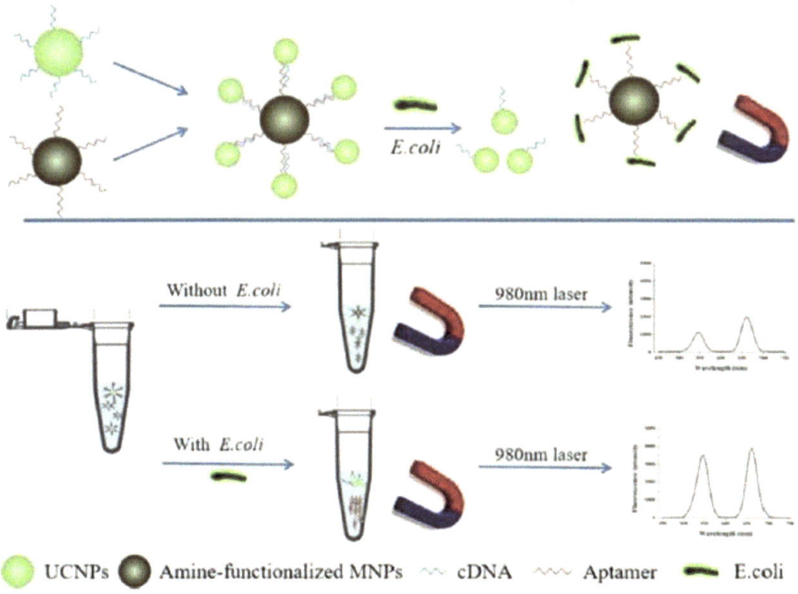

FIGURE 15.1 Flowsheet for magnetic nanoparticles-based estimation of *E. coli* in food samples.

Source: Reprinted with permission from: Ref. [62], ©2020 Elsevier.

refractometer, surface plasma resonance (SPR)). These authors favor use of electrochemical biosensors over optical and mass-based biosensors (such as food matrices), since the former is able to work even with sample turbidity with appreciable sensitivity. Optical and electrochemical biosensors have been promoted in low target analytes such for biologicals [89]. Few authors report employment of electrochemical sensors designed with nanocomposites (nanometal oxide and graphene) for detection of amino acids in food samples [93].

While working with colorimetric techniques, they require enrichment step for quantifiable detection of pathogens. However, the following technique waive that step also. The technique capitalizes on the mutual attraction between positive polyethyleneimine coated gold nanoparticles with negatively charged enzyme (β-galactosidase), and their interaction with the bacteria due to competitive binding. More the bacteria is bound to the aforesaid unit, more the enzyme is released from the adduct that hydrolyzes the subsequently added substrate chlorophenol red β D-glucopyranoside. This forms a yellow to red color in 10 min after 2–3 h of incubation. This method is simple compared to its contemporaries and has been successful in detecting Gram positive and Gram-negative bacteria in potable water samples [103].

In another technique, surface enhanced Raman scattering (SERS) has been used with nanoparticles for detection of foodborne pathogens. Using a sandwich model, investigators have shown possibility of detection of *Escherichia coli* in food products with spherical gold nanoparticles, gold nanorods and SERS. The LOD was 10 CFU/ml [12]. In fact, the combination of gold nanoparticles with aptameters that has been termed as 'spiny' has been used for detection of *Salmonella typhimurium.* The Raman scattering resulted in effective detection of bacteria with LOD of 4 CFU/ml [68].

The lanthanide labeled fluorescent nanoparticle immune-chromatographic strips have exceled at sensing *Escherichia coli* in food samples. Color change is observable in 15 min, even with the naked eyes. Visual LOD was 10^4 CFU/ml and when viewed through strip reader it was 5×10^2 CFU/ml, without any cross reactivity. The strip has been tested on six strains of *Escherichia coli* and it had good consistency in reporting the bacterial concentration. For *Escherichia coli* O157:H7 specifically, LOD is 1 CFU/ml in beef, pork, and chicken samples (8 h incubation), and in milk samples (6 h incubation). Average recovery of the microorganism from product matrix was >85% [109].

15.4.5 ADVANCED CHIP AND COMPUTER-BASED MONITORING OF FOODBORNE PATHOGENS

Visual detection of microorganisms appears to be a long shot, but it may be a reality soon. Considering the potential of microfluidic chip, this feat is very much possible and has been accomplished with *Escherichia coli* in food products by Jin et al. [49]. This group created a self-priming compartmentalization (SPC), microdevice with polymethyl methacrylate (PMMA), and tagged this to loop-mediated isothermal amplification (LAMP). The device is a self-filling microarray that can fill in the sample through a pump. Being a closed device, there is minimal risk of cross contamination. This set up has been examined for assessment of many foodborne microorganisms such as *Salmonella typhimurium, Bacillus cereus, Staphylococcus aureus, Yersinia enterocolitica, Vibrio parahaemolyticus* and *Listeria monocytogenes*. The analytical data is generated within an hour, even when multiple strains are present together. The assay procedure is rather simple, and the setup is portable, making it a better choice over existing detecting methods.

In microfluidic chip-based system for pathogen identification, the microorganisms grow in a single drop of the analytical mix that also contains resazurin, resulting in fluorescent-based detection of bacteria within 5 h. The LOD was found to be 50 CFU/ml and analysis time just 5 h comparer to overnight incubation in classical plate count technique. This setup has been examined for efficiency with milk [3].

A recent publication suggests quick detection (extraction) of foodborne via on-site capture and detection of pathogenic DNA effected by zinc oxide-based sub-micro-pillar arrays. Zinc oxide coated pillars offer high surface area and adhere firmly to bacteria, resulting in enhanced capture of cells. This is because the array surface is positive charged that has enhanced its capture capacity for bacteria in the said setup. The study employed numerical analyzes to understand the finer details of the capture, and mode of operation of the device. The reaction time was <10 min. The LOD for *Escherichia coli* within 10^6–10^1 CFU/ml is observed. This setup could also differentiate between live and dead bacteria using PCR assay (genetic analysis). Favorable aspects of the device included combination of complex structure, high depletion force and high surface electrostatics, as reported [60].

Pseudomonas aeruginosa is another potential pathogen that contaminates food. Polydopamine-polyethyleneimine co-polymer dots have been developed for the said purpose. The arrangement has dual *Pseudomonas aeruginosa* aptameters attached to the dots for improved sensitivity, through labeling. The sensor is capable of detecting the bacteria in a concentration range from 10^1–10^7 CFU/ml, with LOD of 1 CFU/ml, in just 1.5 h. Trials with milk, orange juice and popsicle samples have been a success. The said technique is simple in operation and does not require sophisticated instruments, and organic chemicals [127].

Apart from using technology effectively for sensing microorganism, researchers have also desired making them user-friendly and precise with a smartphone-based method. The smartphone integrated paper sensing approach for a dual approach has been worked out. In this the method paper as well as colorimetric screening of microbes in food products has been designed. The Cu^{2+} triggered oxidation of OPD has been capitalized as the reaction concept. Herein, the capacity of *Escherichia coli* to reduce Cu^{2+} to Cu^{+}. This phenomenon has been made useful to mitigate the initial color of the reaction complex comprising of visible pale-yellow color and orange-yellow fluorescence. Next, a filter paper based visual sensor has been constructed that could employ OPD-Cu^{2+} system with assistance of a UV lamp. The color of sample zones on the paper change from green to dark blue with increasing *Escherichia coli* concentration, when measured by UV lamp at 302 nm The resultant color of the mix could be made useful to detect *Escherichia coli*, through a smartphone-based color scanning application, recording the RGB intensity values. The entire system is sensitive to detect *Escherichia coli* between 10^2 and 10^6 CFU/ml. The fluorescent detection system has LOD 100 CFU/ml and colorimetric method's LOD is 44 CFU/ml. However, visible color change is evident even to the naked eye, especially under daylight that has UV rays. The system has been testified with tap water and milk [108].

Molecular docking-based technique has also revealed considerable potential of computerized simulations to detect pathogens in food. Protease and human serum albumin have been cited as useful domains of the future. They can be used to design specific peptides and protein bodies to target pathogenic microorganisms surface protein to trap them. This could be made useful for enumeration of such microbodies in any food sample as well [101].

15.5 PROCESSES TO ENSURE FOOD SAFETY FROM MICROORGANISMS

Food can be considered the most vulnerable asset for microbial spoilage. Except for few low moisture foods, high salt and oil content foods, most of which are deliberately processes, and all other types of solid, liquid, and semi-solid foods with intermediate and high moisture are vulnerable to spoilage. The challenge is to not only to design steps for elimination of contamination, but also ensure minimal to no loss of nutritive value, texture, and sensory appeal. This target could be difficult to achieve at the outset, but many modern technologies have overcome the bottleneck of food spoilage with novel means. The following sections under this topic provide brief overview of the scientific advancements.

15.5.1 THERMAL METHODS FOR AVERTING MICROBIAL SPOILAGE IN FOODS

The physical methods discussed in this section involve processes that employ non-chemical and non-biological means to avert microbial growth in food products. The first choice that is most common is hot air drying. Hot air drying could be considered one of the basic drying techniques, but it can also limit bacterial growth. According to Zhou et al., hot air drying of basil leaves reduced *Salmonella senfetenberg* significantly by 4 log reductions at 60°C in 90 min; at 80°C in 40 min; and at 100°C in 20 min. Upscaling to pilot plant scale using surrogate strains confirmed the decontamination effect of the treatment on bacteria. A higher vulnerability of *Escherichia coli* P1 over *Enterococcus faecium* was also observed in the said treatment [128]. These findings have significant importance since *Salmonella senfetenberg* has been documented to affect herbs (such as: basil leaves) and is more thermo-resistant compared to *Salmonella enteritis* PT30 and *Escherichia coli* O157:H7 [32, 84].

Escherichia coli ATCC 25922 in red pepper powder by hot air drying (thermal inactivation) has also been witnessed by Zhang and co-workers. Besides thermal effects, the authors have also included effects of water activity, temperature, and particle size in adjudging influence of temperature. For best inactivation, a lower particle size with a comparatively higher water activity, at low temperature has accrued higher bacterial death [120]. The results highlight the fine interplay of independent process variables in ensuring decontamination of pepper. The 5-log reduction in *Listeria monocytogenes* in non-fat spray dried milk powder (water activity 0.3) has been achieved by its heat treatment at 80°C for 60 min. This is a rather encouraging find since the said microorganism is known to endure low moisture content for a long time, but still be viable [6]. Thermal inactivation of *Escherichia coli* O157:H7 and *Salmonella* spp. in ground lean camel meat has also been possible. The D-values observed were 57.5–65°C, in an exposure time of 1.8 to 0.067 min for *Escherichia coli* [79].

Besides regular thermal heating, radio frequency heating has also been engaged for microbial sterilization of *Salmonella typhimurium* in red pepper powder having <0.4–0.7 water activity. The treatment consisted of radio frequency exposure causing a temperature rise to 70°C for 3 min. The bacteria suffered sub-lethal injury when water activity was only >0.50. The authors recommend high ramp in heating procedure so that the bacterium cannot synthesize heat shock protein, thus facilitating the cidal activity [123]. This achievement can prove to be very useful for low moisture foods.

15.5.2 NON-THERMAL PROCESSING METHODS

Although thermal processes have been classical and many new findings also attest their performance, yet they do no justice to volatiles and nutraceuticals that may be affected as a result of the inflicted thermal stress. Though thermal techniques are not completely discounted, the migration of interest to milder, colder, and more benign processes has begun. The following section provides an account of developments recorded in the non-thermal processing methods to arrest microbial growth in food products. They have been divided into radiation based and non-radiation-based methods.

Radiation processing of food has been in practice for quite some time; however, time and again food safety issues are raised since after all the radiation is lethal. The concern arises from the mode of action (MoA) of these rays, but their exposure can be regulated to ward off risks of unwanted and uncontrolled exposure. Irradiation has been debatable choice of sterilization of fresh and processed food in the past, but now modern technique renders controlled exposure for a controlled dose of irradiation that assures food safety both from pathogenic microorganisms as well as over-exposure hazards. Forbye radiation, non-thermal methods also include modern techniques of cold plasma, high pressure processing that have been worked upon. Radiation and non-radiation physical processes for limiting microbial contamination in food products have been enlisted in Table 15.1.

15.5.3 CHEMICAL, BIOLOGICAL, AND NATURAL RESOURCES FOR RESTRICTION OF FOODBORNE PATHOGENS

Chemical and biological treatment method comprising of chemical interventions to control microbial load are also an integral part of the defense against foodborne pathogenic microorganisms. These have been elaborated in Table 15.2. Since chemical treatment, even if not eradicated, is still debated; hence, nowadays natural antimicrobial agents have also become relevant. These natural sources for limiting microorganisms have been enlisted in Table 15.3.

TABLE 15.1 Radiation and Non-Radiation Methods (Physical Methods) to Control Foodborne Pathogens

Technique	Effect	References
Cold plasma processing.	Cold plasma treatment inactivated gas producing yeasts but was benign on lactic acid bacteria in fermented radish paocai.	[125]
High pressure processing.	Decontamination of deboned dry cured ham was brought about by 600 MPa for 5 min, resulting in 2 log units at surface and 3 log units inside the ham for *Listeria monocytogenes*.	[82]
Krypton-chlorine excilamp ($\lambda_{222\ nm}$) and UV-C LED ($\lambda_{280\ nm}$)	Detrimental effect on *Salmonella typhimurium* and *Listeria monocytogenes* by both; but UV-C had higher pathogenicity in flatfish compared to oysters (both model test food products), and vice-versa, due to lower turbidity in flat fish.	[52]
LED ($\lambda_{460\text{–}470\ nm}$) illumination.	Cellular injury to *Listeria monocytogenes* and *Pseudomonas fluorescens* at 4°C in cheese.	[47]
Pulsed light emitting diode ($\lambda_{395\ nm}$) treatment.	2.9 log reduction in *Salmonella* spp. by UV-A radiation in wheat flour.	[30]
UV-C radiation processing.	Decontamination of dried whole black peppercorn by UV-C radiation caused total log reduction of 1.92 CFU *Staphylococcus aureus* and 3.60 CFU in *E. coli*.	[37]
X-ray irradiation.	Arresting growth of *Escherichia coli* and *Salmonella typhimurium* in edible bird's nest by low intensity X-rays.	[122]

TABLE 15.2 Chemical and Biological Aversion Techniques for Aversion of Microbial Contamination in Food Products

Chemical/ Biological Source	Effects	References
	Chemical Sources	
Benzyl isothiocynate.	Inhibitory effect on *Escherichia coli* O157:H7 and *Salmonella enterica* caused cell membrane disruption resulting in cellular lysis; besides, 32% decrement in Shiga toxin produced by *E. coli*.	[81]
Chlorine (ClO_2) produced by sodium chlorite-acid reaction.	>4.0 log CFU/g reduction in *E. coli*, *Listeria monocytogenes* and *Salmonella* spp. was achieved by cumulative exposure of ClO_2 at 300, 1,300, 1,600 ppm-h in baby cut carrots, tomatoes, and blue berries (best results in carrots).	[15]

TABLE 15.2 *(Continued)*

Chemical/ Biological Source	Effects	References
Conenzyme Q0.	Coenzyme Q0 inactivated the microorganism by extending its lag phase and causing cellular damage to *Vibrio parahaemolyticus* when tested on infected fresh shrimps (MIC: 4–16 μg/ml).	[126]
Triglycerol monolaurate.	Detrimental effect on *Cronobacter sakazakii* in affecting reconstituted infant formulation.	[124]
	Biological Sources	
Antimicrobial peptide.	Antimicrobial peptide 1018-K6 eradicates enterotoxic *Staphylococcus aureus*, MRSA *Staphylococcus aureus* and *Listeria monocytogenes* biofilm with cidal MBC of 10 μM and biofilm preventing efficacy at 50 μM.	[22]
Cuticular wax in leaves.	Leaf cuticular wax in lettuce averts growth of *Salmonella* spp. on lettuce leaves at harvest and also post-harvest.	[57]
Narrow spectrum bacteriophage.	Narrow spectrum phage LSE7621 inhibited growth of *Salmonella enterica* serovar Enteridis in lettuce (0.86 log CFU/ml in 6 h) and tofu (3.55 log CFU/ml in 4 h).	[65]
Non-pathogenic surrogates of *Listeria monocytogenes*.	The non-pathogenic surrogate of *Listeria monocytogenes*, *Enterococcus faecium* and *Listeria innocua* were able to target pathogenic bacteria affecting apples.	[94]
Polyvalent bacteriophage.	Polyvalent phage PS5 was able to restrict *Salmonella enteritidis, Salmonella typhimurium,* and *E. coli* O157:H7 in food products.	[31]

TABLE 15.3 Natural Resources for Microbial Abatement in Food Products

Type of Natural Source	Effects	References
	Herbal Essential Oils	
Combination of thyme and oregano essential oils along with bacteriophage.	Treatment of chicken with plant essential oils (1.6%, 3 min) and bacteriophage (Salmonella) at 1.1 × 10^8 PFU/ml for 3 min restricted growth of *Salmonella typhimurium* by 1.9–2.0 CFU/g of antibacterial mix.	[75]
Lemongrass essential oil (citral-rich) and thyme essential oil (carvacrol-rich).	Detrimental effect of lemongrass and thyme essential oils combination was observed in *Cronobacter sakazakii* CICC 21544 through disrupted proton motive force and *psp* regulon.	[14]

TABLE 15.3 *(Continued)*

Type of Natural Source	Effects	References
Mentha haplocalyx and *Schizonepeta tenuifolia* essential oils.	Antifungal effect of essential oils of herbs on *Fusarium oxysporum, Fusarium solani, Cylindrocarpon destructans, Pythium aphanidermatum, Botrytis cinerea, Colletotrichum gloeosporioides* and *Rhizoctonia solani*; affecting cell membrane permeability, reduced mycelia weight, affected spore germination and yield.	[16]
Oregano essential oil (carvacrol-rich).	The essential oil disrupted caused membrane permeability, depolarization, and change in cellular respiration in *Listeria monocytogenes*, resulting in cidal effect. This mechanism was reflected in infected sliced bologna sausages.	[20]
Rose (geraniol-rich) and oregano (carvacrol-rich) essential oils.	The essential oils arrested growth of *Escherichia coli* MTCC 443 and *Bacillus cereus* MTCC 430 in goat meat when maintained at 4°C.	[100]
Thyme essential oil (thymol-rich).	Thyme essential oil had cidal effect on *Listeria monocytogenes* and β-lactum resistant *Escherichia coli*. Listeria biofilm was disrupted at MIC 100 μg/ml, and overall, MIC at 250–800 μg/ml when tested in piglet meat. Planktonic biofilm in bacteria could also be targeted.	[24]
Thyme essential oil (thymol-rich), oregano (carvacrol-rich) essential oil and cinnamon essential oil.	Inhibition of *Listeria monocytogenes* on surface of radish sprouts with MIC for anti-listetrial activity at 0.078 μl/ml, best results with combination of essential oils with a synergistic cidal effect.	[18]
Thyme essential oil and cinnamon essential oil.	Combination of thyme and cinnamon essential oil inhibited growth of *Salmonella* spp. in Tahini, an indigenous ready to food eat product of Middle-East, when tested at 10°C and 37°C.	[2]
	Spice Extracts/Oils	
Clove oil (eugenol-rich).	No biofilm of *Escherichia coli* O157:H7 formed on vegetables (cucumber and lettuce) treated with clove oil solid liposomes.	[23]

TABLE 15.3 *(Continued)*

Type of Natural Source	Effects	References
Photosensitizer curcumin.	Antimicrobial activity of photosensitizer curcumin in chicken skin to inhibit *Listeria moncytogenes* (MIC 10 ppm, MBC 20 ppm) and *Salmonella* spp. (200 ppm, 3.6 log reduction).	[38]
Photosensitizer curcumin with photodynamic inactivation.	Photodynamic inactivated photosensitizer curcumin at 0.1 μM completely inhibited *Listeria monocytogenes* within 5 min of exposure and showed cidal effect on planktonic microbiota of the organism at > 4 log reduction (>99% elimination) in 5 min at 0.2 μM.	[45]
	Plant/Fruit Extracts	
Blueberry extract.	Antimicrobial and membrane deformation effect of blueberry extract on *Vibrio parahaemolyticus* was achieved in salmon with < 25 mg/ml MIC and < 50 mg/ml MBC values for the microbial serovars ATCC 17802, ATCC 33847, and F 13.	[99]
Pecan shell extract.	Inhibitory effect of pecan shell extract was observed with on *Listeria monocytogenes* (1.25–5.0 mg/ml) > *Salmonella enterica* (2.5–5.0 mg/ml) > (>0.5 mg/ml) *Escherichia coli* O157:H7; in fresh-cut cantaloupes and thawed catfish filets effect was best pronounced in *Listeria monocytogenes* with 0.4–4.0 log reductions.	[118]
Silkworm thorn leaf extract.	Antimicrobial effect on *Listeria monocytogenes* as a rinsing medium for fresh cut romaine lettuce and kale (2 log reduction).	[111]

15.6 ROLE OF PROBIOTICS IN AIDING FOOD SUPPLEMENTATION

Microorganism have classically been villains in the food safety. However, a breed of their race is actually helpful to us nutritionally. The salubrious microorganisms which are injected as live (probiotics) have been recommended in our diets as well, since lack of adequate gut bacteria can lead to indigestion, and related complication of the digestive system. Probiotics, as is well known, consist of viable microorganisms that are healthy. Also, as mentioned before, there are certain food products that employ microorganisms as an ingredient, particularly those involving fermentation for products such as yogurt, curd, bread, soya sauce and likewise.

Considering the current shift in lifestyle and habitats that makes individuals indulge in pharmaceuticals (antibiotics), cigarette smoking, lack of physical activity, unbalanced food habits, population of the gut microbiota has taken a toll, and so has our assimilation capacity [8, 21, 91]. Strains such as *Lactobacillus casei* and *Bifidobacterium animalis* are few well known probiotics in food industry as on date. The last decade or so has seen considerable growth in consumer interest in probiotics. Though they have been rendering considerable benefits to all age group consumers, their mechanism of action is yet to be fully understood. Thankfully, many enteric diseases have been controlled by use of probiotic along with classical drugs, which highlights their restorative nature. Lactic acid bacteria (LAB) have averted activity of enteric pathogens in milk and milk products effectively. According to Bisht and Garg, the pathogenicity induced by a host of pathogens including *Bacillus subtilis, Bacillus cereus, Salmonella enteric, Shigella flexneri, Streptococcus pneumonia, Staphylococcus aureus, Staphylococcus epidermidis, Pseudomonas aeruginosa, Escherichia coli, Clostridium perfringens, Listeria monocytogens* and yeast *Candida albicans* could be averted by LAB, corroborating its choice as a biopreservative in food products [10]. But to reap benefits from probiotics, one needs to have a daily dose of 10^8–10^9 CFU in gastro-intestinal tract.

In probiotics, issues of monostrain or multistrain probiotics is a topic of deliberation among experts, with few opining a mix of probiotics to be better suited to meet nutritional needs than the individual strains. The generation of probiotics has also become relevant in the present day. First generation probiotics have differential endurance to human acidic-alkaline medium in gastro-intestinal tract. For example, *Bifidobacterium bifidum* survives 30% and *Lactobacillus acidophilus* survives 10% of human enteric acid-alkaline environment. Also, the bacteria in first generation have a limited shelf life, for example 14 days for *Bifidobacterium longum* [8]. In second generation probiotics, lyophilization has been conducted in polymeric capsules using natural, semi-natural or synthetic fillers. This has ramped up the survival of bacteria by 80%, whereas 90% gain in viability has been achieved with sodium alginate, along with achieving a shelf life of six months. The challenge that remained was the sudden and occurred before arriving in colon. Also, salutary components in bacteria were affected [54, 87]. The third generation further improved on sustainability of the bacteria with encapsulation in polymeric matrices and offering a targeted release. In fourth generation, bacteria are protected from environmental factors by a being a part of a biofilm. This film provides necessary protection against digestive tract pH, a considerably shelf stability and delivery in colon [87]. The generational timeline depicts the improvements made at each level that has been a boon to health supplementation through diet. Medical benefits have also been reaped from probiotics. LAB has also been attributed with medical benefits such as deliverance interleukin 10 to human gut [70], promoting anti-inflammatory response [110].

Besides bacteria, yeasts are also a source of probiotics. One of the best characterized probiotic yeasts has been *Saccharomyces boulardii* because it is better tolerant to heat and pH than *Saccharomyces cerevisiae*. The said strain also has a biopharmaceutical application [92]. The antibiotic effect of the bacterium on *Clostridium difficile* has also been demonstrated [55].

At pH 2, the survivability of *Saccharomyces boulardii* is up to 75% compared to *Saccharomyces cerevisiae* at 30% [41]. *Saccharomyces boulardii* is susceptible to bile at 0.1% [33]. This suggests great ability to withstand physiological conditions and still be bioactive. Also, *Saccharomyces boulardii* is able to confer medical benefits through immune-modulatory functions through facilitating short chain fatty acids (SCFs) production, and in maintenance of butyrate concentration in the blood [90], arresting proliferation caused by *Clostridium difficle* and enact antibacterial effects on *Bacillus* spp. [77]. *Saccharomyces boulardii* also aids restoration of tight junctions via E-cadherin and p120 catenin expression in colonic explants from inflammatory bowel syndrome patients [102]. Arguably, these features contribute to its high preference as a probiotic.

Aspergillus awamori when fed to broiler chickens caused weight gain compared to control group chicken [88]; whereas when the fungi *Chrysonilia crassa* was added to the diet of broiler chickens, immunity as well as body mass increased [98]. Similar improved body mass in chickens has been observed with co-administration of *Saccharomyces boulardii* and *Pichia pastoris* to chicken feed [25].

15.7 MODERN FRONTIERS AND ADVANCES IN FOOD ANALYSIS

15.7.1 SURFACE ENHANCED RAMAN SCATTERING (SERS) BASED TECHNIQUES

There have been many modern techniques also that have been recently developed or being employed for detection of food spoilage. Few the notable techniques worked upon lately have been mentioned in this section. In an advancement under non-destructive food testing SERS has been found to be much useful. Literature indicates multimodal application of SERS with emergent flexible SERS substrates. These authors promoted swab sampling has been favored for collection of target compounds directly from the sample; and *in situ* detection has been found to be best for easy, accurate, and non-destructive means of analysis. Primer method of infiltration has been advocated prior to swab sampling. This is thus a tripartite analytical process. Such a modern technique has been envisaged to be much useful in monitoring proposes on toxins, adulterants, illegal additives, foodborne pathogens, etc. [121]. This system of SERS analysis has been verified by another group using milk and chicken as food sample, targeting on *Staphylococcus aureus* contamination in specific. Using artificial peroxidase enzyme as a trigger for SERS on *S. aureus* has also been testified successfully by other researchers, with good agreement with classical plate count method. This similarity in findings establish the reliability of SERS, according to these researchers [66].

15.7.2 NUCLEIC ACID-BASED METHODS

Nucleic acid based analytical procedures have also been employed for testing for food pathogens. Such practices when incorporated in sensors have produced appreciable results food adulterants/contaminants, biotoxins., pathogenic contaminates, heavy metal toxicity, etc. Moreover, the experts suggest sandwiching the said nucleic acid detection with optical, electrochemical, and mass based sensing mechanisms. Aptamers that are short nucleic acids with specific selectivity and affinity for a particular target have been recommended for the said application. These aptamers have high efficiency and low cost, favoring their use. Other nucleic acids such as ribozymes and DNAses have also been cited as important contributors to such sensing [46]. Few other experts have the opinion that when using nucleic acid-based processes/sensors, sample pretreatment should be done to avoid noise in signals. These pretreatments are centrifugation, filtration, immunogenic separation or their combinations [51].

15.7.3 MACHINE LEARNING-BASED TECHNIQUES

Machine learning has been instrumental in modern day automation of various processes; hence, food pathogen and toxin detection has also benefitted from the same. A report on machine learning based paper chromogenic identification of spoilage microorganisms has been published. The process has been lauded for its ability to sense multiplexed samples with viable pathogens in food. The machine learning network consisted of multi-layer neural networks that provided accuracy in detection up to 95%. It could distinguish between strains of microorganisms contaminating a food, such as *L. monocytogenes* and *E. coli* when tested on fresh lettuce leaves. The process bypasses otherwise necessary steps of enrichment, culturing, incubation, and other sample preparation steps [115]. Tandem method of Raman spectroscopy (RS) with machine learning has also been verified by Yan et al. These authors achieved serotype level differentiation amongst pathogenic strains of microorganisms by matching them with their unique Raman spectra (23 common strains from 7 genera), with accuracy best at 95%. These authors employed decision tree algorithm to differentiate bacterial strains at serotype level [113].

15.7.4 LOOP-MEDIATED ISOTHERMAL AMPLIFICATION (LAMP)

The LAMP method has been quite useful in surveillance on the entire class of pathogens such as bacteria, virus, other microbial contaminants, as well as food allergens. Though conceptually similar to PCR, LAMP offers few unique advantages such as rapidity, high sensitivity and specificity, and lower cost. LAMP has also been reported to offer fluorescence, turbidity, and electrochemical sensing abilities. To provide better automation, accuracy, and speed, integration of LAMP with electronics-based system of complementary metal oxide semiconductor (CMOS) has been recommended [58]. However, with LAMP, the sample preparation needs to be appropriate warding off analytical barriers such as polysaccharides and polyphenols that are abundant in vegetables, animal samples and fruits. The authors have laid special emphasis on detection inhibitors that hinder appropriate analysis [76].

15.7.5 MICROFLUIDICS-BASED TECHNIQUES

Use of microfluidic based devices for detection of microorganisms, using the microfluidic chip offers rapid, portable, inexpensive, and immediate detection of pathogens in edible products. Further sample analysis by amplification, immunoassay, surface plasmon resonance and impedance spectroscopy could be further carrier out in this procedure [43]. Microfluidics chip-based samples could also be subjected to detection by alternative methods such as biosensors that comprise of amperometric, voltametric, fluorescence, colorimetric, impedance, and RS [36].

15.7.6 SMARTPHONE-BASED DETECTION

Detection of contaminates under the concept of Point of care tests. The method is modern and involves sensors material consisting of metal nanoparticles, carbon, and quantum dots, and analytical procedure of lateral flow assay, paper-based method, and microfluidic chips, etc. Typically, they have been useful in detection of contaminates/heavy metals in environmental water. The electronics integration is such that the results are displayed directly in a smartphone [95].

15.8 FUTURE SCOPE AND CHALLENGES

Considering the increase in global population, the demand to supply chain of food would plausibly be strained. In order to ensure greater quality of health of our successors, we need to ensure our health first, and that would be with the quality of food we consume. We need to foresee the employability of microorganisms constructively, and need to develop techniques that allow better control at various check points in between instead of a beginning and end control. Intelligent packing is one of the coming technologies that would be smart enough to detect the changes in pH of the packaged food and alert based on color change in a sensor on the packaging. Development of resistance in pathogenic microorganisms affecting food is also concerning and we may need specialized genomic techniques to counter them, since the endurance renders these microbes' ability to sustain physical and chemical attenuation. More use of irradiation could be useful under optimized appropriate doses.

Besides, abattoirs should be hygienic and so should be food warehouses and silos. With advanced instrumentation, nanotechnology, and artificial intelligence (AI), the future appears to be bright and with many constructive possibilities in ensuring food safety from foodborne pathogens. Moreover, the global rising trend of probiotics in duet would provide a better means of supplementation our diet with salutary ingredients. Non-thermal methods of obtaining active constituents also needs to be ramped. Few of the recent findings suggest strong influence of these techniques on efficacy of natural antimicrobial agents. The delicate extraction of active constituents by non-thermal methods such as such as supercritical carbon-dioxide extraction (SC-CO_2) even from matrices as flowers for antimicrobial effects attest to their preference in the modern industry [39]. SC-CO_2 is so effective yet benign on active compounds that even mathematical scale up feasibility by experimental modeling has been successful [40]. More useful research is warranted in areas exploiting the natural resources, but in a way to best preserve the potency of the vulnerable and/or volatile fractions, which are otherwise very potent antimicrobial agents.

15.9 SUMMARY

Foodborne pathogens have been a nuisance since man started consuming food. They have been plaguing food products and many a times are difficult to identify and contain. The concept of farm to fork would be best propagated of the farm produce is safe from pathogenic microorganism. This entails care from the farm produce, to warehouse to supply chain, production facility, then again finished product supply chain, to the store/grocery and finally to the consumer. Given the virulence of microbes that affect, and their intensity, the global food security issue would be incomplete with adequate measure to contain these pathogenic microbiotas. Also, the salubrious probiotics need to be enhanced in our diet to supplement necessary microflora that aids digestion and other medical benefits. As an epilog, it is fair to summarize that the two facets of microorganisms are both obfuscating (in being a friend and foe) yet intriguing, and warrant attention by researchers for designing a safe, secure, and sustainable food program for the world in the future.

ACKNOWLEDGMENT

The authors thank their parent organization and its staff for providing adequate encouragement, funding, and facilities for drafting the chapter.

KEYWORDS

- **detection**
- **foodborne pathogens**
- **loop-mediated isothermal amplification**
- **microorganisms**
- **nanotechnology**
- **non-thermal processing**
- **pathogens**
- **probiotics**
- **spoilage**

REFERENCES

1. Algaba, I. G., Verhaegen, B., Jennes, M., Rahman, M., Coucke, W., Cox, E., Dorny, P., et al., (2018). Pork as a source of transmission of *Toxoplasma gondii* to humans: A parasite burden study in pig tissues after infection with different strains of *Toxoplasma gondii* as a function of time and different parasite stages. *International Journal for Parasitology, 48*(7), 555–560.
2. Al-Nabuslsi, A. A., Osaili, T. M., Olaimat, A. N., Almasri, W. E., Ayyash, M., Al-Holy, M. A., Jaradat, Z. W., et al., (2020). Inactivation of *Salmonella* spp. in tahini using plant essential oil extracts. *Food Microbiology, 86*, E-article ID 103338.
3. An, X., Zuo, P., & Ye, B. C., (2020). A single cell droplet microfluidic system for quantitative determination of foodborne pathogens. *Talanta, 209*, E-article ID 120571.
4. Andritsos, N. D., Tzimotoudis, N., & Mataragas, M., (2020). Estimating the performance of four culture media used for enumeration and detection of *Campylobacter* species in chicken meat. *LWT-Food Science and Technology, 118*, E-article ID 108808.
5. Aragrande, M., & Canali, M., (2020). Integrating epidemiological and economic models to identify the cost of foodborne diseases. *Experimental Parasitology, 210*, E-article ID 107832.
6. Ballom, K. F., Tsai, H. C., Taylor, M., Tang, J., & Zhu, M. J., (2020). Stability of *Listeria monocytogenes* in non-fat dry milk powder during isothermal treatment and storage. *Food Microbiology, 87*, E-article ID 103376.
7. Barrere, V., Tompkins, E., Armstrong, M., Bird, P., Bastin, B., & Goodridge, L., (2020). Optimization of *Salmonella* detection in garlic, onion, cinnamon, red chili pepper powders and green tea. *International Journal of Food Microbiology, 316*, E-article ID 108440.
8. Biedermann, L., Zeitz, J., Mwinyi, J., Sutter-Minder, E., Rehman, A., Ott, S. J., Stey, C. S., et al., (2013). Smoking cessation induces profound changes in the composition of the intestinal microbiota in humans. *PLoS One, 8*(3), E-article ID e59260.

9. Binh, D. X., Minh, N. N., & Nguyet, D. T., (2017). Prevalence of *Listeria monocytogenes, E. coli, Salmonella* Spp. and *Staphylococcus aureus* bacteria contamination on meat at public market in the north of Vietnam. *SOJ Microbiology and Infectious Diseases, 5*(5), 1–22.
10. Bisht, N., & Garg, A., (2019). Antagonistic activity of lactic acid bacteria (LAB) against common enteric pathogens isolated from milk and milk products and evaluation of their probiotic attributes. *Bioscience Biotechnology Research Communications, 12*(4), 1173–1184.
11. Boatemaa, S., Barney, M., Drimie, S., Harper, J., Korsten, L., & Pereira, L., (2019). Awakening from the listeriosis crisis: Food safety challenges, practices, and governance in the food retail sector in South Africa. *Food Control, 104*(3), 333–342.
12. Bozkurt, A. G., Buyukgoz, G. G., Soforoglu, M., Tamer, U., Suludere, Z., & Boyaci, I. H., (2018). Alkaline phosphatase labeled SERS active sandwich immunoassay for detection of *Escherichia coli. Spectrochimica Acta Part A: Molecular and Biomolecular Spectroscopy, 194*, 8–13.
13. Cacciòm, S. M., Chalmers, R. M., Dorny, P., & Robertson, L. J., (2018). Foodborne parasites: Outbreaks and outbreak investigations. A meeting report from the European network for foodborne parasites (Euro-FBP). *Food and Waterborne Parasitology, 10*(1), 1–5.
14. Cao, Y., Zhou, A., Zhou, D., Xiao, X., Yu, Y., & Li, X., (2020). *Cronobacter sakazakii* CICC 21544 responds to the combination of carvacrol and citral by regulating proton motive force. *LWT-Food Science and Technology, 122*(1), E-article ID 109040.
15. Chai, H. E., Hwang, C. A., Huang, L., Wu, V. C., & Sheen, L. Y., (2020). Feasibility and efficacy of using gaseous chlorine dioxide generated by sodium chlorite-acid reaction for decontamination of foodborne pathogens on produce. *Food Control, 108*, E-article ID 106839.
16. Chen, C. J., Li, Q. Q., Zeng, Z. Y., Duan, S. S., Wang, W., Xu, F. R., Cheng, Y. X., & Dong, X., (2020). Efficacy and mechanism of *Mentha haplocalyx* and *Schizonepeta tenuifolia* essential oils on the inhibition of *Panaxnoto ginseng* pathogens. *Industrial Crops and Products, 145*(1), E-article ID 112073.
17. Chen, Z. G., Zhong, H. X., Luo, H., Zhang, R. Y., & Huang, J. R., (2019). Recombinase polymerase amplification combined with unmodified gold nanoparticles for *Salmonella* detection in milk. *Food Analytical Methods, 12*(1), 190–197.
18. Cho, Y., Kim, H., Beuchat, L. R., & Ryu, J. H., (2020). Synergistic activities of gaseous oregano and thyme thymol essential oils against *Listeria monocytogenes* on surfaces of a laboratory medium and radish sprouts. *Food Microbiology, 86*, E-article ID 103357.
19. Christidis, T., Hurst, M., Rudnick, W., Pintar, K. D., & Pollari, F., (2020). A comparative exposure assessment of foodborne, animal contact and waterborne transmission routes of *Salmonella* in Canada. *Food Control, 109*(6), E-article ID 106899.
20. Churklam, W., Chaturongakul, S., Ngamwongsatit, B., & Aunpad, R., (2020). The mechanisms of action of carvacrol and its synergism with nisin against *Listeria monocytogenes* on sliced bologna sausage. *Food Control, 108*, E-article ID 106864.
21. Clarke, S. F., Murphy, E. F., O'Sullivan, O., Lucey, A. J., Humphreys, M., Hogan, A., Hayes, P., et al., (2014). Exercise and associated dietary extremes impact on gut microbial diversity. *Gut, 63*(12), 1913–1920.
22. Colagiorgi, A., Festa, R., Di Ciccio, P. A., Gogliettino, M., Balestrieri, M., Palmieri, G., Anastasio A., & Ianieri, A., (2020). Rapid biofilm eradication of the antimicrobial

peptide 1018-K6 against *Staphylococcus aureus*: A new potential tool to fight bacterial biofilms. *Food Control, 107*(5), E-article ID 106815.

23. Cui, H., Zhang, C., Li, C., & Lin, L., (2020). Inhibition of *Escherichia coli* O157: H7 biofilm on vegetable surface by solid liposomes of clove oil. *LWT-Food Science and Technology, 117*(1, 2), E-article ID 108656.
24. Cusimano, M. G., Di Stefano, V., La Giglia, M., Di Marco, L., Presti, V., Schillaci, D., Pomilio, F., & Vitale, M., (2020). Control of growth and persistence of *Listeria monocytogenes* and β-lactam-resistant *Escherichia coli* by thymol in food processing settings. *Molecules, 25*(2), E-article ID E383.
25. De Los Santos, D. G., De Los Santos, J. R. G., Gil-Turnes, C., Gaboardi, G., Silva, L. F., Franca, R., Fernandes, C. G., & Conceicao, F. R., (2018). Probiotic effect of *Pichia pastoris* X-33 produced in parboiled rice effluent and YPD medium on broiler chickens. *PLoS One 13*(2), E-article ID 0192904.
26. Denyes, J. M., Dunne, M., Steiner, S., Mittelviefhaus, M., Weiss, A., Schmidt, H., Klumpp, J., & Loessner, M., (2017). Modified bacteriophage S16 long tail fiber proteins for rapid and specific immobilization and detection of *Salmonella* cells. *Applied and Environmental Microbiology, 83*(12), E-article ID 00277.
27. Draz, M. S., & Shafiee, H., (2018). Applications of gold nanoparticles in virus detection. *Theranostics, 8*(7), 1985–2017.
28. Du, J., Singh, H., Dong, W. J., Bai, Y. H., & Yi, T. H., (2018). Colorimetric detection of *Listeria monocytogenes* using one-pot biosynthesized flower-shaped gold nanoparticles. *Sensors and Actuators B: Chemical, 265*, 285–292.
29. Du, J., Wu, S., Niu, L., Li, J., Zhao, D., & Bai, Y., (2020). A gold nanoparticles-assisted multiplex PCR assay for simultaneous detection of *Salmonella typhimurium*, *Listeria monocytogenes* and *Escherichia coli* O157: H7. *Analytical Methods, 12*(2), 212–217.
30. Du, L., Prasad, A. J., Gänzle, M., & Roopesh, M., (2020). Inactivation of *Salmonella* spp. in wheat flour by 395 nm pulsed light emitting diode (LED) treatment and the related functional and structural changes of gluten. *Food Research International, 127*(8), E-article ID 108716.
31. Duc, H. M., Son, H. M., Yi, H. P. S., Sato, J., Ngan, P. H., Masuda, Y., Honjoh, K., & Miyamoto, T., (2020). Isolation, characterization, and application of a polyvalent phage capable of controlling *Salmonella* and *Escherichia coli* O157: H7 in different food matrices. *Food Research International, 131*, E-article ID 108977.
32. Elviss, N., Little, C., Hucklesby, L., Sagoo, S., Surman-Lee, S., De Pinna, E., & Threlfall, E. J., (2009). Microbiological study of fresh herbs from retail premises uncovers an international outbreak of salmonellosis. *International Journal of Food Microbiology, 134*(1, 2), 83–88.
33. Fietto, J. L., Araújo, R. S., Valadão, F. N., Fietto, L. G., Brandão, R. L., Neves, M. J., Gomes, F. C. O., et al., (2004). Molecular and physiological comparisons between *Saccharomyces cerevisiae* and *Saccharomyces boulardii*. *Canadian Journal of Microbiology, 50*(8), 615–621.
34. Fletcher, S., Boonwaat, L., Moore, T., Chavada, R., & Conaty, S., (2015). Investigating an outbreak of staphylococcal food poisoning among travelers across two Australian states. *Western Pacific Surveillance and Response Journal, 6*(2), 17–21.
35. Friker, B., Morach, M., Püntener, S., Cernela, N., Horlbog, J., & Stephan, R., (2020). Assessing the microbiological quality of raw goats' and ewes' tank milk samples in Switzerland. *International Dairy Journal, 102*, E-article ID 104609.

36. Fu, X., Sun, J., Liang, R., Guo, H., Wang, L., & Sun, X., (2021). Application progress of microfluidics-integrated biosensing platforms in the detection of foodborne pathogens. *Trends in Food Science & Technology, 116*(4), 115–129.
37. Gabriel, A. A., David, M. M. C., Elpa, M. S. C., & Michelena, J. C. D., (2020). Decontamination of dried whole black peppercorns using ultraviolet-c irradiation. *Food Microbiology, 88*(s8), E-article ID 103401.
38. Gao, J., & Matthews, K. R., (2020). Effects of the photosensitizer curcumin in inactivating foodborne pathogens on chicken skin. *Food Control, 109*(5), E-article ID 106959.
39. Ghosh, P. K., Bhattacharjee, P., & Das, S., (2014). Antimicrobial activity of supercritical carbon dioxide extracts of tuberose (*Polianthes tuberosa* Linn.) flowers against common pathogens. *International Journal of Pharmaceutical Sciences and Research, 5*(4), 1279–1289.
40. Ghosh, P. K., & Bhattacharjee, P., (2016). Mathematical modeling of supercritical carbon dioxide extraction of methyl eugenol from tuberose flowers. *Korean Journal of Chemical Engineering, 33*(5), 1681–1691.
41. Graff, S., Chaumeil, J. C., Boy, P., Lai-Kuen, R., & Charrueau, C., (2008). Influence of pH conditions on the viability of *Saccharomyces boulardii* yeast. *The Journal of General and Applied Microbiology, 54*(4), 221–227.
42. Gutiérrez, D., Rodríguez-Rubio, L., Martínez, B., Rodríguez, A., & García, P., (2016). Bacteriophages as weapons against bacterial biofilms in the food industry. *Frontiers in Microbiology, 7*, E-article ID 1379.
43. Han, X., Liu, Y., Yin, J., Yue, M., & Mu, Y., (2021). Microfluidic devices for multiplexed detection of foodborne pathogens. *Food Research International, 143*(4), E-article ID 110246.
44. He, Y., Ren, Y., Guo, B., Yang, Y., Ji, Y., Zhang, D., Wang, J., et al., (2020). Development of a specific nanobody and its application in rapid and selective determination of *Salmonella enteritidis* in milk. *Food Chemistry, 310*, E-article ID 125942.
45. Huang, J., Chen, B., Li, H., Zeng, Q. H., Wang, J. J., Liu, H., Pan, Y., & Zhao, Y., (2020). Enhanced antibacterial and antibiofilm functions of the curcumin-mediated photodynamic inactivation against *Listeria monocytogenes*. *Food Control, 108*, E-article ID 106886.
46. Huo, B., Hu, Y., Gao, Z., & Li, G., (2021). Recent advances on functional nucleic acid-based biosensors for detection of food contaminants. *Talanta, 222*(1), E-article ID 121565.
47. Hyun, J. E., & Lee, S. Y., (2020). Antibacterial effect and mechanisms of action of 460–470 nm light-emitting diode against *Listeria monocytogenes* and *Pseudomonas fluorescens* on the surface of packaged sliced cheese. *Food Microbiology, 86*, E-article ID 103314.
48. Ishikawa, K., (1990). *Introduction to Quality Control* (3rd edn., p. 435), 3A Corporation, Tokyo, Japan.
49. Jin, Z., Ding, G., Li, G., Yang, G., Han, Y., Hao, N., Deng, J., et al., (2020). Rapid detection of foodborne bacterial pathogens using visual high-throughput microfluidic chip. *Journal of Chemical Technology & Biotechnology, 95*(5), E-article ID 6331.
50. Kim, J. H., Jung, S., & Oh, S. W., (2020). Combination of bacteria concentration and DNA concentration for rapid detection of *E. coli* O157: H7, *L. monocytogenes*, and

S. Typhimurium without microbial enrichment. *LWT-Food Science and Technology, 117*(4), E-article ID 108609.
51. Kim, J. H., & Oh, S. W., (2021). Pretreatment methods for nucleic acid-based rapid detection of pathogens in food: A review. *Food Control, 121*(12), E-article ID 107575.
52. Kim, S. S., Shin, M., Kang, J. W., Kim, D. K., & Kang, D. H., (2020). Application of the 222nm krypton-chlorine lamp and 280nm UVC light-emitting diode for the inactivation of *Listeria monocytogenes* and *Salmonella typhimurium* in water with various turbidities. *LWT-Food Science and Technology, 117*(12), E-article ID 108458.
53. Kim, Y. J., Kim, H. S., Chon, J. W., Kim, D. H., Hyeon, J. Y., & Seo, K. H., (2018). New colorimetric aptasensor for rapid on-site detection of *Campylobacter jejuni* and *Campylobacter coli* in chicken carcass samples. *Analytica Chimica Acta, 1029*, 78–85.
54. Klayraung, S., Viernstein, H., & Okonogi, S., (2009). Development of tablets containing probiotics: Effects of formulation and processing parameters on bacterial viability. *International Journal of Pharmaceutics, 370*(1, 2), 54–60.
55. Koon, H. W., Su, B., Xu, C., Mussatto, C. C., Tran, D. H. N., Lee, E. C., Ortiz, C., et al., (2016). Probiotic *Saccharomyces boulardii* CNCM I-745 prevents outbreak-associated *Clostridium difficile*-associated cecal inflammation in hamsters. *American Journal of Physiology-Gastrointestinal and Liver Physiology, 311*(4), G610-G623.
56. Kretzer, J. W., Lehmann, R., Schmelcher, M., Banz, M., Kim, K. P., Korn, C., & Loessner, M. J., (2007). Use of high-affinity cell wall-binding domains of bacteriophage endolysins for immobilization and separation of bacterial cells. *Applied and Environmental Microbiology, 73*(6), 1992–2000.
57. Ku, K. M., Chiu, Y. C., Shen, C., & Jenks, M., (2020). Leaf cuticular waxes of lettuce are associated with reduced attachment of the foodborne pathogen *Salmonella* spp. at harvest and after postharvest storage. *LWT-Food Science and Technology, 117*(2), E-article ID 108657.
58. Lakshmi, B. A., & Kim, S., (2021). Recent trends in the utilization of LAMP for the diagnosis of viruses, bacteria, and allergens in food. In: Viswanath, B., (ed.), *Recent Developments in Applied Microbiology and Biochemistry* (Vol. 2, pp. 291–297). Elsevier, Inc., San Dego – USA.
59. Ledlod, S., Areekit, S., Santiwatanakul, S., & Chansiri, K., (2020). Colorimetric aptasensor for detecting *Salmonella spp.*, *Listeria monocytogenes* and *Escherichia coli* in meat samples. *Food Science and Technology International, 26*(5), 430–443.
60. Lee, K. S., Song, Y., Kim, C. H., Kim, Y. T., Kang, T., Lee, S. J., Choi, B. G., & Lee, K. G., (2020). Development of zinc oxide-based sub-micro pillar arrays for on-site capture and DNA detection of foodborne pathogen. *Journal of Colloid and Interface Science, 563*, 54–61.
61. Lewis, E., Hudson, J., Cook, N., Barnes, J., & Haynes, E., (2020). Next-generation sequencing as a screening tool for foodborne pathogens in fresh produce. *Journal of Microbiological Methods, 171*, E-article ID 105840.
62. Li, H., Ahmad, W., Rong, Y., Chen, Q., Zuo, M., Ouyang, Q., & Guo, Z., (2020). Designing an aptamer based magnetic and up conversion nanoparticles conjugated fluorescence sensor for screening *Escherichia coli* in food. *Food Control, 107*, E-article ID 106761.
63. Li, J., Ma, B., Fang, J., Zhi, A., Chen, E., Xu, Y., Yu, X., Sun, C., & Zhang, M., (2020). Recombinase polymerase amplification (RPA) combined with lateral flow immunoassay for rapid detection of *Salmonella* in food. *Foods, 9*(1), E-article ID 9010027.

64. Ling, N., Shen, J., Guo, J., Zeng, D., Ren, J., Sun, L., Jiang, Y., et al., (2020). Rapid and accurate detection of viable *Vibrio parahaemolyticus* by sodium deoxycholate-propidium monoazide-qPCR in shrimp. *Food Control, 109*, E-article ID 106883.
65. Liu, A., Liu, Y., Peng, L., Cai, X., Shen, L., Duan, M., Ning, Y., et al., (2020). Characterization of the narrow-spectrum bacteriophage LSE7621 towards *Salmonella enteritidis* and its biocontrol potential on lettuce and tofu. *LWT-Food Science and Technology, 118*(10), E-article ID 108791.
66. Liu, S., Li, H., Hassan, M. M., Ali, S., & Chen, Q., (2021). SERS based artificial peroxidase enzyme regulated multiple signal amplified system for quantitative detection of foodborne pathogens. *Food Control, 123*(1, 2), E-article ID 107733.
67. Ma, X., Song, L., Zhou, N., Xia, Y., & Wang, Z., (2017). A novel aptasensor for the colorimetric detection of *S. typhimurium* based on gold nanoparticles. *International Journal of Food Microbiology, 245*, 1–5.
68. Ma, X., Xu, X., Xia, Y., & Wang, Z., (2018). SERS aptasensor for *Salmonella typhimurium* detection based on spiny gold nanoparticles. *Food Control, 84*, 232–237.
69. Mark, C. M., Zuchel, J., Schaffner, D. W. S., & Strawn, L. K., (2020). Growth and survival of *Listeria monocytogenes* on intact fruit and vegetable surfaces during postharvest handling: A systematic literature review. *Journal of Food Protection, 83*(1), 108–128.
70. Martín, R., Chain, F., Miquel, S., Natividad, J. M., & Sokol, H., (2014). Effects in the use of a genetically engineered strain of *Lactococcus lactis* delivering *in situ* IL-10 as a therapy to treat low-grade colon inflammation. *Human Vaccines & Immunotherapeutics, 10*(6), 1611–1621.
71. Martinović, T., Andjelković, U., Gajdošik, M. Š., Rešetar, D., & Josić, D., (2016). Foodborne pathogens and their toxins. *Journal of Proteomics, 147*, 226–235.
72. McCollum, J. T., Cronquist, A. B., Silk, B. J., Jackson, K. A., O'connor, K. A., Cosgrove, S., Gosack, J. P., et al., (2013). Multistate outbreak of listeriosis associated with cantaloupe. *The New England Journal of Medicine, 369*(10), 944–953.
73. Michael, M., Acuff, J., Lopez, K., Vega, D., Phebus, R., Thippareddi, H., & Channaiah, L. H., (2020). Comparison of survival and heat resistance of *Escherichia coli* O121 and *Salmonella* in muffins. *International Journal of Food Microbiology, 317*, E-article ID 108422.
74. Momin, K. M., Milton, A. A. P., Ghatak, S., Thomas, S. C., Priya, G. B., Das, S., Shakutala, I., et al., (2020). Development of a novel and rapid polymerase spiral reaction (PSR) assay to detect *Salmonella* in pork and pork products. *Molecular and Cellular Probes, 50*, E-article ID 101510.
75. Moon, S. H., Waite-Cusic, J., & Huang, E., (2020). Control of *Salmonella* in chicken meat using a combination of a commercial bacteriophage and plant-based essential oils. *Food Control, 110*(14), E-article ID 106984.
76. Moon, Y. J., Lee, S. Y., & Oh, S. W., (2022). A review of isothermal amplification methods and food-origin inhibitors against detecting foodborne pathogens. *Foods, 11*(3), E-article ID 11030322.
77. Naimah, A. K., Al-Manhel, A. J. A., & Al-Shawi, M. J., (2018). Isolation, purification, and characterization of antimicrobial peptides produced from *Saccharomyces boulardii*. *International Journal of Peptide Research and Therapeutics, 24*(3), 455–461.
78. Nerín, C., Aznar, M., & Carrizo, D., (2016). Food contamination during food process. *Trends in Food Science & Technology, 48*, 63–68.

79. Osaili, T. M., Al-Nabulsi, A. A., Dhanasekaran, D. K., Hasan, F., Rao, S., Fatima, H., Ayyash, M., et al., (2020). Growth behavior and thermal inactivation of *E. coli* O157: H7 and *Salmonella* spp. in ground lean camel meat. *International Journal of Food Microbiology, 316*, E-article ID 108423.
80. O'Sullivan, L., Bolton, D., McAuliffe, O., & Coffey, A., (2020). The use of bacteriophages to control and detect pathogens in the dairy industry. *International Journal of Dairy Technology, 73*(1), 1–11.
81. Patel, J., Yin, H. B., Bauchan, G., & Mowery, J., (2020). Inhibition of *Escherichia coli* O157: H7 and *Salmonella enterica* virulence factors by benzyl isothiocyanate. *Food Microbiology, 86*, E-article ID 103303.
82. Pérez-Baltar, A., Serrano, A., Montiel, R., & Medina, M., (2020). *Listeria monocytogenes* inactivation in deboned dry-cured hams by high pressure processing. *Meat Science, 160*, E-article ID 107960.
83. Pissuwan, D., Gazzana, C., Mongkolsuk, S., & Cortie, M. B., (2020). Single and multiple detections of foodborne pathogens by gold nanoparticle assays. *Wiley Interdisciplinary Reviews: Nanomedicine and Nanobiotechnology, 12*(1), E-article ID e1584.
84. Podolak, R., Lucore, L., & Harris, L. J., (2017). Heat resistance of *Salmonella* and other bacterial pathogens in low-moisture foods. In: Podolak, R., & Black, D. G., (eds.), *Control of Salmonella and Other Bacterial Pathogens in Low-Moisture Foods* (pp. 121–148). Chichester – UK: John Wiley & Sons, Inc.
85. Poshtiban, S., Javed, M. A., Arutyunov, D., Singh, A., Banting, G., Szymanski, C. M., & Evoy, S., (2013). Phage receptor binding protein-based magnetic enrichment method as an aid for real time PCR detection of foodborne bacteria. *Analyst, 138*(19), 5619–5626.
86. Rodrigues, C., Da Silva, A. L. B. R., & Dunn, L. L., (2020). Factors impacting the prevalence of foodborne pathogens in agricultural water sources in the southeastern United States. *Water, 12*(1), 51–58.
87. Salas-Jara, M. J., Ilabaca, A., Vega, M., & García, A., (2016). Biofilm forming *Lactobacillus*: New challenges for the development of probiotics. *Microorganisms, 4*(3), E-article ID 4030035.
88. Saleh, A., Hayashi, K., Ijiri, D., & Ohtsuka, A., (2014). Beneficial effects of *Aspergillus awamori* in broiler nutrition. *World's Poultry Science Journal, 70*(4), 857–864.
89. Saravanan, A., Kumar, P. S., Hemavathy, R., Jeevanantham, S., Kamalesh, R., Sneha, S., & Yaashika, P. R., (2021). Methods of detection of foodborne pathogens: A review. *Environmental Chemistry Letters, 19*(1), 189–207.
90. Schneider, S. M., Girard-Pipau, F., Filippi, J., Hébuterne, X., Moyse, D., Hinojosa, G. C., Pompei, A., & Rampal, P., (2005). Effects of *Saccharomyces boulardii* on fecal short-chain fatty acids and microflora in patients on long-term total enteral nutrition. *World Journal of Gastroenterology, 11*(39), 6165–6169.
91. Scott, K. P., Gratz, S. W., Sheridan, P. O., Flint, H. J., & Duncan, S. H., (2013). The influence of diet on the gut microbiota. *Pharmacological Research, 69*(1), 52–60.
92. Sen, S., & Mansell, T. J., (2020). Yeasts as probiotics: Mechanisms, outcomes, and future potential. *Fungal Genetics and Biology, 137*(Suppl 1), E-article ID 103333.
93. Shenashen, M. A., Emran, M. Y., El Sabagh, A., Selim, M. M., El Marakbi, A., & El-Safty, S. A., (2022). Progress in sensory devices of pesticides, pathogens, coronavirus, and chemical additives and hazards in food assessment: Food safety concerns. *Progress in Materials Science, 124*(2020), E-article ID 100866.

94. Sheng, L., Shen, X., & Zhu, M. J., (2020). Screening of non-pathogenic surrogates of *Listeria monocytogenes* applicable for chemical antimicrobial interventions of fresh apples. *Food Control, 110*, E-article ID 106977.
95. Sivakumar, R., & Lee, N. Y., (2021). Recent progress in smartphone-based techniques for food safety and the detection of heavy metal ions in environmental water. *Chemosphere, 275*(1), E-article ID 130096.
96. Slot, E., Zaaijer, H. L., Molier, M., Van, D. H. K., Prinsze, F., & Hogema, B. M., (2017). Meat consumption is a major risk factor for hepatitis E virus infection. *PLoS One, 12*(4), E-article ID 0176414.
97. Soon, J. M., Brazier, A. K., & Wallace, C. A., (2020). Determining common contributory factors in food safety incidents–a review of global outbreaks and recalls 2008–2018. *Trends in Food Science & Technology, 97*(7), 76–87.
98. Sugiharto, S., Yudiarti, T., Isroli, I., Widiastuti, E., Wahyuni, H. I., & Sartono, T. A., (2018). The effect of fungi-origin probiotic *Chrysonilia crassa* in comparison to selected commercially used feed additives on broiler chicken performance, intestinal microbiology, and blood indices. *Journal of Advanced Veterinary and Animal Research, 5*(3), 332–342.
99. Sun, X. H., Hao, L. R., Xie, Q. C., Lan, W. Q., Zhao, Y., Pan, Y. J., & Wu, V. C. H., (2020). Antimicrobial effects and membrane damage mechanism of blueberry (*Vaccinium corymbosum* L.) extract against *Vibrio parahaemolyticus*. *Food Control, 111*(8), E-article ID 107020.
100. Syed, I., Banerjee, P., & Sarkar, P., (2020). Oil-in-water emulsions of geraniol and carvacrol improve the antibacterial activity of these compounds on raw goat meat surface during extended storage at 4°C. *Food Control, 107*(1), E-article ID 106757.
101. Tao, X., Huang, Y., Wang, C., Chen, F., Yang, L., Ling, L., Che, Z., & Chen, X., (2020). Recent developments in molecular docking technology applied in food science: A review. *International Journal of Food Science & Technology, 55*(1), 33–45.
102. Terciolo, C., Dobric, A., Ouaissi, M., Siret, C., Breuzard, G., Silvy, F., Marchiori, B., et al., (2017). *Saccharomyces boulardii* CNCM I-745 restores intestinal barrier integrity by regulation of E-cadherin recycling. *Journal of Crohn's and Colitis, 11*(8), 999–1010.
103. Thiramanas, R., & Laocharoensuk, R., (2016). Competitive binding of polyethyleneimine-coated gold nanoparticles to enzymes and bacteria: A key mechanism for low-level colorimetric detection of gram-positive and gram-negative bacteria. *Microchimica Acta, 183*(1), 389–396.
104. Torregrosa, L., Echeverria, G., Illa, J., Torres, R., & Giné-Bordonaba, J., (2020). Spatial distribution of flavor components and antioxidants in the flesh of 'Conference' pears and its relationship with postharvest pathogens susceptibility. *Postharvest Biology and Technology, 159*, E-article ID 111004.
105. Trnčíková, T., Hrušková, V., Oravcová, K., Pangallo, D., & Kaclíková, E., (2009). Rapid and sensitive detection of *Staphylococcus aureus* in food using selective enrichment and real-time PCR targeting a new gene marker. *Food Analytical Methods, 2*(4), 241–248.
106. Tsaregorodtseva, E., Kabanova, T., Okhotnikov, S., & Dolgorukova, M., (2020). Approaches to safety and quality of foodstuffs in the European Union and Russia. *IOP Conference Series: Earth and Environmental Science, 421*(3), Article ID 032003.
107. Wachiralurpan, S., Chansiri, K., & Lieberzeit, P. A., (2020). Direct detection of *Listeria monocytogenes* DNA amplification products with quartz crystal microbalances at

elevated temperatures. *Sensors and Actuators B: Chemical, 308*(2009), E-article ID 127678.
108. Wang, C., Gao, X., Wang, S., & Liu, Y., (2020). A smartphone-integrated paper sensing system for fluorescent and colorimetric dual-channel detection of foodborne pathogenic bacteria. *Analytical and Bioanalytical Chemistry, 412*(3), 611–620.
109. Wang, Q., Long, M., Lv, C., Xin, S., Han, X., & Jiang, W., (2020). Lanthanide-labeled fluorescent-nanoparticle immunochromatographic strips enable rapid and quantitative detection of *Escherichia coli* O157: H7 in food samples. *Food Control, 109*, E-article ID 106894.
110. Wells, J. M., (2011). Immunomodulatory mechanisms of lactobacilli. *Microbial Cell Factories, 10*(Suppl 1), S1–S17.
111. Woo, H. J., Kang, J. H., Lee, C. H., & Song, K. B., (2020). Application of *Cudrania tricuspidata* leaf extract as a washing agent to inactivate *Listeria monocytogenes* on fresh-cut romaine lettuce and kale. *International Journal of Food Science & Technology, 55*(1), 276–282.
112. Wu, B., Yang, T., Zou, D., Jin, L., Liang, X., Li, T., Huang, G., & Zhang, J., (2020). Nuclear magnetic resonance biosensor based on streptavidin–biotin system and poly-l-lysine macromolecular targeted gadolinium probe for rapid detection of *Salmonella* in milk. *International Dairy Journal, 102*, E-article ID 104594.
113. Yan, S., Wang, S., Qiu, J., Li, M., Li, D., Xu, D., Li, D., & Liu, Q., (2021). Raman spectroscopy combined with machine learning for rapid detection of foodborne pathogens at the single-cell level. *Talanta, 226*, E-article ID 122195.
114. Yan, W., Qian, J., Ge, Y., Ye, K., Zhou, C., & Zhang, H., (2020). Principal component analysis of MALDI-TOF MS of whole-cell foodborne pathogenic bacteria. *Analytical Biochemistry, 592*, E-article ID 113582.
115. Yang, M., Liu, X., Luo, Y., Pearlstein, A. J., Wang, S., Dillow, H., Reed, K., et al., (2021). Machine learning-enabled non-destructive paper chromogenic array detection of multiplexed viable pathogens on food. *Nature Food, 2*(2), 110–117.
116. Yang, X., Wu, Q., Huang, J., Wu, S., Zhang, J., Chen, L., Wei, X., et al., (2020). Prevalence and characterization of *Salmonella* isolated from raw vegetables in China. *Food Control, 109*, E-article ID 106915.
117. Yao, S., Zhao, C., Liu, Y., Nie, H., Xi, G., Cao, X., Li, Z., et al., (2020). Colorimetric immunoassay for the detection of *Staphylococcus aureus* by using magnetic carbon dots and sliver nanoclusters as o-phenylenediamine-oxidase mimetics. *Food Analytical Methods, 13*(4), 833–838.
118. Yemmireddy, V. K., Cason, C., Moreira, J., & Adhikari, A., (2020). Effect of pecan variety and the method of extraction on the antimicrobial activity of pecan shell extracts against different foodborne pathogens and their efficacy on food matrices. *Food Control, 112*, E-article ID 107098.
119. Yeni, F., Yavaş, S., Alpas, H., & Soyer, Y., (2016). Most common foodborne pathogens and mycotoxins on fresh produce: A review of recent outbreaks. *Critical Reviews in Food Science and Nutrition, 56*(9), 1532–1544.
120. Zhang, B., Zhang, L., Cheng, T., Guan, X., & Wang, S., (2020). Effects of water activity, temperature and particle size on thermal inactivation of *Escherichia coli* ATCC 25922 in red pepper powder. *Food Control, 107*, E-article ID 106817.
121. Zhang, D., Pu, H., Huang, L., & Sun, D. W., (2021). Advances in flexible surface-enhanced Raman scattering (SERS) substrates for nondestructive food detection:

Fundamentals and recent applications. *Trends in Food Science & Technology, 109*, 690–701.
122. Zhang, H., Ha, T. M. H., Seck, H. L., & Zhou, W., (2020). Inactivation of *Escherichia coli* O157: H7 and *Salmonella typhimurium* in edible bird's nest by low-energy x-ray irradiation. *Food Control, 110*(3), E-article ID 107031.
123. Zhang, H., Zhao, Y., Gong, C., & Jiao, S., (2020). Effect of radio frequency heating stress on sublethal injury of *Salmonella typhimurium* in red pepper powder. *LWT-Food Science and Technology, 117*(1), E-article ID 108700.
124. Zhang, S., Xiong, J., Lou, W., Ning, Z., Zhang, D., & Yang, J., (2020). Inhibition of *Cronobacter sakazakii* in reconstituted infant formula using triglycerol monolaurate and its effect on the sensory properties of infant formula. *International Journal of Food Microbiology, 320*, E-article ID 108518.
125. Zhao, N., Ge, L., Huang, Y., Wang, Y., Wang, Y., Lai, H., Wang, Y., et al., (2020). Impact of cold plasma processing on quality parameters of packaged fermented vegetable (*Radish paocai*) in comparison with pasteurization processing: Insight into safety and storage stability of products. *Innovative Food Science & Emerging Technologies, 60*(2), E-article ID 102300.
126. Zheng, X., Guo, J., Rao, H., Guo, D., Huang, Y., Xu, Y., Liang, S., et al., (2020). Antibacterial and antibiofilm activity of coenzyme-Q_0 against *Vibrio parahaemolyticus*. *Food Control, 109*, E-article ID 106955.
127. Zhong, Z., Gao, R., Chen, Q., & Jia, L., (2020). Dual-aptamers labeled polydopamine-polyethyleneimine copolymer dots assisted engineering a fluorescence biosensor for sensitive detection of *Pseudomonas aeruginosa* in food samples. *Spectrochimica Acta Part A: Molecular and Biomolecular Spectroscopy, 224*, E-article ID 117417.
128. Zhou, Z., Zuber, S., Campagnoli, M., Moser, M., Devlieghere, F., & Uyttendaele, M., (2020). Decontamination effect of hot-air drying against bacterial pathogen and surrogate strains on basil leaves, from laboratory to pilot scale settings. *LWT-Food Science and Technology, 122*(1), E-article ID 109035.

CHAPTER 16

Sensory Analysis and Consumer Acceptance of Legume-Based Gluten-Free Bakery Products

Edibe S. Erten and Tayyibe Erten

ABSTRACT

The replacement of gluten is essential for the development of gluten-free foods. To achieve this, variety of flours (such as corn flour, chestnut flour, and buckwheat flour) have been investigated by food scientists. Legume flours seem preferable to other flours among these gluten-alternative sources due to their nutritional quality. In this chapter, recent developments in legume-based gluten-free bakery products were investigated. Due to their high protein content, legumes have numerous applications in the gluten-free food industry. However, undesirable flavor formation limits their widespread usage. For this reason, the effects of replacing gluten-containing ingredients with legumes on the sensory properties and consumer acceptance were discussed as the main focus of this chapter.

16.1 INTRODUCTION

In recent years, dietary patterns and lifestyles have started to change globally with the increasing rate of chronic diseases. Therefore, the new popular diets have been introduced by understanding the relationship between health and food consumption. A gluten-free diet (GFD) is one of these popular diets, which is crucial for persons with coeliac disease (CD) or other gluten-related diseases.

The Functional Foods: Nutrient and Health Benefits.
Megh R. Goyal, Junaid Ahmad Malik, and Anu Kumari (Eds.)

A GFD has become popular worldwide since some consumers see it as a 'healthier option.' However, it has attracted more attention from many healthy populations by increasing health concerns and consumer awareness of food. Consequently, this attitude has expanded the gluten-free market. However, it is three times more expensive than its gluten-containing counterparts; and owing to the high demand for gluten-free products, its market share worldwide is increasing every year [35].

Gluten-free food products have been gaining popularity with an increase in their varieties [59]. In the past, gluten-free products were only offered to people with health problems caused by gluten. Since they were only a tiny percentage compared to the total food consumers' market, there were not many brands offering gluten-free choices. Since the avoidance of gluten is no longer only a health-related issue (as discussed in Section 16.3 of this chapter), more consumers prefer to use gluten-free products. As a result, more variety has been sought by consumers. Studies on different ingredients have also been increased for nutritional value, texture, and sensory properties with increased demand in recent years. Sensory properties of a product and consumer acceptance are among the essential factors of new products in the market.

In this chapter, replacing gluten-containing ingredients with legume-based ingredients in bakery products was investigated, and their effects on sensory properties and acceptance were discussed in detail.

16.2 GLUTEN AND ITS FUNCTIONS

Gluten is described as a remaining viscoelastic part when the other water-soluble compounds are washed out in grain doughs. It is found in cereals' endosperm with starch, and they form 70–80% of cereal proteins, while the remaining proteins consist of albumins, globulins, peptides, and amino acids.

Gluten proteins are the primary storage and structural complex protein in some cereals, and they are mostly found in wheat. Since they include prolamin and glutamine amino acids in their structures, they are also called prolamins. In terms of solubility, prolamins can be classified into two groups in wheat: gliadins and glutenins. The glutenins usually are insoluble in water, salt, and alcohol but partially soluble in a dilute acid solution. However, when the disulfide bonds disintegrate by reducing agent-alcohol mixture at increased temperature, it forms into monomeric subunits; thus, it becomes alcohol-soluble like prolamins [20, 48].

Moreover, the total molecular weight of glutenins ranges from 30 to 140 kDa, including high molecular weight (HMW ~ 90–140 kDa) and low molecular weight fractions (LMW ~ 30–75 kDa) by SDS-PAGE (sodium dodecyl polyacrylamide gel electrophoresis) method. Owing to the disulfide bonds of cysteine amino acids in HMW fractions, the viscoelasticity of dough is an advantage. Therefore, the bread volume is one of the quality parameters generated by this subunit [7].

On the other hand, gliadins consist of heterogeneous associations of single-chain polypeptides. They are only soluble in aqueous alcohol solution (70%), including a high amount of glutamine and proline amino acids. Their ratio in gluten proteins is approximately 50%, and according to their electrophoretic mobility in the A-PAGE method, they are classified as four subunits; α, β, γ, and ω gliadins. The α and γ gliadins are rich in methionine and cysteine residues which contribute to the intrachain disulfide bonds and the structure of proteins. On the contrary, ω gliadin has less sulfur-containing amino acids, but they are rich in glutamine, proline, and phenylalanine. However, in terms of toxicity, the α-gliadin is found to be the most harmful gliadin as β, γ, and ω gliadins are less deleterious effects compared to α-gliadin [7, 8].

The molecular weight of gliadins is between 30 and 75 kDa, and they produce a network with glutenins through non-covalent hydrophobic interaction and hydrogen bonding. Furthermore, these also generate the viscosity via non-covalent interactions, hydrogen bonding and van der Waal's forces; thus, the viscous characteristics of dough can occur owing to this characteristic [7].

Although known as a wheat protein, gluten can also be found in rye, barley, and some hybridized varieties. However, this structural protein is called differently in other cereals, such as; hordeins in barley and secalins in the rye. Sometimes oats can also be considered gluten-containing because of possible cross-contamination with other cereals during the harvest [48].

The dough formation of wheat flour in many bakery products relies on the gluten proteins; especially quantity and quality of gluten proteins will determine the dough and quality of end products [7]. When the wheat flour is hydrated and mixed with water, gluten proteins form a three-dimensional (3D) network in dough systems with starch molecules, especially in hard wheat products, such as bread, pasta, and pizza. Therefore, the continuation of this network depends on the inter and intrachain interactions of glutenin and gliadin fractions, which confer unique rheological properties such as viscosity and elasticity [7, 15].

Furthermore, the dough strength and extensibility are affected by the gliadin/glutenin ratio since gliadin plays an important role in viscosity while glutenin provides elasticity on hydrated gluten. Although some conflicting ideas about the effect of gliadin on bread's loaf volume, yet the positive effect is mostly mentioned in the literature. Most of the studies pointed out that γ-gliadins showed a positive correlation between bread-making and dough quality as ω-gliadins have less positive effect on bread volume due to lacking cysteine amino acid [7]. On the other hand, while gliadins provide viscosity and extensity, glutenins contribute toughness and elasticity to dough systems when completely hydrated. Similarly, tit was also found that HMW glutenins with different molecular weights promote viscoelasticity by creating β-sheet protein structures [41].

16.3 GLUTEN-FREE DIET (GFD)

Gluten proteins are usually digested into the polypeptides, di-tripeptides, and amino acids by gastric, pancreatic, and intestinal enzymes in healthy individuals. However, gluten proteins cannot degrade in the intestine in wheat hypersensitivity cases because of enzyme-resistant proline-rich sequences. Since pepsins and chymotrypsin hardly break down the peptide bonds, the long peptides head to lamina propria, where the disease-specific immune response is generated by T-cells and antibody response [48]. Therefore, in this case, the necessity of a GFD emerges for individuals susceptible to gluten hypersensitivity, such as CD, wheat allergy (WA), and gluten ataxia (GA). In this diet, the cereals containing gliadin (such as wheat, rye, and barley and their products) should be omitted in the diet. Moreover, the products contaminated with gliadin should also be avoided for the well-being of patients. Since gluten damages the small intestine and causes nutrient malabsorption, therefore the GFD is particularly important for coeliac patients.

The fruits, vegetables, seafood, dairy products, legumes, nuts, and poultry can be used for preparations of gluten-free foods. However, some of these products may contaminate gluten or gluten-containing products; thus, the hidden gluten may be found even in gluten-free foods. Therefore, to avoid cross-contamination, gluten-free foods can be prepared in particular section of the production area. The probability of the contamination is needed to specify in the product label and the ingredient list [15].

GFD is also the only nutritional therapy for individuals, who suffer from CD and other autoimmune gluten-related disorders. This diet needs to be strictly followed to avoid any undesirable consequences. However, since most staple foods include wheat or cross-contaminated gluten, therefore it becomes difficult in practice.

16.3.1 GLUTEN RELATED HEALTH DISORDERS

16.3.1.1 COELIAC DISEASE (CD)

Coeliac disease (CD) is extensively researched and one of the well-known gluten-related disorders. It is a gluten enteropathy, affecting approximately 1% of the world's population. This disease is triggered by consuming wheat, rye, and barley proteins and foods contaminated with these cereals [30]. After consuming gluten-containing foods by genetically susceptible individuals who carry HLA type II DQ2/DQ8 alleles, the T-cell occurs the immune response to transglutaminase. As a result, the mucosal tissue of the small intestine is damaged, and the villous atrophy emerges, which leads to digestion problems and malabsorption of nutrients [4, 16].

The symptoms of CD can be observed as both intestinal and extra-intestinal, including chronic diarrhea, malnutrition of certain nutrients (e.g., calcium, iron, vitamins), weight loss, anemia, growth retardation in children, hepatic dysfunction, anemia, constipation, and chronic fatigue [16, 54, 56]. Moreover, this disease is diagnosed with serological testing, duodenal biopsy, genetic testing, and screening the first-degree coeliac relatives. Type I *diabetes mellitus* patients and individuals with down syndrome are the high-risk groups in terms of CD, where studies presented a high prevalence [16]. Moreover, anti-tissue transglutaminase (Anti-tTG), anti-endomysial antibody (EMA), anti-deamidated gliadin peptides (Anti-DGP), IgA (immunoglobulin A) and IgG (immunoglobulin G) antibodies are used for detection of serological markers. However, Anti-tTG is considered the most specified antibody with high sensitivity to CD diagnosis [54].

The only efficient treatment for CD is dietary therapy, and the GFD. Avoidance of gluten-containing cereals and foods is strictly recommended for these patients since it enhances the clinical manifestations and nutritional status and heals intestinal damage. Nevertheless, due to the low content of nutrients (such as fiber, minerals, vitamins, and proteins in gluten-free products), constipation, and micro-macronutrient malnutrition are observed in the GFD patients [44]. Therefore, the fortifications of gluten-free products with high nutritional quality ingredients have become a necessity for this diet.

16.3.1.2 WHEAT ALLERGY (WA)

Wheat allergy (WA) is considered one of the most severe and common food allergies. In opposition to CD, the water-soluble globulin and albumin also induce WA and gluten proteins. The allergy is mediated by either IgE (immunoglobulin E) or non-IgE. When the individual is exposed to wheat via skin contact or inhalation, the T helper type 2 cells activate, and IgE is produced. After binding with gluten, IgE generates histamine production and release, which causes clinical symptoms such as skin rash, itching, sneezing, coughing, abdominal pain, diarrhea, and headache. It sometimes causes severe effects such as anaphylaxis and death [54, 56]. Baker's asthma is this type of allergy, and it is observed in bakery workers after inhalation of the cereal dust. However, most symptoms like coughing, sneezing, tearing, and itching are not experienced when the wheat is cooked unless it is contaminated with raw wheat products [44, 56].

On the other hand, the symptoms of non-IgE-mediated allergy appear after 2 hours or more than 2 h of wheat intake. Indigestion, vomiting, headaches, and arthralgia are the main signs of the allergy since this type of allergy is related to eosinophilic esophagitis and eosinophilic gastritis [44, 54].

16.3.1.3 DERMATITIS HERPETIFORMIS (DH)

Dermatitis herpetiformis (DH) is an autoimmune, chronic disease often associated with CD since similar serological precursors are present in patients [44, 56]. However, EMA, DGP, and Anti-tTG antibodies are also recognized as specific serologic markers for DH [6]. The recurring intestinal damage develops when the genetically susceptible individuals are exposed to gluten proteins. The prevalence of this disease is higher in males than females, with a ratio of 3:2, and it is prevalent in northern Europe and the UK [54].

In this disease, the skin is affected by blistering rashes; thus, the symptoms include polymorphic lesions with erythema, urticarial plaques, and hyperpigmentation on the extensor surface of elbows, knees, and shoulders, are observed. Unlike CD, gastrointestinal (GI) symptoms rarely occur, but villous atrophy presents in approximately 70% of patients when they encounter gluten. Moreover, DH patients have identical HLA alleles (DQ2 and DQ8) as CD; hence, GFD has also been applied to DH patients for nutritional treatment [54, 56].

16.3.1.4 GLUTEN ATAXIA (GA)

Gluten ataxia (GA) is another autoimmune disease, and cerebellar ataxia occurs when genetically susceptible individuals digest gluten. The disorder affects Purkinje cells and causes neurological consequences such as cerebellar damage. The mechanism of occurrence is still unclear; however, it is presumed that the villous injury in CD causes vitamin E and Vitamin B1 deficiencies, leading to pino-cerebellar degeneration. This disease is diagnosed with anti-tTG, anti-gliadin, and anti-TG6 (anti-transglutaminase 6) antibodies [44, 56].

16.3.1.5 NON-CELIAC GLUTEN/WHEAT SENSITIVITY

Non-coeliac gluten sensitivity (NCGS) is a novel disease among gluten-related disorders. It is partly the reason for the rising popularity of GFD among individuals who are not formally diagnosed with CD or WA. It is mainly seen in adult females with a 0.6–13% prevalence. According to recent studies, the symptoms are not triggered only by gluten. However, other wheat compounds such as low fermentable carbohydrates and wheat amylase-trypsin can also play a role to appear of them. Hence it is also called wheat sensitivity instead of gluten sensitivity. The symptoms of this disorder are similar to CD and WA, such as bloating, abdominal pain, weight loss, and fatigue. However, these manifestations are not a result of an allergic or immune response [54].

Moreover, distinctively from CD, persistent intestinal damage is not observed in NCGS, and no specific serological marker exists for NCGS. Thus, it is determined by screening the patients and their reaction to wheat/gluten consumption. In most cases, it has been seen that after elimination of gluten or wheat in diet, the symptoms are disappeared in patients, and the general health improves; however, the symptoms reoccur within a few hours or days of consumption of gluten/gluten-containing foods [24]. Therefore, GFD is also an important treatment option for these novel disorders like other gluten-related disorders.

16.3.2 LIFE CHOICE

Apart from having one gluten-related disorder, the number of individuals who adopted a gluten-free lifestyle and diet has increased. For instance,

25% of the population choose GFD products in USA. According to research, people choose this diet since they thought it as a 'healthy option,' 'health benefits,' and 'healthy for digestion.' It is also stated that the GFD food is mostly bought by people, who are not associated with coeliac or other related disorders due to the inaccurate assumption that GFD is healthier than a gluten-containing diet [42]. Furthermore, although the controlled human research on the weight-loss feature of GFD is limited, yet some studies endorsed the obesogenic effect of gluten in animal models [19]. Therefore, GFD is considered one of the popular diets to reduce body weight.

Athletes also prefer GFD because of 'enhancing body composition' and 'improving sports performance' expectations despite not being formally diagnosed with gluten-related diseases [35]. However, short-term application of GFD is reported with no effect on well-being and calorie loss for non-coeliac athletes [19]. Moreover, this diet is applied to non-gluten-related GI diseases such as irritable bowel disease to soothe the pain, decrease the bloating or improve the health status.

Contrary to the health benefit beliefs about GFD, it is reported that most gluten-free products have low quantities of micronutrients (MNs), fibers, and excess amounts of sugar, salt, and saturated fats. Mainly, processed gluten-free foods include high fats, sugar, and energy content. This situation has raised the risk of obesity and other obesity-related health problems for coeliac patients. It now has become a threat to the healthy people who apply a GFD diet and have an obesogenic environment. Although gluten-free vegetables and fruits increase diets' nutritional quality, they are not enough to assume that GFD is a good diet for healthy subjects. Therefore, it is mostly not recommended for people without gluten-related diseases.

The market share of these products has boosted in recent years to meet consumer expectations. According to economic studies, despite the high cost of gluten-free products, the followers of this diet, especially young adults, are willing to pay more. Though the relation of this diet with weight loss, health improvement, and decreasing the gluten sensitivity is still unclear, yet the number of people on this diet as a lifestyle is foreseen to increase soon [35].

16.4 GLUTEN ALTERNATIVES IN BAKERY PRODUCTS

Since bakery products consisting of wheat flour are mostly the staple foods of many nations, the GFD challenges the patients. The increasing number of patients with gluten-related disorders and the low quality of gluten-free products lead to the necessity of high nutritional product quality for patients. New formulations and ingredients have emerged to produce gluten-free products as wheat-based bakery products [34]. Therefore, pseudo-cereals, legumes, and vegetable flours are examined and are being used as alternative gluten sources [26].

Replacing gluten is challenging since gluten proteins play an essential role in bakery products. It makes the dough extensible, cohesive, and easily shaped while providing consistent gas retention in dough while processing. Hence, the gluten replacement additives are expected to present these features. In case of lack of gluten, the dough becomes liquidized leading to distorted interior structure in bread and disintegration of the pasta dough because of the unstabilized protein network [41].

Gluten-free products are mostly low in volume, crumb softness, and unable to retain the gas during processing. Considering the necessity of a gluten-network structure in many bakery products, especially in bread, the ingredients used as gluten alternatives are expected to provide this critical role. The hydrocolloids such as xanthan, hydroxypropyl methylcellulose, carboxymethylcellulose, and emulsifiers are mostly added to ensure this essential function [9].

Another issue with gluten-free bakery products is low nutritional quality. Most of the selected base flour is rich in starch, which provides filling ingredients, but the quality of protein, minerals, unsaturated fatty acids, and vitamins is low. These lead to a high glycemic index (GI) containing food and insufficient nutritional quality in gluten-free products, which contributes to an unhealthy diet for individuals applying GFD. The sensory quality is also an important factor for application to diet and consumer acceptability of gluten-free products. The chosen flour is expected to respond to sensory quality parameters such as color, flavor, taste, mouthfeel, texture, and smell. Therefore, there are a variety of flour from different sources to replace wheat flour for gluten-free baking. Each one of them improves the products in different aspects [41].

16.4.1 RICE FLOUR

Rice (*Oryza sativa* L.) is one of the major cereals in the world and the main ingredient for half of the world's population, particularly in Far Eastern Asia and the Indian subcontinent. China is the leading producer, and since rice is the staple food, it is used for multiple products such as rice cakes, rice noodles, and rice bread. Therefore, for the gluten-free bakery formulations, rice is mainly added to the formulations due to its neutral taste and aroma and its affordable market availability [34]. It is also non-allergic, easily digested, and has a mild flavor. Moreover, the high starch content helps gelatinization and the structure of the many bakery products. Thus, rice flour is extensively used to reform many gluten-free products due to these properties. Although it has these benefits, the functionality of rice proteins is low due to hydrophobic characteristics, which inhibit the viscoelasticity of dough. Furthermore, the nutritional quality of rice flour is also poor in terms of amino acid composition, minerals, vitamins, and lipids compared to wheat. For this reason, it can be blended with other gluten-free flours (legumes and pseudo-cereals) to overcome the disadvantages of nutritional and functional quality [33].

16.4.2 CORN FLOUR

Corn or maize (*Zea mays*) is considered a safe cereal in terms of gluten, and it is mostly consumed in the USA. It widely participates in gluten-free formulations such as corn flour or corn starch because of its high starch content. It is the main ingredient in gluten-free products and is used as a starch source. The maize flour is high in dietary fiber, unsaturated fatty acids, and Vitamin B6, while protein and mineral amounts are low. It is generally evaluated as a base flour with rice flour or alone in gluten-free bakery products. Corn starch, the byproduct of oil production, is also preferred in formulations. According to the pre-treatments to starch before usage, the quality and functional properties can change in products. Corn flour and starch are preferable for pasta products since they provide high viscosity resulting more firm and structured dough. When the starch is pregelatinized, the rheological properties and networked structure in pasta improve; thus, cooking quality and overall acceptability also enhance. However, corn flour is poor in protein, mineral, and vitamin quality like rice flour. Therefore, in bakery products, the corn flour is mostly enriched with hydrocolloids, legume flours, protein isolates, and other flours [17, 21].

16.4.3 PSEUDO-CEREAL FLOURS

The pseudo-cereals have been evaluated to improve the quality and functionality of gluten-free products due to their nutritional and phytochemicals.

In nutritional aspects, besides being rich sources of starch, fiber, vitamin, unsaturated fatty acids, minerals, and high-quality essential amino acids, they are also rich in saponins, phytosterols, polyphenols, and betalains, which are mostly lacking in these products. Among these seeds, amaranth, quinoa, flaxseed, buckwheat, and teff are the most studied and considered good functional ingredients in gluten-free bakery products. Products containing amaranth, quinoa, and buckwheat have a high amount of protein, fiber, and minerals, making them an alternative fortification agent to rice and corn-based products with better consumer acceptance [8].

Especially in bread, amaranth, quinoa, and buckwheat presented an improvement in crumb texture and color, and viscoelasticity of dough compared with starch-based gluten-free bread formulations while enhancing the nutritional quality in terms of fiber, protein, minerals (iron and calcium) and vitamin [1, 57]. Moreover, owing to their high dietary fiber content, the swelling ratio of starch can be adjusted, which leads to higher gas and water retention capacity with increased viscosity of doughs. In this way, the crumb softness and bread loaf volume improve as gluten-free bread gains more acceptability. In addition to this, whole grain pseudo-cereals also include natural emulsifiers that chelate with amylose; thus, the starch retrogradation and bread stealing throughout storage reduce by limiting the swelling of starch [9].

Marti et al. [31] reported that fermented teff enhanced the physical features and nutritional quality of gluten-free bread, and decreased the staling compared to control and non-fermented teff gluten-free bread. The hardness of bread was lower than control corn-based bread, while the stable protein network was created with teff fermentation.

In bread formulation, pseudo-cereals are also mostly used in other bakery products such as pasta and cookies. D'amico et al. [13] showed that textural features of pasta consist of amaranth, quinoa, and buckwheat enhanced with higher drying temperature, which enabled the structural integrity and decreased cooking loss. On the other hand, enrichment of buckwheat with rice flour in biscuits was found that the phenolic and mineral content are higher than the rice flour-based controls besides having better sensory qualities [45].

16.4.4 OTHER FLOURS

Studies have presented other candidate flours as a replacement and fortification agent for gluten-free products. Chia seed (*Salvia hispanica*) has recently been used in food formulations due to its natural hydrocolloid properties appearing with hydration. Hence its usage in gluten-free formulations is evaluated to strengthen the structure of dough as it behaves as a hydrocolloid while increasing the nutritional quality of products. When used with other fortification flours, it enhances the loaf volume of bread and the nutritional quality of products [57].

Acorn flour is considered a novel ingredient for gluten-free bakery products although it has been used for many years in different world regions, including Japan and African countries. Depending on species, the nutritional quality of the flour's changes but amino acid composition, including essential amino acids, starch amount and unsaturated fatty acid composition, significantly increase the nutritional quality of foods. Furthermore, acorn flour improves the gluten-free bread quality by enhancing the bread volume and crumb texture and reducing the staling through retrogradation ratio [27].

Chestnut flour (*Castanea* spp.) is another alternative flour source for gluten-free products. It participates in gluten-free cookie, wafer, and bread formulations. It helps to increase cookie sensory quality when used with legume flours by masking the flavor. It also provides elasticity to the dough, positively affects the rheological characteristics by binding water with its fiber and retards the staling by reducing the moisture in gluten-free-bread [55, 57].

Sorghum flour (*Sorghum bicolor* L.) is a safe cereal for people with gluten-related disorders due to having a similar protein to corn. However, the improvement is not observed when used as the only fortification flour. When used with other alternative flours such as potato and corn starch, it amends the functions, or the pre-treatment, including heat treatment, is applied [41].

16.5 LEGUME-BASED INGREDIENTS FOR GLUTEN-FREE PRODUCTS

Legumes are an alternative plant protein source for those unable to consume animal protein sources due to their nutritional quality. They are rich in starch, proteins with essential amino acids, vitamins, minerals, and dietary fiber. Although chickpea, lupin, and soya include high lipid content, this family contains a low amount of fat. Thanks to these nutritional and chemical content, they are highly preferred in gluten-free baking as a main ingredient or for fortification purposes. Moreover, it has been used because of its functions in bakery goods. Legume proteins exhibit high solubility, gelation, water-holding, and fat absorption capacity, and emulsifying and foaming characteristics; thus, they improve the gel-forming, viscosity of dough, bread loaf volume, texture, and dough structure [18].

Chickpea (*Cicer arietinum* L.) is an essential legume with its high-quality chemical composition. It is composed of starch, phenolic contents, protein composing eight essential amino acids, minerals, and a high amount of fiber. In addition to lowering the glycemic index (GI) of products by high dietary fiber composition, the gelation capability and high water-fat binding ability of chickpea fiber lead to improved texture, sensory properties and shelf-life of products. The protein in chickpea also shows emulsifier characteristics that affect and results in higher loaf volume in bread. Therefore, it has been widely studied in many gluten-free bakery products, including mostly bread, pasta, cookies, muffins, and noodles. In these products, the enhancement is provided by increasing dough viscoelasticity and batter viscosity when used with other flour or hydrocolloids because only chickpea flour does not form the dough. Even mixing is limited by a ratio of 20% since the dough becomes sticky and hard to process after this rate [32].

Soybean (*Glycine max* L.) is preferred for both gluten-free and vegetarian/vegan diets for protein and fat quality. It is considered an alternative protein source by its complementary amino acid content; thus, it is used to promote the nutritional quality of gluten-free products. It is added into the formulas in different forms, including soy protein, okara, and fatted/defatted soy flour. Since the protein content of defatted soy flour is eight times higher than wheat proteins, the addition of gluten-free formulas boosts the nutritional quality. Moreover, the amino acids in soybean protein contribute to the interactions with other compounds in dough systems; they bind and detain the water well. Hence the staling of bread reduces, and the storage time extends. The dietary fiber is also rich in soybean, which allows gel formation, increment at water holding capacity, thickening of the dough and improvement in structure besides increasing the dietary fiber content of the gluten-free goods. Furthermore, the gluten-free formulations enriched with other legume flours, enzymes, and hydrocolloids exhibit better functional, nutritional, and structural quality. Combining soybean with chickpea presented a more synergistic effect on bakery products than using them alone. On the other hand, the flavorless or beany flavor and low crust and crumb quality have limited the addition of the soybean in products [32, 52].

Bean (*Phaseolus vulgaris* L.) is another fortification legume that raises dietary fiber and protein content and mineral and vitamin substances in gluten-free products. Owing to rich zinc and better iron absorption in the body than other beans, their usage increases the nutritional quality of gluten-free bakery products. Like chickpea, the addition of bean flour also enhances the viscosity of the mixture and bread loaf volume in bread. It increases water absorption capacity while reducing cooking loss in pasta. Hence it improves the structure and acceptance of the products [18, 22]. Fava bean (*Vicia fava*) is also preferred in gluten-free cookies, cakes, and pasta because of their functional properties like other legume flours. Recent studies showed that the addition of fava bean flour into formulations enhanced the water holding and emulsifying capacity. However, it is noted that the pre-treatment and enzyme addition are essential factors in the improvement of functional properties in pasta. At the same time, hydrocolloids are more critical to the textural quality of cakes [5, 43].

Lentils (*Lens culinaris Medikus*) are found in different colors (yellow, red, green) and chemical compositions depend on the characteristics of the plant sources. The protein, dietary fiber and carbohydrates are higher than in other legumes, and saturated fat with sugar are lower than in other legumes. Fortification of lentils in gluten-free cake and bread increases consumer acceptability by improvement of sensory characteristics along with increment in foaming, fat/water binding and fat absorption. However, the product color might change, reducing overall acceptability [18, 40].

Similar results can be observed when the cowpea is added to the formulations. The cowpea (*Vigna unguiculate* L.) is rich in amylose content leading to an increment of viscosity in cake batter; however, the cowpea protein limits the starch gelation when used in gluten-free products. The protein and mineral content are also higher in cowpea fortified gluten-free products. However, the color is darker due to Maillard and caramelization reactions during baking when the ratio of cowpea increases in the gluten-free flour mixture [25].

Carob (*Ceratonia siliqua* L.) participates in gluten-free formulation owing to its protein and carbohydrates functions. Besides having proteins with similar functional characteristics to wheat gluten, the soluble and insoluble dietary fiber of germ also contributes to the structure of gluten-free bakery products. This soluble fiber is widely used as a hydrocolloid in many food products; thus, dough's rheological, structural, and textural quality through its utilization. It helps increase the dough's viscoelasticity, viscosity, and water holding capacity, resulting in better porosity and crumb softness [57].

16.6 SENSORY ASPECTS OF LEGUME-BASED GLUTEN-FREE PRODUCTS

Legumes have different compositions and substituting the wheat flour in gluten-free bakery products with legume-based ingredients changes the sensory aspect of products, such as texture, appearance, taste, and aroma. For example, among legumes, lentils, and chickpea flours had the most odor-active compounds when baked caused by the Maillard reaction [28].

There are several sensory analyzes to investigate the quality properties and consumer acceptance of a product (Figure 16.1). In panel tests, trained panelists describe the product and investigate quality parameters. In a consumer test, untrained potential consumers are used as test subjects. In the case of legume-based products, consumer acceptance tests are critical because of their undesirable beany and grassy flavors. Both analyzes were used to describe and investigate the acceptability of gluten-free legume-based bakery products.

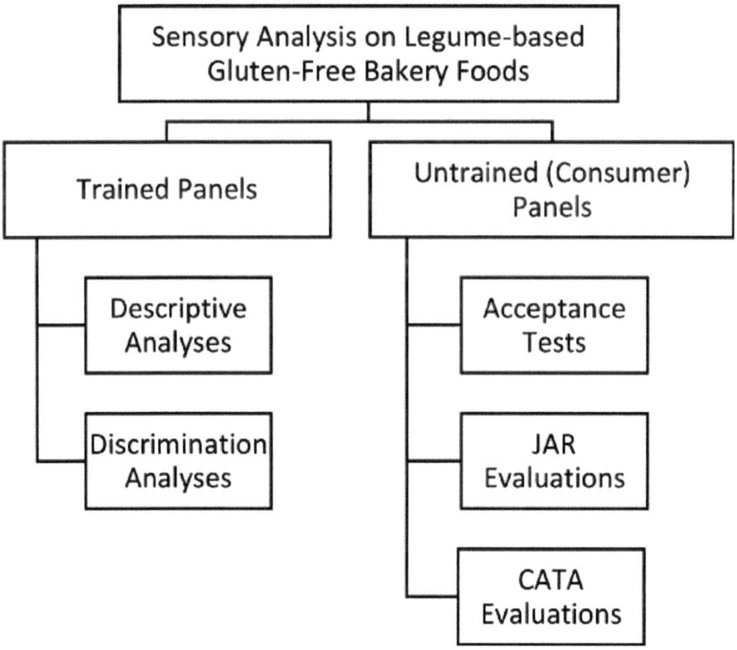

FIGURE 16.1 Classification of sensory analyzes on legume-based gluten-free bakery products.

Legend: CATA: Check-all-that-apply; and JAR: Just about right.

Furthermore, the studies are divided into two categories: (i) the legumes were used alone as an ingredient for replacing the wheat-based product; and (ii) legume-based ingredients were used with other gluten-free ingredients.

16.6.1 PANEL TESTS ON SENSORY PROPERTIES

This section will discuss sensory evaluations with the trained panel on legume-based gluten-free bakery products. There have been several studies in recent years that investigated the sensory properties of these products. Most of the studies focused on chickpeas and other more traditional and local legume-based flours. Even though gluten-free bread was more extensively studied in the past, there has been more variety of bakery products, from muffins to cookies, pancakes, and brownie samples. Descriptors specific for each type of product are decided or generated in sensory panels.

Olojede et al. [36] analyzed legume flours (chickpea and cowpea) and added sorghum-based sourdough with trained 10 panelists. A 9-point hedonic scale was used to evaluate appearance, taste, texture, aroma, and overall acceptability. The control sample only contained sorghum flour and scored significantly less than other samples for all sensory parameters. Compared, chickpea flour added samples were liked better than samples with cowpea flours with scores higher than 7.

In a study by Ouazib et al. [38], wheat flour was replaced 100% by raw, germinated, toasted, and roasted chickpea flours in bread samples. Nine panelists used descriptive terms such as appearance, flavor, color, and taste and evaluated samples based on a 9-point scale. All scores were below the sensory acceptance level. There was no significant difference among samples (except for appearance); however, bread samples with raw and germinated chickpeas scored higher than other samples.

In another gluten-free bread-related study [53], bread samples were prepared with soy flour and fava bean flour (fermented and unfermented) and evaluated in a descriptive panel with five trained panelists. Bread samples were evaluated based on evenness, crumbliness, softness, springiness, toughness, and intensity of the color. Samples with soybean flours had a significantly lower score in appearance terms. Samples with fava beans had better scores on a 5-point scale and the sample with unfermented fava beans had the highest score in terms of softness. Fava beans were also investigated in cookie samples.

Schmelter et al. [49] analyzed cookies made with six types of Vicio fava bean flours in a Flash profiling analysis. Around 17 trained panelists evaluated the samples and they found that all samples were in an acceptable range, except one variety (Fuego). They suggested that Vicio fava beans could be used in gluten-free cookies after flour treatment and mild-flavored ingredients.

Brownies were the product of interest by Pio Ávila et al. [40], in which, to obtain a gluten-free brownie, rice flour was mixed with either white bean flour or lentil flour. Other ingredients such as sugar, eggs, cocoa powder, amaranth, and Brazil nuts were added to all samples. While regular 50% chocolate was added to the control sample with wheat flour, gluten-free chocolate was used in gluten-free brownies. Descriptive analysis with 20 trained panelists was conducted in which 20 descriptive terms were generated in the first part. These terms, such as firm, aerated, nice, sweet, bitter, crumbly, crispy, tasty, nutty, soft, compact, light-colored, chocolate, and dark, describe the properties of samples. In the second part, after training, panelists evaluated samples based on these terms with a 9-cm scale by comparing three brownie samples (100% wheat flour, 6:1 rice flour and lentil flour and 6:1 rice flour and white bean flour). Gluten-free brownies showed a more aerated texture, lighter color and more bitter taste. Since these were undesirable properties in brownie products, it was suggested that it was possible to enhance the sensory properties of brownies with flour treatment.

Akshata et al. [3] formulated gluten-free pancakes with rice flour along with xanthan gum, whey protein, mung bean flour and dietary fiber. The control sample had only wheat flour and another sample contained only rice flour. Rice flour and mung bean flour were mixed in different ratios in other experimental samples. They conducted a descriptive analysis with trained 18 panelists. Seven attributes were decided (color, surface, shape, handfeel, flavor, mouthfeel, and taste) and evaluated on a 10-cm scale. Overall quality was evaluated separately. Samples with additional ingredients (xanthan gum, whey protein and dietary fiber) had high scores in sensory analysis; however, the samples with mung bean flours were not scored within an acceptable range. Mung beans were also used in another study.

Jeong & Chung [25] evaluated muffins with three mung bean varieties and three cowpea varieties mixed with waxy rice flour. Waxy rice flour and individual legume flour were mixed 1:1 ratio in each sample. Sensory evaluation was conducted by the response of 24 trained panelists in a 7-point acceptance test. All samples had increased nutritional value compared to wheat flour-based muffins and all samples received acceptable scores from the panelist. Quality parameters were better in samples with cowpea flour, which had higher scores in sensory evaluations than the samples with mung bean flour. Muffins with chickpea flour were also subjected to quantitative descriptive analysis (QDA®) with trained eight panelists [23]. Aside from chickpea flour, different ingredients such as whey protein, xanthan gum and inulin are added to muffins. Three descriptors for appearance (general, height, color, and sponginess), four descriptors for manual texture (springiness, cohesiveness, and stickiness), five descriptors for oral texture (hardness, adhesiveness, moisture, difficulty to chew and difficulty in swallowing) and three descriptors for taste attributes (typical muffin taste, chickpea taste and aftertaste) were evaluated. The scores of panelists varied from each other. A significant consensus was not achieved; however, the sample with chickpea flour and xanthan gum had better sensory scores than other muffin samples. Furthermore, it was concluded that chickpeas' taste did not result in a disliking by panelists.

16.6.2 CONSUMER ACCEPTANCE TESTS

This chapter has investigated consumer studies on legume-based gluten-free bakery products. There have been more studies with consumers than trained panels in recent years on these products, presumably because of the commercial importance of gluten-free products in the global market.

In bread-related studies, rice-based gluten-free products are generally compared with legume-based products, or legume-based ingredients were added to rice-based products to improve their nutritional and texture properties and decrease glycemic indices. For instance, Santos et al. [46] compared the bread containing chickpea flour with or without psyllium powder and the bread containing rice flour with cassava starch as a control sample. They conducted a consumer acceptance test with 65 untrained panelists and found that overall liking of bread with chickpea flour was scored significantly higher than bread with rice flour. They have also indicated that the addition of psyllium did not affect the acceptability scores of chickpea-containing breads while improving the texture properties. In another study, bread with rice flour and bean flours were compared individually and in mixed samples [2]. With 64 consumer panelists, aroma, texture, flavor, and overall liking with consumer acceptance test based on a 9-point hedonic scale, softness with just about right (JAR) test based on a 5-point scale and emotional response to bread with check-all-that-apply (CATA) test base on 33 facial emojis were evaluated. Rather than individual ingredients with rice and bean flour, mixed-flour samples scored better in the acceptance test and JAR evaluation. According to CATA evaluation, using mixture of rice flour and bean flour in bread samples correlated positive emotions with texture and acceptance.

Mixing rice flour with legume-based ingredients did not always result in the best results. Lian et al. [29] used chickpea and soybean okara to use them as thickeners in gluten-free bread samples made with rice flour and corn starch. Okaras were mixed corn starch and rice flour with a ratio of 50:50. Around 16 untrained panelists evaluated breads on a 9-point hedonic scale based on taste, appearance, moistness, and overall liking. Although soybean okara samples were scored higher than chickpea samples, scores of both samples were below the sensory acceptance level.

Other than rice-based products, corn-based bread samples were also compared with legume-based ingredients. Cunha et al. [12] fortified maize bread with legumes such as pea, chickpea, fava bean and lentil. Overall liking was graded by 60 untrained tasters on a 9-point hedonic scale. There was no significant difference among the samples; however, the bread sample with chickpea flour had the closest score to the control sample.

There are also reports in which wheat bread samples were used as control. For example, chickpea bread samples were evaluated by adding cassava starch, maize starch, potato starch and rice flour [47]. Untrained 53 panelists evaluated bread samples. They used a 10-cm scale for grading appearance, color, aroma, taste, texture, and overall liking. Adding potato starch and cassava starch to chickpea flour resulted in the highest acceptability scores, and these scores were similar to wheat bread and gluten-free white bread. In another study, wheat bread was used as a control and compared with samples with extruded broken bean flour along with rice flour and corn starch [22]. Based on the results of 240 consumer panelists, the control sample had the highest score in texture; the consumer acceptability score for all gluten-free samples was similar to the control sample, except in the sample with 60% broken bean flour in terms of texture.

Cookies and biscuits were other products of interest in consumer tests. There was more variety of legume-based and other gluten-free ingredients in these products than in bread-related studies. Simons & Hall III [51] analyzed cookies with pinto beans in raw, cooked, germinated, and germinated/steam blanched forms. Oats, rice, and tapioca were also added to the bean flours. A 9-point hedonic scale acceptance test was conducted with 31 untrained panelists. All samples were scored higher than 6-points and there was no significant difference among samples. They concluded that as long as the optimum level of pinto beans (40%) was added to the flour mix for cookies, the heat treatment of beans, which was applied to reduce undesirable 'beany' flavor and requires more energy and cost, was not necessary to produce a gluten-free product.

Ostermann-Porcel et al. [37] added soybean okara and manioc flour mixture to produce gluten-free cookies. Around 70 untrained consumers evaluated cookies based on a 9-point hedonic scale for the consumer acceptance test. Samples with 15% and 30% okara were more acceptable than higher additions. Therefore, soybean okara addition at high concentrations affected consumer acceptance negatively, as it was the situation in the bread samples above [29]. However, peanut okara might not have as much negative effect on cookies as soybean okara.

Cardillo Diniz et al. [10] modified the formula of traditional Argentinian Alfajour cookies using different mixtures of peanut okara with rice and sorghum flours instead of wheat flour. Samples with higher percentages of peanut okara and sorghum had higher acceptance levels by 100 untrained consumer subjects. Other than cookies, biscuits were also evaluated.

Wesley et al. [58] conducted a consumer acceptance analysis on biscuits made with cooked and raw rice and white bean flours. The added bean flour enhanced 'rice biscuits' nutritional value and fiber content. Around 121 untrained panelists evaluated samples and they found that biscuits with cooked bean flours were more acceptable than raw bean flours. Consumers' purchase intent was also investigated, and products with cooked bean flours were more prone to be purchased.

Di Cairano et al. [14] used buckwheat (50%) and sorghum (30%) flours as base ingredients and added chickpea and lentil flours (20%) in gluten-free biscuits. They used a 100% wheat flour sample as a control, and one of the biscuit samples did not contain any legumes. Around 76 consumers evaluated samples with a 100-point likeness scale. Lentil-containing samples scored higher than the samples with chickpea. The control sample and the sample without any legume scored higher than samples with legume because of the 'sandy' mouthfeel of legume-containing samples.

Consumers also evaluated cake-style bakery products. Pio Ávila et al. [40] investigated the gluten-free brownies with a consumer acceptance, JAR and CATA tests with 100 untrained panelists. The acceptance test was scored based on a 9-point hedonic scale. Around 20 attributes from descriptive analysis (Section 16.6.1) were used in JAR and CATA tests. All samples were scored more than 6-point; however, consumers liked the sample with lentil flourless. They suggested that when formulating a gluten-free brownie, lentil, and white bean flours were viable alternatives to using with color and texture modification.

Furthermore, Sciammaro et al. [50] analyzed muffins with carob beans, rice, and corn flours. They conducted a consumer acceptance test (9-point hedonic scale) with untrained 76 panelists and found that muffins with 66% carob flour had the best score. Andrade et al. [5] mixed fava bean flour with galactomannan and xanthan gum in cake samples and analyzed six combinations with a preference ordering test and consequently an acceptance test with selected cakes. Around 70 untrained consumers evaluated and compared fava bean-containing samples with corn flour-containing samples. Samples with fava beans scored higher than samples with corn flour. They concluded that adding hydrocolloids to fava bean-containing samples did not affect the sensory scores and acceptance.

Chompoorat et al. [11] prepared cupcakes with red kidney bean flour with the addition of rice flour from 5% to 25%. Around 20 untrained panelists evaluated cupcakes with a 9-point hedonic scale and samples with 25% rice flour had the highest score for all attributes (overall quality, flavor, texture, and taste).

Pasqualone et al. [39] analyzed pizza samples with lentil, rice, and corn flours. Two types of lentil flours were used (untreated and extruded-cooked flours). Around 70 untrained consumers evaluated these samples based on a 5-point likeness scale; and samples with extruded-cooked lentil flour scored significantly higher than those with untreated lentil flour.

16.7 SUMMARY

Several legumes have been added to bakery formulas to produce gluten-free products in recent years. Replacing wheat flour with legume-based flours increased the nutritional quality of bakery products; however, sensory properties were generally negatively affected because of changes in texture, appearance, flavor, and taste. Mixing legume flours with other gluten-free flours enhanced all sensory properties compared to using only legume or non-legume-based gluten-free ingredients. To enhance the sensory properties of gluten-free bakery goods, mixing of different gluten-free flour sources, the addition of texture enhancing ingredient and flour treatment was suggested by researchers.

KEYWORDS

- **algae**
- **biological activities**
- **extraction**
- **farming**
- **nutraceuticals**
- **pharmaceuticals**
- **secondary metabolites**

REFERENCES

1. Aguiar, E. V., Santos, F. G., Centeno, A. C. L., & Capriles, V. D., (2021). Influence of pseudo-cereals on gluten-free bread quality: A study integrating dough rheology, bread physical properties and acceptability. *Food Research International, 150*(Part A), E-article ID 110762.
2. Aguiar, E. V., Santos, F. G., Faggian, L., Da Silveira, A. M. B., Araújo, V. A., Conti, A. C., & Capriles, V. D., (2022). An integrated instrumental and sensory techniques for assessing liking, softness and emotional related of gluten-free bread based on blended rice and bean flour. *Food Research International, 154*, E-article ID 110999.
3. Akshata, B., Indrani, D., & Prabhasankar, P., (2019). Effects of ingredients and certain additives on rheological and sensory characteristics of gluten-free eggless pancake. *Journal of Food Processing and Preservation, 43*(10), E-article ID 14129.
4. Allen, B., & Orfila, C., (2018). The availability and nutritional adequacy of gluten-free bread and pasta. *Nutrients, 10*(10), 1370–1377.
5. Andrade, F. J. E. T., Albuquerque, P. B. S., Moraes, G. M. D., Farias, M. D. P., Teixeira-Sá, D. M. A., Vicente, A. A., & Carneiro-da-Cunha, M. G., (2018). Influence of hydrocolloids (galactomannan and xanthan gum) on the physicochemical and sensory characteristics of gluten-free cakes based on fava beans (*Phaseolus lunatus*). *Food & Function, 9*(12), 6369–6379.
6. Antiga, E., & Caproni, M., (2015). The diagnosis and treatment of dermatitis herpetiformis. *Clinical, Cosmetic and Investigational Dermatology, 8*, 257–265.
7. Barak, S., Mudgil, D., & Khatkar, B. S., (2015). Biochemical and functional properties of wheat gliadins: A review. *Critical Reviews in Food Science and Nutrition, 55*(3), 357–368.
8. Bascuñán, K. A., Vespa, M. C., & Araya, M., (2017). Celiac disease: Understanding the gluten-free diet. *European Journal of Nutrition, 56*(2), 449–459.
9. Brites, L. T., Schmiele, M., & Steel, C. J., (2018). Gluten-Free bakery and pasta products. In: Holban, A. M., & Grumezescu, A. M., (eds.), *Alternative and Replacement Foods* (pp. 385–410). Burlington – ON, Canada: Apple Academic Press.

10. Cardillo, D. R., Morcatti, C. F., & Ferreira, R. J., (2021). Effect of different gluten-free flours on the sensory characteristics of a vegan alfajor: Vegan gluten-free Alfajor development. *Food Science and Technology International, 27*(2), 145–150.
11. Chompoorat, P., Kantanet, N., Hernández, E. Z. J., & Rayas-Duarte, P., (2020). Physical and dynamic oscillatory shear properties of gluten-free red kidney bean batter and cupcakes affected by rice flour addition. *Foods, 9*(5), 616–621.
12. Cunha, L. M., Fonseca, S. C., Lima, R. C., Loureiro, J., Pinto, A. S., Vaz, P. M. C., & Brites, C., (2019). Consumer-driven improvement of maize bread formulations with legume fortification. *Foods, 8*(7), 235–241.
13. D'Amico, S., Mäschle, J., Jekle, M., Tömösközi, S., Langó, B., & Schoenlechner, R., (2015). Effect of high temperature drying on gluten-free pasta properties. *LWT-Food Science and Technology, 63*(1), 391–399.
14. Di Cairano, M., Condelli, N., Galgano, F., & Caruso, M. C., (2022). Experimental gluten-free biscuits with underexploited flours versus commercial products: Preference pattern and sensory characterization by Check All That Apply Questionnaire. *International Journal of Food Science & Technology, 57*(4), 1936–1944.
15. El-Khoury, D., Balfour-Ducharme, S., & Joye, I. J., (2018). A review on the gluten-free diet: Technological and nutritional challenges. *Nutrients, 10*(10), 1410–1418.
16. Elli, L., Branchi, F., Tomba, C., Villalta, D., Norsa, L., Ferretti, F., Roncoroni, L., & Bardella, M. T., (2015). Diagnosis of gluten related disorders: Celiac disease, wheat allergy and non-celiac gluten sensitivity. *World Journal of Gastroenterology: WJG, 21*(23), 7110–7119.
17. Ferreira, S. M. R., De Mello, A. P., Dos Anjos, M. D. C. R., Krüger, C. C. H., Azoubel, P. M., & De Oliveira, A. M. A., (2016). Utilization of sorghum, rice, corn flours with potato starch for the preparation of gluten-free pasta. *Food Chemistry, 191*, 147–151.
18. Foschia, M., Horstmann, S. W., Arendt, E. K., & Zannini, E., (2017). Legumes as functional ingredients in gluten-free bakery and pasta products. *Annual Review of Food Science and Technology, 8*(1), 75–96.
19. Freire, R., (2020). Scientific evidence of diets for weight loss: Different macronutrient composition, intermittent fasting, and popular diets. *Nutrition, 69*, E-article ID 110549.
20. Gabler, A. M., & Scherf, K. A., (2020). Comparative characterization of gluten and hydrolyzed wheat proteins. *Biomolecules, 10*(9), 1227–1236.
21. Gao, Y., Janes, M. E., Chaiya, B., Brennan, M. A., Brennan, C. S., & Prinyawiwatkul, W., (2018). Gluten-free bakery and pasta products: Prevalence and quality improvement. *International Journal of Food Science & Technology, 53*(1), 19–32.
22. Gomes, L. D. O. F., Santiago, R. D. A. C., Carvalho, A. V., Carvalho, R. N., Oliveira, I. G. D., & Bassinello, P. Z., (2015). Application of extruded broken bean flour for formulation of gluten-free cake blends. *Food Science and Technology, 35*(2), 307–313.
23. Herranz, B., Canet, W., Jiménez, M. J., Fuentes, R., & Alvarez, M. D., (2016). Characterization of chickpea flour-based gluten-free batters and muffins with added biopolymers: Rheological, physical, and sensory properties. *International Journal of Food Science & Technology, 51*(5), 1087–1098.
24. Hollon, J., Puppa, E. L., Greenwald, B., Goldberg, E., Guerrerio, A., & Fasano, A., (2015). Effect of gliadin on permeability of intestinal biopsy explants from celiac disease patients and patients with non-celiac gluten sensitivity. *Nutrients, 7*(3), 1565–1576.

25. Jeong, D., & Chung, H. J., (2019). Physical, textural, and sensory characteristics of legume-based gluten-free muffin enriched with waxy rice flour. *Food Science and Biotechnology, 28*(1), 87–97.
26. Jnawali, P., Kumar, V., & Tanwar, B., (2016). Celiac disease: Overview and considerations for development of gluten-free foods. *Food Science and Human Wellness, 5*(4), 169–176.
27. Korus, J., Witczak, M., Ziobro, R., & Juszczak, L., (2015). The influence of corn flour on rheological properties of gluten-free dough and physical characteristics of the bread. *European Food Research and Technology, 240*(6), 1135–1143.
28. Krause, S., Keller, S., Hashemi, A., Descharles, N., Bonazzi, C., & Rega, B., (2022). From flours to cakes: Reactivity potential of pulse ingredients to generate volatile compounds impacting the quality of processed foods. *Food Chemistry, 371*, E-article ID 131379.
29. Lian, H., Luo, K., Gong, Y., Zhang, S., & Serventi, L., (2020). Okara flours from chickpea and soy are thickeners: Increased dough viscosity and moisture content in gluten-free bread. *International Journal of Food Science & Technology, 55*(2), 805–812.
30. Lionetti, E., Gatti, S., Pulvirenti, A., & Catassi, C., (2015). Celiac disease from a global perspective. *Best Practice & Research Clinical Gastroenterology, 29*(3), 365–379.
31. Marti, A., Marengo, M., Bonomi, F., Casiraghi, M. C., Franzetti, L., Pagani, M. A., & Iametti, S., (2017). Molecular features of fermented teff flour relate to its suitability for the production of enriched gluten-free bread. *LWT-Food Science and Technology, 78*(1), 296–302.
32. Melini, F., Melini, V., Luziatelli, F., & Ruzzi, M., (2017). Current and forward-looking approaches to technological and nutritional improvements of gluten-free bread with legume flours: A critical review. *Comprehensive Reviews in Food Science and Food Safety, 16*(5), 1101–1122.
33. Mert, S., Sahin, S., & Sumnu, G., (2015). Development of gluten-free wafer sheet formulations. *LWT-Food Science and Technology, 63*(2), 1121–1127.
34. Nespeca, L. D. S., Da Silva, P. H. F., Barlati, V. Da. S. T., Bona, E., Leimann, F. V., Marques, L. L. M., Cardoso, F. A. R., et al., (2021). How does the replacement of rice flour with flours of higher nutritional quality impact the texture and sensory profile and acceptance of gluten-free chocolate cakes? *International Journal of Food Science & Technology, 56*(4), 2019–2029.
35. Newberry, C., McKnight, L., Sarav, M., & Pickett-Blakely, O., (2017). Going gluten free: The history and nutritional implications of today's most popular diet. *Current Gastroenterology Reports, 19*(11), 1–8.
36. Olojede, A. O., Sanni, A. I., & Banwo, K., (2020). Effect of legume addition on the physiochemical and sensorial attributes of sorghum-based sourdough bread. *LWT-Food Science and Technology, 118*, E-article ID 108769.
37. Ostermann-Porcel, M. V., Quiroga-Panelo, N., Rinaldoni, A. N., & Campderrós, M. E., (2017). Incorporation of okara into gluten-free cookies with high quality and nutritional value. *Journal of Food Quality, 2017*(23), 88–97.
38. Ouazib, M., Garzon, R., Zaidi, F., & Rosell, C. M., (2016). Germinated, toasted, and cooked chickpea as ingredients for breadmaking. *Journal of Food Science and Technology, 53*(6), 2664–2672.
39. Pasqualone, A., Costantini, M., Faccia, M., Difonzo, G., Caponio, F., & Summo, C., (2022). The effectiveness of extruded-cooked lentil flour in preparing a gluten-free pizza with improved nutritional features and a good sensory quality. *Foods, 11*(3), 482–489.

40. Pio Ávila, B., Cardozo, L. O., Alves, G. D., Gularte, M. A., Monks, J., & Elias, M. C., (2019). Consumers' sensory perception of food attributes: Identifying the ideal formulation of gluten-and lactose-free brownie using sensory methodologies. *Journal of Food Science, 84*(12), 3707–3716.
41. Rai, S., Kaur, A., & Chopra, C. S., (2018). Gluten-free products for celiac susceptible people. *Frontiers in Nutrition, 5,* 116–123.
42. Reilly, N. R., (2016). The gluten-free diet: Recognizing facts, fiction, and fad. *The Journal of Pediatrics, 175*, 206–210.
43. Rosa-Sibakov, N., Heiniö, R. L., Cassan, D., Holopainen-Mantila, U., Micard, V., Lantto, R., & Sozer, N., (2016). Effect of bioprocessing and fractionation on the structural, textural, and sensory properties of gluten-free fava bean pasta. *LWT-Food Science and Technology, 67*, 27–36.
44. Sabença, C., Ribeiro, M., Sousa, T. D., Poeta, P., Bagulho, A. S., & Igrejas, G., (2021). Wheat/gluten-related disorders and gluten-free diet misconceptions: A review. *Foods, 10*(8), 1765–1772.
45. Sakač, M., Pestorić, M., Mišan, A., Nedeljković, N., Jambrec, D., Jovanov, P., Banjac, V., et al., (2015). Antioxidant capacity, mineral content and sensory properties of gluten-free rice and buckwheat cookies. *Food Technology and Biotechnology, 53*(1), 38–46.
46. Santos, F. G., Aguiar, E. V., Rosell, C. M., & Capriles, V. D., (2021). Potential of chickpea and psyllium in gluten-free breadmaking: Assessing bread's quality, sensory acceptability, and glycemic and satiety indexes. *Food Hydrocolloids, 113*, E-article ID 106487.
47. Santos, F. G., Fratelli, C., Muniz, D. G., & Capriles, V. D., (2018). Mixture design applied to the development of chickpea-based gluten-free bread with attractive technological, sensory, and nutritional quality. *Journal of Food Science, 83*(1), 188–197.
48. Scherf, K. A., Koehler, P., & Wieser, H., (2016). Gluten and wheat sensitivities: An overview. *Journal of Cereal Science, 67*, 2–11.
49. Schmelter, L., Rohm, H., & Struck, S., (2021). Gluten-free bakery products: Cookies made from different *Vicia Faba* bean varieties. *Future Foods, 4*, E-article ID 100038.
50. Sciammaro, L. P., Ferrero, C., & Puppo, M. C., (2018). Gluten-free baked muffins developed with *Prosopis alba* flour. *LWT Food Science & Technology, 98*(1), 568–576.
51. Simons, C. W., & Hall, C., (2018). Consumer acceptability of gluten-free cookies containing raw cooked and germinated pinto bean flours. *Food Science & Nutrition, 6*(1), 77–84.
52. Šmídová, Z., & Rysová, J., (2022). Gluten-free bread and bakery products technology. *Foods, 11*(3), 480–487.
53. Sozer, N., Melama, L., Silbir, S., Rizzello, C. G., Flander, L., & Poutanen, K., (2019). Lactic acid fermentation as a pre-treatment process for Faba bean flour and its effect on textural, structural, and nutritional properties of protein-enriched gluten-free Faba bean breads. *Foods, 8*(10), 431–437.
54. Taraghikhah, N., Ashtari, S., Asri, N., Shahbazkhani, B., Al-Dulaimi, D., Rostami-Nejad, M., Rezaei-Tavirani, M., et al., (2020). An updated overview of spectrum of gluten-related disorders: Clinical and diagnostic aspects. *BMC Gastroenterology, 20*(1), 1–12.
55. Torra, M., Belorio, M., Ayuso, M., Carocho, M., Ferreira, I. C., Barros, L., & Gómez, M., (2021). Chickpea and chestnut flours as non-gluten alternatives in cookies. *Foods, 10*(5), 911–918.

56. Tovoli, F., Masi, C., Guidetti, E., Negrini, G., Paterini, P., & Bolondi, L., (2015). Clinical and diagnostic aspects of gluten related disorders. *World Journal of Clinical Cases: WJCC, 3*(3), 275–286.
57. Tsatsaragkou, K., Protonotariou, S., & Mandala, I., (2016). Structural role of fiber addition to increase knowledge of non-gluten bread. *Journal of Cereal Science, 67*, 58–67.
58. Wesley, S. D., André, B. H. M., & Clerici, M. T. P. S., (2021). Gluten-free rice & bean biscuit: Characterization of a new food product. *Heliyon, 7*(1), E-article ID 05956.
59. Wunsch, N., (2020). Gluten-free products launches worldwide 2008–2018. https://www.statista.com/statistics/1028534/product-launch-gluten-free-products-worldwide/ (accessed on 2 November 2023).

Index

H

Q

R

T